Basic Topology 2

Professor Mahima Ranjan Adhikari (1944–2021)

Avishek Adhikari · Mahima Ranjan Adhikari

Basic Topology 2

Topological Groups, Topology of Manifolds and Lie Groups

Avishek Adhikari
Department of Mathematics
Presidency University
Kolkata, West Bengal, India

Mahima Ranjan Adhikari
Institute for Mathematics, Bioinformatics, Information Technology and Computer Science (IMBIC)
Kolkata, West Bengal, India

Professor Mahima Ranjan Adhikari is deceased.

ISBN 978-981-16-6579-0 ISBN 978-981-16-6577-6 (eBook)
https://doi.org/10.1007/978-981-16-6577-6

Mathematics Subject Classification: 22-XX, 51Hxx, 55-XX, 55Pxx, 14F45

This Springer imprint is published by the registered company Springer Nature Singapore Pte Ltd.
The registered company address is: 152 Beach Road, #21-01/04 Gateway East, Singapore 189721, Singapore

Dedicated to

my grandparents Naba Kumar and Snehalata Adhikari,
my mother Minati Adhikari
who created my interest in mathematics
at my very early childhood:

—Avishek Adhikari

Dedicated to

my mother Snehalata Adhikari (1921–2007),
on occasion of her birth centenary, who gave
me the first lesson of mathematics:

—Mahima Ranjan Adhikari

Preface

The series of three books on *Basic Topology* is a project book funded by the Government of West Bengal. It is designed to introduce many variants of a basic course in topology through the study of point-set topology, topological groups, topological vector spaces, manifolds, Lie groups, homotopy and homology theories with an emphasis on their applications in modern analysis, geometry, algebra, and theory of numbers. Topics in topology are vast. The range of its basic topics is distributed among different topological subfields such as general topology, topological algebra, differential topology, combinatorial topology, algebraic topology, and geometric topology. Each volume of the present book is considered a separate textbook that promotes active learning of the subject highlighting the elegance, beauty, scope, and power of topology. Volume 1 studies metric spaces and general topology. It considers the general properties of topological spaces and their mappings. Volume 2 considers additional structures other than topological structures studied in Volume 1. The contents of Volume 2 are expanded in five chapters. The chapterwise description of Volume 2 runs as follows:

Chapter 1 provides a background on algebra, topology and analysis for a smooth study of this volume. Chapter 2 studies topological groups and topological vector spaces using topology and algebra. It provides many interesting geometrical objects of interest and relates algebra with geometry and analysis. Chapter 3 studies topological problems of the theory of differentiable manifolds and differentiable functions on differentiable manifolds. The concepts of diffeomorphisms, embeddings and bundles play important roles in differential topology. The main aim of this chapter is to convey an elementary approach to differential topology avoiding algebraic topology, which is studied in Volume 3. Chapter 3 starts with topological manifolds followed by smooth manifolds and smooth mappings leading to differential topology. The most basic results embedded in the theory of differential manifolds is essentially due to the work of a single man, H. Whitney (1907–1989). Whitney's embedding theorems say that every manifold can be embedded in a Euclidean space as a closed subspace. It implies that any manifold may be considered as a submanifold of a Euclidean space. Sard's theorem is studied, and it is applied to prove the fundamental theorem of algebra, and the Brouwer fixed-point theorem without using the tools of algebraic topology.

Chapter 4 starts with Hilbert's fifth problem and discusses a study of Lie groups and Lie algebra. Chapter 5 presents a brief history of the emergence of the concepts leading to the development of topological groups, and also Lie groups as mathematical topics with their motivations.

This volume is complete in itself and deals with the introductory concepts of topological algebra, differential topology and Lie groups and discusses certain classical problems such as the fundamental theorem of algebra, Brouwer fixed-point theorem, classification of closed surfaces and the Jordan curve theorem without using the formal techniques of algebraic topology, which are studied in Volume 3.

Anyone who will study Volume 2 and solve exercises in the book will earn a thorough knowledge in topological groups and vector spaces (which form an integral part of topological algebra general topology) and also knowledge in topology of manifolds and Lie groups (which form an integral part of differential topology).

The book is a clear exposition of the basic ideas of topology and conveys a straightforward discussion of the basic topics of topology and avoids unnecessary definitions and terminologies. Each chapter starts with highlighting the main results of the chapter with motivation and is split into several sections which discuss related topics with some degree of thoroughness and ends with exercises of varying degrees of difficulties, which not only impart additional information about the text covered previously but also introduce a variety of ideas not treated in the earlier texts with certain references to the interested readers for more study. All these constitute the basic organizational units of the book.

The authors acknowledge Higher Education Department of Government of West Bengal for sanction of financial support to the Institute for Mathematics, Bioinformatics and Computer Science (IMBIC) toward writing this book vide order no. 432(Sanc)/ EH/P/SE/ SE/1G-17/07 dated August 29, 2017, and also to IMBIC, University of Calcutta, Presidency University, Kolkata, India, and Moulana Abul Kalam Azad University of Technology, West Bengal, for providing infrastructure toward implementing the scheme.

The authors are indebted to the authors of the books and research papers listed in the bibliographies at the end of each chapter and are very thankful to Profs. P. Stavrions (Greece), Constantine Udriste (Romania), Akira Asada (Japan) and also to the reviewers of the manuscript for their scholarly suggestions for the improvement of the book. We are thankful to Md. Kutubuddin Sardar for his cooperation towards the typesetting of the manuscript and to many UG and PG students of Presidency University and Calcutta University, and to many other individuals who have helped in proofreading the book. The authors' thanks are also due to IMBIC, Kolkata, for providing the authors with a library and other facilities toward the manuscript development work of this book. To those whose names have been inadvertently not entered, the authors apologize. Finally, the authors acknowledge, with heartfelt thanks, the patience and sacrifice of the long-suffering family of the authors, especially Dr. Shibopriya Mitra Adhikari and Master Avipriyo Adhikari.

Kolkata, India
June 2021

Avishek Adhikari
Mahima Ranjan Adhikari

A Note on Basic Topology—Volumes 1–3

The topic "topology" has become one of the most exciting and influential fields of study in modern mathematics because of its beauty and scope. This subject aims to make a qualitative study of geometry in the sense that if one geometric object is continuously deformed into another geometrical object, then these two geometric objects are considered topologically equivalent, called *homeomorphic*. Topology starts where sets have some cohesive properties, leading to defining the continuity of functions.

The series of three books on *Basic Topology* is a project book funded by the Government of West Bengal, which is designed to introduce many variants of a basic course in topology through the study of point-set topology, topological groups, topological vector spaces, manifolds, Lie groups, homotopy and homology theories with an emphasis on their applications in modern analysis, geometry, algebra and theory of numbers.

The topics in topology are vast. The range of its basic topics is distributed among different topological subfields such as general topology, topological algebra, differential topology, combinatorial topology, algebraic topology and geometric topology. Each volume of the present book is considered as a separate textbook that promotes active learning of the subject highlighting elegance, beauty, scope and power of topology.

Basic Topology—Volume 1: Metric Spaces and General Topology

This volume majorly studies metric spaces and general topology. It considers the general properties of topological spaces and their mappings. The special structure of a metric space induces a topology having many applications of topology in modern analysis, geometry and algebra. The texts of Volume 1 are expanded into eight chapters.

Basic Topology—Volume 2: Topological Groups, Topology of Manifolds and Lie Groups

This volume considers additional structures other than topological structures studied in Volume 1 and links topological structure with other structures in a compatible way to study topological groups, topological vector spaces, topological and smooth manifolds, Lie groups and Lie algebra and also gives a complete classification of closed surfaces without using the formal techniques of homology theory. Volume 2 contains five chapters.

Basic Topology—Volume 3: Algebraic Topology and Topology of Fiber Bundles

This volume mainly discusses algebraic topology and topology of fiber bundles. The main aim of topology is to classify topological spaces up to homeomorphism. To achieve this goal, algebraic topology constructs algebraic invariants and studies topological problems by using these algebraic invariants. Because of its beauty and scope, algebraic topology has become an essential branch of topology. Algebraic topology is an important branch of topology that utilizes algebraic tools to study topological problems. Its basic aim is to construct algebraic invariants that classify topological spaces up to homeomorphism. It is found that this classification, usually in most cases, is up to homotopy equivalence.

This volume conveys a coherent introduction to algebraic topology formally inaugurated by H. Poincaré (1854–1912) in his land-marking *Analysis situs*, Paris, 1895, through his invention of fundamental group and homology theory, which are topological invariants. It studies Euler characteristic, the Betti number and also certain classic problems such as the Jordan curve theorem. It considers Higher homotopy groups and establishes links between homotopy and homology theories, axiomatic approach to homology and cohomology inaugurated by Eilenberg and Steenrod. It studies the problems of converting topological and geometrical problems to algebraic ones in a functorial way for better chances for solutions.

This volume also studies geometric topology and manifolds by using algebraic topology. The contents of Volume 3 are expanded in seven chapters.

Just after the concept of homeomorphisms is clearly defined, the subject of topology begins to study those properties of geometric figures which are preserved by homeomorphisms with an eye to classify topological spaces up to homeomorphism, which stands as the ultimate problem in topology, where a geometric figure is considered to be a point set in the Euclidean space $\mathbf{R}^n$. But this undertaking becomes hopeless, when there exists no homeomorphism between two given topological spaces. The concepts of topological properties and topological invariants play key tools in such problems:

(a) The concept of "topological property" such as compactness and connectedness, which is introduced in general topology, solves this problem in a very few cases, which are studied in Volume 1. A study of the subspaces of the Euclidean plane $\mathbf{R}^2$ gives an obvious example.

(b) On the other hand, the subjects algebraic topology and differential topology (studied in Volume 2) were born to solve the problems of impossibility in many cases with a shift of the problem by associating *invariant objects* in the sense that homeomorphic spaces have the same object (up to equivalence), called *topological invariants*. Initially, these objects were integers, and subsequent research reveals that more fruitful and interesting results can be obtained from the algebraic invariant structures such as groups and rings. For example, homology and homotopy groups are very important algebraic invariants which provide strong tools to study the structure of topological spaces.

Contents

About the Authors

Avishek Adhikari, Ph.D., M.Sc. (Gold Medalist), is Professor at the Department of Mathematics, Presidency University, Kolkata, India. A recipient of the President of India Medal, ISCA Young Scientist Award, and the NANUM Fund by International Mathematical Union, Professor Adhikari did his Ph.D. from the Indian Statistical Institute, India, under the guidance of Professor Bimal Roy, the former Director of the Indian Statistical Institute. Earlier, Professor Adhikari was a faculty member at the Department of Pure Mathematics, University of Calcutta, Kolkata, from July 2006–January 2019. He is the founder Secretary of the Institute for Mathematics, Bioinformatics, Information Technology & Computer Science (IMBIC), India, having branches in Sweden and Japan, and the Treasurer of the Cryptology Research Society of India. He was the former Eastern Zonal coordinator of the M.Sc. and Ph.D. scholarship examinations by the National Board of Higher Mathematics (NBHM). He was a Post-Doctoral fellow at INRIA-Rocquencourt, France, and Visiting Scientist at Linkoping University, Sweden; Indian Statistical Institute, Kolkata. Professor Adhikari delivered invited talks at various universities and institutions abroad.

He has authored four textbooks on mathematics including *Basic Modern Algebra with Applications* (Springer, 2014) and edited two volumes including Mathematical and Statistical Applications in Life Sciences and Engineering (Springer, 2017). His research papers have been published in reputed international journals, conference proceedings and contributed volumes. Professor Adhikari is one of the investigators of the twelve sponsored research projects funded by agencies like DRDO, WESEE (the Ministry of Defense), DST, DIT, NBHM of the Government of India, including two international collaborative projects supported by DST-JSPS and DST-JST (both Indo-Japan projects). Five of his Ph.D. students have already been awarded degrees, and two of the students have submitted their theses and, currently, three Ph.D. scholars are conducting research under his supervision.

Mahima Ranjan Adhikari, Ph.D., M.Sc. (Gold Medalist), is the founder president of the Institute for Mathematics, Bioinformatics and Computer Science (IMBIC), Kolkata, India. He is a former professor at the Department of Pure Mathematics, University of Calcutta, India. His research papers are published in national and international journals of repute, including the *Proceedings of American Mathematical Society*. He has authored nine textbooks and is the editor of two, including: *Basic Modern Algebra with Applications* (Springer, 2014), *Basic Algebraic Topology and Applications* (Springer, 2016), and *Mathematical and Statistical Applications in Life Sciences and Engineering* (Springer, 2017).

Twelve students have been awarded Ph.D. degree under his guidance on various topics such as algebra, algebraic topology, category theory, geometry, analysis, graph theory, knot theory and history of mathematics. He has visited several universities and research institutions in India, USA, UK, Japan, China, Greece, Sweden, Switzerland, Italy, and many other countries on invitation. A member of the American Mathematical Society, Prof. Adhikari is on the editorial board of several journals of repute. He was elected as the president of the Mathematical Sciences Section (including Statistics) of the 95th Indian Science Congress, 2008. He has successfully completed research projects funded by the Government of India.

Chapter 1
Background on Algebra, Topology and Analysis

This chapter communicates a few basic concepts of algebra, analysis and topology for the smooth study of **Basic Topology, Volume 2** of the present series of books. A detailed study of the topological concepts and results is available in Volume 1 of the present series. Moreover, the books Adhikari (2016, 2022); Adhikari and Adhikari (2003, 2006, 2014, 2022); Alexandrov (1979); Borisovich et al. (1985); Bredon (1983); Dugundji (1966); Hu (1969); MacLane (1971); Simmons (1963); Williard (1970) and some other references are given in Bibliography.

1.1 Some Basic Concepts on Algebraic Structures and Their Homomorphisms

A group, ring, vector space, module or any algebraic system is defined as a nonempty set endowed with special algebraic structures. A homomorphism (transformation) is a function preserving the specific structures of the algebraic system. Their detailed study is available in the book [Adhikari and Adhikari, Springer, 2014] and also in many other books.

1.1.1 Groups and Fundamental Homomorphism Theorem

This subsection presents some basic concepts of group theory which are subsequently used. Historically, the earlier definition of a group as the set of permutations (i.e., bijections) on a nonempty set X with the property that a combination (called composition) of two permutations is also a permutation on X is generalized to the present definition of an abstract group by a set of axioms.

A. Adhikari and M. R. Adhikari, *Basic Topology 2*,
https://doi.org/10.1007/978-981-16-6577-6_1

Definition 1.1.1 A nonempty set G together with a binary operation

$$m : G \times G \to G,$$

the image of (x,y) under m, denoted by xy, usually called a multiplication, is called a group (abstract) if the algebraic system (G, m) has the following properties:

G(1) $xy(z) = x(yz)$ for all x, y, z in G (associative law);
G(2) there exists an element e in G such that $xe = ex = x$ for all x in G (existence of identity);
G(3) for each x in G, there is an element x' in G such that $xx' = x'x = e$ (existence of inverse).

Remark 1.1.2 In a group G, the identity element e is unique and for every x in G; the element x' in G(3) is also unique. The element x' abbreviated as x^{-1} is called the inverse of x in G. In additive notation, xy is written as $x + y$, e as 0 (zero) and x^{-1} as $-x$.

Definition 1.1.3 A group G is said to be commutative if the binary operation "m" on G is commutative in the sense that $xy = yx$ for all x, y in G. The term "abelian group" is usually used in honor of N. H. Abel (1802–1829) when the composition law is in additive notation. A finite group G is a group having its underlying set finite; otherwise, a group G is said to be infinite.

Example 1.1.4 Under usual addition, $\mathbf{Z}$ forms an infinite group; on the other hand, $\mathbf{Z}_n$ forms a finite group.

Example 1.1.5 (*Permutation group*) Let X be a nonempty set X and $\mathcal{P}(X)$ be the set of all permutations (bijective mappings) on X. Then under the usual composition of mappings, $\mathcal{P}(X)$ forms a group, called permutation group on X. In particular, if X consists of only n elements, then $\mathcal{P}(X)$ is called the symmetric group on n elements, denoted by S_n.

Example 1.1.6 (*Circle group*) The set $S^1 = \{z \in \mathbf{C} : |z| = 1\}$ forms a group under the usual multiplication of complex numbers, called the circle group in $\mathbf{C}$.

Example 1.1.7 (*General linear group*) The set $GL(n, \mathbf{R})$ ($GL(n, \mathbf{C})$) of all invertible (nonsingular) $n \times n$ real (complex) matrices forms a group under usual multiplication of matrices, called the general linear group of order n over $\mathbf{R}(\mathbf{C})$.

A subset of a group forming a group, called a subgroup, creates interest in many situations.

Proposition 1.1.8 *Let G be a group and H be a nonempty subset of G. Then H is a subgroup of G if and only if $xy^{-1} \in H$ for all $x, y \in H$.*

Example 1.1.9 Let G be a group and $Z(G)$ be defined by $Z(G) = \{g \in G : gx = xg \text{ for all } x \in G\}$. Then $Z(G)$ forms a subgroup of G, called the **center** of G. This subgroup is also sometimes written as $C(G)$.

Definition 1.1.10 Let G be a group and H be a subgroup of G. For every $x \in G$,

(i) the set

$$xH = \{xh : h \in H\}$$

is called a left coset of H in G and

(ii) the right coset Hx is defined by the set

$$Hx = \{hx : h \in H\}.$$

Definition 1.1.11 Let G be group. A subgroup H is said to be a normal subgroup of G, if

$$xH = Hx, \forall x \in G;$$

equivalently, H is a normal subgroup of G if

$$xHx^{-1} = \{xhx^{-1} : h \in H\} \subset H, \forall x \in G.$$

Theorem 1.1.12 *Let G be a group and N be a normal subgroup of G. If G/N is the set of all cosets of N in G, then it is a group under the composition*

$$xN \circ yN = xyN, \forall x, y \in G$$

*called the **factor group (or quotient group)** of G by N, denoted by G/N.*

Definition 1.1.13 Given an additive abelian group G, and a normal subgroup N of G, the factor group G/N is defined under the group operation

$$(x + N) + (y + N) = (x + y) + N, \forall x, y \in G$$

and this group is sometimes called difference group.

Definition 1.1.14 Let G and K be two groups. A map

$$f : G \to K$$

is said to be a homomorphism (of groups) if

$$f(xy) = f(x)f(y) \forall x, y \in G.$$

Remark 1.1.15 A homomorphism of groups maps identity element into identity element and inverse element into the inverse element.

Definition 1.1.16 A homomorphism $f : G \to K$ of groups is said to be

(i) a monomorphism if f is injective;
(ii) an embedding if f is a monomorphism;

(iii) an epimorphism if f is surjective;
(iv) an isomorphism if f is bijective.

In particular, an isomorphism of G onto itself is called an automorphism of G, and a homomorphism of G into itself is called an endomorphism.

Remark 1.1.17 Two isomorphic groups are considered as replicas of each other, because they have identical algebraic properties. So, two isomorphic groups are identified in group theory and are considered the same group (up to isomorphism).

Definition 1.1.18 The kernel of a homomorphism $f : G \to K$ of groups, denoted by $kerf$, is defined by

$$kerf = \{x \in G : f(x) = e_K\},$$

where e_K is the identity element of K.

Theorem 1.1.19 *Let $f : G \to K$ be a group homomorphism. Then $kerf$ is a normal subgroup of G. Conversely, if N is a normal subgroup of G, then the map*

$$\pi : G \to K/N, \ x \mapsto xN$$

is an epimorphism with N as its kernel.

Corollary 1.1.20 *Let N be a subgroup of a given group G. Then N is a normal subgroup of G iff it is the kernel of some homomorphism.*

Theorem 1.1.21 (First Isomorphism Theorem) *Let $f : G \to K$ be a group homomorphism. Then f induces an isomorphism*

$$\tilde{f} : G/kerf \to Imf, \ x\ kerf \mapsto f(x), \ for\ all\ x \in G.$$

Corollary 1.1.22 *Let $f : G \to K$ be an epimorphism of groups. Then the groups $G/kerf$ and K are isomorphic.*

Example 1.1.23 Let $G = GL(n, \mathbf{R})$ be the general linear group over $\mathbf{R}$ and $\mathbf{R}^*$ be the multiplicative group of nonzero real numbers. If

$$det : G \to \mathbf{R}^*, \ M \mapsto det\ M$$

is the determinant function and

$$N = \{M \in GL(n, \mathbf{R}) : detM = 1\},$$

then N is a normal subgroup of G such that the groups $GL(n, \mathbf{R})/N$ and $\mathbf{R}^*$ are isomorphic by Corollary 1.1.22.

Definition 1.1.24 Given a group G and a pair of elements $g, h \in G$, the commutator of g and h is the element $ghg^{-1}h^{-1}$, denoted by $[g, h]$. The subgroup K of G generated by the set $S = \{ghg^{-1}h^{-1} : g, h \in G\}$ is called the **commutator subgroup** of G and it consists of all finite products of commutators of G.

Theorem 1.1.25 *Let G be a group and K be the commutator subgroup of K. Then*

(i) *K is a normal subgroup of G;*
(ii) *the quotient group G/K is always commutative;*
(iii) *the group G is commutative iff its commutator subgroup K is the trivial group.*

1.1.2 Concept of Group Actions

The concept of **group actions** unifies the historical concept of the group of transformations and the axiomatic concept of a group used today.

Definition 1.1.26 Let G be a group with identity element e and X be a nonempty set. Then G is said to act on X from the left if there is map

$$\mu : G \times X \to X,$$

the image $\mu(g, x)$, denoted by $g(x)$ or gx such that

(i) $e(x) = x$, for all $x \in X$;
(ii) $(gk)(x) = g(k(x))$, for all $x \in X$, and $g, k \in G$.

The set X endowed with a left G-action on X is said to be a left G-set.

A right G-set is defined in an analogous way. The ordered triple (X, G, μ) is sometimes called a **transformation group**.

Remark 1.1.27 There is a one-to-one correspondence between the left and right G-set structures on X. So it is sufficient to study only one of them according to the situation.

Example 1.1.28 The Euclidean n-space $\mathbf{R}^n$ is a left $G(n, \mathbf{R})$-set under usual multiplication of matrices.

Example 1.1.29 Let G be a group and H be a subgroup of G.

(i) The action $H \times G \to G, x \mapsto hx$ (left **translation** by h) is a group action of H on G;
(ii) The action $H \times G \to G, x \mapsto hxh^{-1}$ (**conjugation** by h) is also an action of H on G;
(iii) If H is normal in G, then the action $G \times H \to H, h \mapsto xhx^{-1}$ is an action of G on H.

Remark 1.1.30 There may exist different actions of a group on a set.

Proposition 1.1.31 *Let X be a left G-set and $g \in G$ be an arbitrary point. Then the map*

$$\psi_g : X \to X, x \mapsto g(x)$$

is a bijection, i.e., a permutation on X.

Proof ψ_g and ψ_g^{-1} are two maps such that $\psi_g \circ \psi_g^{-1} = I_X = \psi_g^{-1} \circ \psi_g$ and hence ψ_g is a bijection. ❑

Remark 1.1.32 The concept of a G-set X is equivalent to the concept of a representation of the group G by the permutations on X.

Definition 1.1.33 Let X be a left G-set. Two given elements x, y in X are called G-equivalent, if $y = g(x)$ for some $g \in G$. The relation of being G-equivalent, being an equivalence relation $\sim$ on X, the quotient set $X/\sim$, denoted by $X mod\ G$ is called the **orbit set** of X, obtained by the action of G on X. For an element $x \in X$, the set

$$G(x)$$

is called the **orbit** of x, which is also denoted by $orb\ (x)$ and the subgroup G_x of G defined by

$$G_x = \{g \in G : g(x) = x\}$$

is called the **stabilizer or isotropy group** at x.

1.1.3 Ring of Real-Valued Continuous Functions

The ring $\mathcal{C}([0, 1])$ of real-valued continuous functions on the closed interval $[0, 1]$ is an important ring in mathematics. This ring gives an interplay between real analysis and algebra.

Definition 1.1.34 Let $\mathcal{C}([0, 1])$ be the set of all real-valued continuous functions on the closed interval $[0, 1]$, i.e., $\mathcal{C}([0, 1]) = \{f : [0, 1] \to \mathbf{R}\ such\ that\ f\ is\ continuous\}$. Then $\mathcal{C}([0, 1])$ forms a commutative ring under usual pointwise addition and multiplication:

$$(f + g)(x) = f(x) + g(x),\ and\ (f \cdot g)(x) = f(x)g(x),\ \forall\ f, g \in \mathcal{C}([0, 1])\ and\ \forall x \in [0, 1],$$

where the right-hand addition and multiplication are the usual addition and multiplication of real numbers. The ring $\mathcal{C}([0, 1])$ is called the **ring of real-valued continuous functions on** $[0, 1]$.

Example 1.1.35 The ring $\mathcal{C}([0, 1]) = R$ contains divisors of zero and hence it is not an integral domain. Consider the elements $f, g \in R$

$$f : [0, 1] \to \mathbf{R},\ f(x) = \begin{cases} 0, & \text{if } 0 \leq x \leq 1/2 \\ x - \frac{1}{2}, & \text{if } 1/2 \leq x \leq 1 \end{cases}$$

$$g : [0, 1] \to \mathbf{R},\ f(x) = \begin{cases} \frac{1}{2} - x, & \text{if } 0 \leq x \leq 1/2 \\ 0, & \text{if } 1/2 \leq x \leq 1 \end{cases}.$$

Then $f, g \in \mathcal{C}([0, 1])$ are nonzero elements such that their product $f \cdot g = 0$. This implies that the ring $\mathcal{C}([0, 1])$ cannot be an integral domain.

Theorem 1.1.36 *For each point* $x \in \mathbf{I} = [0, 1]$, *the ideal*

$$M_x = \{f \in \mathcal{C}([0, 1]) : f(x) = 0\}$$

is a maximal ideal of the ring $\mathcal{C}([0, 1])$. *Moreover, if* $\mathcal{M}$ *is the set of all maximal ideals of* $C([0, 1])$, *then the map*

$$\psi : I \to \mathcal{M}, x \mapsto M_x$$

is a bijection.

Proof This is left as an exercise or see Volume 1. ❑

1.1.4 Fundamental Theorem of Algebra

This subsection states the Fundamental Theorem of Algebra, and its proof is given in Chap. 3 of this volume by using the tools of differential topology and also in Chap. 2 of Volume 3 by using the tools of homotopy theory. The fundamental theorem of algebra asserts the completeness of the field **C**.

Definition 1.1.37 A field F is said to algebraically closed or complete if every polynomial ring $F[x]$ of degree n ($n \geq 1$) over F has a root in F.

Example 1.1.38 The field **C** is algebraically closed. It follows from the fundamental theorem of algebra, Theorem 1.1.39.

Theorem 1.1.39 (Fundamental Theorem of Algebra) *Let* **C** *be the field of complex numbers. Every nonconstant polynomial over* **C** *has a root in* **C**.

1.1.5 Vector Space and Linear Transformations

This subsection recalls the concept of vector spaces (linear spaces) which have a very strong algebraic structure to solve many specific problems. Let **F** be the field **F** = **R** or **C**. For an arbitrary field F, the definition of a vector space is similar.

Definition 1.1.40 A vector space or a linear space over a field F (whose elements are called scalars) with identity element e is an additive abelian group V (whose elements are called vectors) together with an external law of composition, called scalar multiplication

$$m : F \times V \to V,$$

the image of (α, v) under m abbreviated as αv, if the following conditions are satisfied for all $x, y \in V$ and $\alpha, \beta \in F$:

(i) $ex = x$;
(ii) $\alpha(x + y) = \alpha x + \alpha y$;
(iii) $(\alpha + \beta)x = \alpha x + \beta x$;
(iv) $(\alpha\beta)x = \alpha(\beta x)$.

Example 1.1.41 F^n is a vector space over any field F. In particular, the n-dimensional Euclidean space $\mathbf{R}^n$ is a vector space over the field **R**.

1.1.6 Basis for a Vector Space

This subsection studies the basis of vector spaces. The n vectors such as $e_1 = (1, 0, \ldots, 0), \ldots, e_n = (0, 0, \ldots, 1) \in \mathbf{R}^n$ determine every vector of $\mathbf{R}^n$ uniquely. This leads to the concept of basis (finite) of a vector space.

Definition 1.1.42 Let B be a nonempty subset (finite or infinite) of a vector space V over F. A vector $v \in V$ is called a linear combination of vectors of B over F if it can be expressed as

$$v = a_1 v_1 + \cdots + a_n v_n, \ a_i \in F; \ v_i \in B; \ i = 1, 2, \ldots, n.$$

Proposition 1.1.43 *Let B be a nonempty subset (finite or infinite) of a vector space V over F. Then the set $\mathbf{L}(B)$ of all linear combinations of vectors of B is a subspace of the vector space V, which is the smallest subspace of V containing B.*

Definition 1.1.44 $\mathbf{L}(B)$ is called the subspace generated or spanned by B in V. In particular, if $V = \mathbf{L}(B)$ for some $B \subset V$, then B is said to be a set of generators for V. It is said to be finitely or infinitely generated according as card B is finite or infinite.

Example 1.1.45 The vector space $\mathbf{C}^n$ is finitely generated. On the other hand, the polynomial ring $\mathbf{R}[x]$ with real coefficients is not finitely generated, because any linear combination of a finite set of polynomials is a polynomial whose degree does not exceed the maximum degree say k of the set of polynomials, but $\mathbf{R}[x]$ has polynomials having degree $> k$.

Definition 1.1.46 A nonempty subset B of a vector V is said to be a basis of V over F if

(i) V is generated by the set B and
(ii) B is linearly independent over F.

If a vector space V is trivial, i.e., $V = \{0\}$, then the empty set $\emptyset$ is taken as its basis.

Theorem 1.1.47 (Existence theorem) *Every vector space has a basis. The cardinality of every basis of a vector space is the same.*

Definition 1.1.48 Given a nonzero vector space V over F, the cardinality of every basis of V being the same, this common value is said to be the dimension of V, abbreviated as dim V. The vector space is said to be finite- or infinite-dimensional if V has a finite or an infinite basis B, i.e., if card B is finite or not. If dim B is n, then V is said to be an n-dimensional vector space over F.

Example 1.1.49 $\mathbf{R}^n$ is an n-dimensional vector space over $\mathbf{R}$. On the other hand, $\mathbf{R}[x]$ is an infinite-dimensional vector space over $\mathbf{R}$.

Example 1.1.50 The real quaternions form a 4-dimensional vector space $\mathcal{H}$ over $\mathbf{R}$, with a basis $\mathbf{B} = \{1, i, j, k\}$.

1.1.7 Algebra over a Field

Definition 1.1.51 An **algebra** $\mathcal{A}$ is a vector space of a field F, with multiplication defined in such a way that $\mathcal{A}$ is also a ring in which the scalar multiplication and ring multiplication are linked by the property

$$\alpha(xy) = (\alpha x)y = x(\alpha y),\ \forall x, y \in \mathcal{A},\ \textit{and}\ \forall \alpha \in F.$$

An algebra $\mathcal{A}$ is said to be commutative if

$$xy = yx,\ \forall x, y \in \mathcal{A}$$

and it is said to have an identity, denoted by 1, if

$$1x = x1,\ \forall x \in \mathcal{A}.$$

(i) An algebra $\mathcal{A}$ over the field $\mathbf{R}$ is said to be real;
(ii) An algebra $\mathcal{A}$ over the field $\mathbf{C}$ is said to be complex.

Example 1.1.52 The vector space $\mathcal{C}([0, 1])$ is a commutative real algebra with identity.

Definition 1.1.53 Let $\mathcal{A}$ be an algebra over a field F. Then a nonempty subset $\mathcal{B}$ is said to be a subalgebra of $\mathcal{A}$ if $\mathcal{B}$ is itself an algebra over F under the same operations defined on $\mathcal{A}$.

Example 1.1.54 Given an algebra $\mathcal{A}$, the subset

$$\mathcal{C} = \{x \in \mathcal{A} : xy = yx, \ \forall\, y \in \mathcal{A}\}$$

of $\mathcal{A}$ is a subalgebra of $\mathcal{A}$, called the center of $\mathcal{A}$.

Definition 1.1.55 A homomorphism $f : \mathcal{A} \to \mathcal{B}$ from an algebra $\mathcal{A}$ to an algebra $\mathcal{B}$ over the same field F is a mapping with the properties

(i) $f(x + y) = f(x) + f(y)$;
(ii) $f(xy) = f(x)f(y)$;
(iii) $f(\alpha x) = \alpha f(x)$,

$\forall\, x, y \in \mathcal{A}$ and $\forall\, \alpha \in F$.

1.2 Some Basic Results of Set Topology

This section recalls some basic topological concepts and results studied in **Basic Topology, Volume 1** of the present series of books to facilitate a smooth study of its Volume 2.

1.2.1 Continuity of Functions

Definition 1.2.1 Let X and Y be topological spaces. A function $f : X \to Y$ is said to be continuous if $f^{-1}(U) \subset X$ is an open set in X for every open set U in Y.

The concept of continuity can also be equally well formulated by closed sets in topological settings given in Theorem 1.2.2.

Theorem 1.2.2 *Let X and Y be topological spaces. A function $f : X \to Y$ is continuous iff $f^{-1}(A) \subset X$ is a closed set in X for every closed set A in Y.*

1.2.2 Identification Topology

Proposition 1.2.3 *Let (X, τ) be a topological space and Y be a nonempty arbitrary set. Given a surjective map $f : X \to Y$, the topology*

$$\tau_f = \{V \subset Y : f^{-1}(V) \in \tau\}$$

is the strongest (largest) topology on Y such that the function $f : (X, \tau) \to (Y, \tau_f)$ is continuous.

Proof Let σ be a topology on Y such that the given function

$$f : (X, \tau) \to (Y, \sigma)$$

is continuous. Let $U \in \sigma$ be an arbitrary open set in Y. Then by continuity of f, it follows that $f^{-1}(U) \in \tau$ and hence $U \in \tau_f$. This implies that $\sigma \subset \tau_f$. ❑

Definition 1.2.4 The topology τ_f defined in Proposition 1.2.3 is called the **identification topology**.

Definition 1.2.5 Let (X, τ) be a topological space and Y be a nonempty set. If $f : X \to Y$ is a surjective map, then a topology σ is defined on Y such that it is the largest topology on Y for which f is continuous by declaring that a subset U of Y is in σ iff $f^{-1}(U) \in \tau$. Then f is said to be an **identification map** with σ its identification topology, and the corresponding topological space (Y, σ) is called its **identification space**.

Example 1.2.6 Let (X, τ) be a topological space and $\mathcal{P}$ be a partition of X into disjoint subsets generated by an equivalence relation $\sim$, which produces the quotient set $Y = X/\sim$. Then a new topological space (Y, σ), called an identification space, is defined if the points of Y are the members of $\mathcal{P}$ and if $p : X \to Y$ maps each point of X to the member of $\mathcal{P}$ containing it, and the topology σ of Y is the largest for which p is continuous. Hence, the map

$$p : (X, \tau) \to (Y, \sigma), \ x \mapsto [x]$$

is an identification map.

Theorem 1.2.7 *Let (X, τ) and (Y, σ) be topological spaces. If $f : X \to Y$ is a continuous surjective map such that f maps open sets of X to open sets of Y or f maps closed sets of X to closed sets of Y, then f is an identification map.*

Proof Let $f : X \to Y$ be a continuous surjective map such that f maps open sets of X to open sets of Y and τ_f be the topology defined in Proposition 1.2.3. To prove the theorem, it is sufficient to prove that $\sigma = \tau_f$. Given any open set $U \in \sigma$, it follows from continuity of f and definition of τ_f that

$$f^{-1}(U) \in \tau \implies U \in \tau_f \implies \sigma \subset \tau_f.$$

Conversely, for any $V \in \tau_f$, by surjectivity of f, it follows that $f(f^{-1}(V)) = V$. Again, by using the hypothesis that f maps open sets of X to open sets of Y, it follows that

$$f^{-1}(V) \in \tau \implies V = f(f^{-1}(V)) \in \sigma \implies \tau_f \subset \sigma.$$

Next, let f map closed sets of X to closed sets of Y and $U \in \sigma$ be an arbitrary open set. Then by continuity of f and definition of τ_f, it follows as discussed above that $\sigma \subset \tau_f$. Conversely, let $U \in \tau_f$. Then

$$f^{-1}(U) \in \tau \implies X - f^{-1}(U) \text{ is closed in } (X, \tau) \implies f(X - f^{-1}(U)) = Y - U$$

is closed in (Y, σ), since

$$f^{-1}(Y - U) = X - f^{-1}(U)$$

and by hypothesis, f maps closed sets into closed sets. Hence, it follows that

$$U \in \sigma \implies \tau_f \subset \sigma.$$

Thus in both cases, $\sigma \subset \tau_f$ and $\tau_f \subset \sigma$ assert that $\tau_f = \sigma$. ❑

1.2.3 *Open Base for a Topology*

Definition 1.2.8 Let (X, τ) be a topological space. A collection $\mathcal{B}$ of open sets $\{U\}$ in X is said to be an open base for the topology τ if every open set in X can be expressed as a union of some members of $\mathcal{B}$.

Remark 1.2.9 An open base of a topological space is characterized in Theorem 1.2.10.

Theorem 1.2.10 *Let (X, τ) be a topological space. Then a given collection of open sets $\mathcal{B}$ forms an open base of the topology τ if and only if for any open set U and any point $x \in U$ there exists a set $V \in \mathcal{B}$ such that*

$$x \in V \subset U.$$

Proof See Chap. 3 of Basic Topology, Volume 1. ❑

Definition 1.2.11 In the set $\mathbf{R}$, a subset $U \subset \mathbf{R}$ is said to be open for a topology τ on $\mathbf{R}$ if for every point $x \in \mathbf{U}$, there are points $a, b \in \mathbf{R}, \ with \ a < b$ such that

$$x \in (a, b) \subset U.$$

The collection $\mathcal{B}$ of such sets $\{U\}$ forms an open base for the topology τ, called Euclidean topology on $\mathbf{R}$.

1.2.4 Product Topology

Definition 1.2.12 Given topological spaces (X, τ) and (Y, σ), let

$$\mathcal{B} = \{U \times V : U \in \tau, \ and \ V \in \sigma\} \subset X \times Y.$$

Then $\mathcal{B}$ is the family of all subsets of $X \times Y$ of the form $U \times V$, where $U \in \tau, \ and \ V \in \sigma$ are such that the intersection of any two elements $U \times V, \ U' \times V' \in \mathcal{B}$ is also in $\mathcal{B}$, because

$$(U \times V) \cap (U' \times V') = (U \cap U') \times (V \times V') \in \mathcal{B}.$$

This asserts that $\mathcal{B}$ forms an open base for a topology $\tau_{\mathcal{B}}$, called the product topology on $X \times Y$ generated by the open base $\mathcal{B}$, and the set $X \times Y$ endowed with the **product topology** $\tau_{\mathcal{B}}$ is called a **product space**.

Remark 1.2.13 Product topology on a finite product of topological spaces is defined in an analogous way.

The product topology on $X \times Y$ is now studied in Theorem 1.2.14 by the projection maps

$$p_1 : X \times Y \to X, \ (x, y) \mapsto x, \ and \ p_2 : X \times Y \to Y, \ (x, y) \mapsto y.$$

Theorem 1.2.14 *Let X and Y be two topological spaces with the product topology* $\tau_{\mathcal{B}}$ *on* $X \times Y$. *Then*

(i) *the projection map*

$$p_1 : X \times Y \to X, \ (x, y) \mapsto x$$

is continuous;

(ii) *the projection map*

$$p_2 : X \times Y \to Y, \ (x, y) \mapsto y$$

is also continuous;

(iii) *the projection maps p_1 and p_2 send open sets to open sets;*
(iv) *the product topology on $X \times Y$ is the smallest topology such that the projection maps p_1 and p_2 are both continuous.*

Proof **(i)** Given any open set U of X, the set $p_1^{-1}(U) = U \times Y \in \tau_{\mathcal{B}} \implies p_1$ is continuous. Given any open set V of Y, the set $p_2^{-1}(V) = X \times V \in \tau_{\mathcal{B}} \implies p_2$ is continuous.

(ii) Let $U \times V \in \tau_{\mathcal{B}}$ be an arbitrary basic open set in the product topology on $X \times Y$. Then other open sets of the product space $X \times Y$ is a union of these basic open sets. So to prove part (ii), it is sufficient to consider the effect of p_1 and p_2 on the basic open set $U \times V$. As $p_1(U \times V) = U$ and $p_2(U \times V) = V$, where U is an open set of X and V is an open set of Y, and it follows that p_1 and p_2 send open sets of the product space $X \times Y$ to open sets of X and Y, respectively.

(iii) Suppose μ is an arbitrary topology on $X \times Y$ such that the projection maps p_1 and p_2 are both continuous. Then for any open set U of X and any open set V of Y, the set

$$p_1^{-1}(U) \cap p_2^{-1}(V) \in \mu \implies U \times V \in \mu \implies$$

the product topology $\tau_{\mathcal{B}}$ is contained in the given topology μ on $X \times Y$ so that p_1 and p_2 are both continuous. This proves (iii). ❑

Example 1.2.15 The Euclidean plane $\mathbf{R}^2 = \mathbf{R} \times \mathbf{R}$ has the usual topology, which is the product topology such that it is the smallest topology with the property that

$$p_1 : \mathbf{R} \times \mathbf{R},\ (x, y) \mapsto x \text{ and } p_2 : \mathbf{R} \times \mathbf{R},\ (x, y) \mapsto y$$

are both continuous. This implies that the map

$$f : \mathbf{R} \times \mathbf{R},\ (x, y) \mapsto x + y$$

is continuous, since $f(x, y) = (p_1 + p_2)(x, y),\ \forall (x, y) \in \mathbf{R} \times \mathbf{R}$ and hence $f = p_1 + p_2$ is the sum of two continuous maps which are continuous.

1.2.5 Connectedness and Path Connectedness

We recall the concepts of connectedness and path connectedness of topological spaces along with their associated results (see Chap. 5 of Basic Topology, Volume 1) for using them in this volume.

Definition 1.2.16 A topological space (X, τ) is said to be **connected** if it is not the union of two disjoint nonempty open subsets of X; equivalently, (X, τ) is connected, whenever $X = A \cup B$, where $A \neq \emptyset$ and $B \neq \emptyset$, then $A \cap \overline{B} \neq \emptyset$ or $\overline{A} \cap B \neq \emptyset$.

Example 1.2.17 **(i)** The real line space $(\mathbf{R}, \sigma)$ with usual topology σ is connected.
(ii) Every open or closed interval of $(\mathbf{R}, \sigma)$ is connected.
(iii) Euclidean n-space $\mathbf{R}^n$ is connected for all $n \geq 1$.

Proposition 1.2.18 *If $f : (X, \tau) \to (Y, \sigma)$ is a continuous map from a connected space X, then $f(X)$ is also connected.*

Corollary 1.2.19 *If $f : (X, \tau) \to (Y, \sigma)$ is a homeomorphism, then X is connected iff Y is connected.*

Remark 1.2.20 Connectedness of a topological space is a topological property by Corollary 1.2.19 in the sense that it is preserved under every homeomorphism.

Definition 1.2.21 Let (X, τ) be a topological space and $x \in X$ be an arbitrary point. Then the largest (maximal) connected set C_x (from the set theoretic viewpoint) containing the point x is called the **connected component or component** of x in X, which is the union of all connected sets in X containing the point x.

Example 1.2.22 The punctured Euclidean plane $\mathbf{R}^* = \mathbf{R} - \{0\}$ has two connected components such as

(i) $A = \{x \in \mathbf{R}^* : x < 0\}$ and
(ii) $B = \{x \in \mathbf{R}^* : x > 0\}$;
On the other hand, $\mathbf{R}$ has only one connected component, which is $\mathbf{R}$ itself.

Recall the concept of path connectedness and its related results from Chap. 5 of Volume 1.

Definition 1.2.23 A topological space X is said to be **path-connected** if for every pair of points $a, b \in X$, there is a path connecting the points a, b in X, i.e., if there exists a continuous map

$$f : \mathbf{I} \to X \text{ such that } f(0) = a, \ f(1) = b \text{ and } f(\mathbf{I}) \subset X,$$

where $\mathbf{I} = [0, 1]$ has the subspace topology inherited from the real line space $\mathbf{R}$.

Example 1.2.24 Every convex set in $\mathbf{R}^n$ is path-connected.

Recall the following results from Chap. 5 of Volume 1.

Theorem 1.2.25 *Let a topological space X be such that every pair of points can be joined by some connected subset of X. Then X is connected.*

Proof Let a topological space X be such that every pair of points can be joined by some connected subset of X in the sense that any path joining these points will entirely lie in a connected subset of X. If X is not connected, then there exist two disjoint open sets U and V such that $X = U \cup V$. Let $a \in U$ and $b \in V$ be two points and $W \subset X$ be a connected set containing the points a and b. Then $U_1 = U \cap W$ and $V_1 = V \cap W$ are two disjoint nonempty open sets in W such that $W = U_1 \cap V_1$. But it contradicts the connectedness of W. ❑

Corollary 1.2.26 *Every path-connected space is connected.*

Proof It follows from Theorem 1.2.25. ❑

1.2.6 Compactness

We now recall the concept of compactness, which was born through the Heine–Borel theorem for bounded closed sets of the Euclidean line.

Definition 1.2.27 Let (X, τ) be a topological space. It is said to be **compact** if every open covering of X has a finite subcover.

Definition 1.2.28 A subset A of the Euclidean n-space $\mathbf{R}^n$ is said to be bounded if the coordinates of every point $x = (x_1, x_2, \ldots, x_n) \in A$ are bounded in the sense that there is a real number $r > 0$ such that

$$|x_i| \leq r, \ \forall i = 1, 2, \ldots, n.$$

Example 1.2.29 **(i)** Every closed interval $[a, b]$ in the Euclidean line $\mathbf{R}$ is compact.
(ii) Every closed and bounded subspace of the Euclidean n-space $\mathbf{R}^n$ is compact.
(iii) No open interval (a, b) in the Euclidean line $\mathbf{R}$ is compact.

Theorem 1.2.30 (Heine–Borel) *A subset A of the Euclidean n-space* $\mathbf{R}^n$ *is compact iff it is closed and bounded.*

Example 1.2.31 Consider the unit 2-sphere S^2 in Euclidean space $\mathbf{R}^3$ and the real projective plane $\mathbf{R}P^2$.

(i) The unit 2-sphere $S^2 = \{(x_1, x_2, x_3) \in \mathbf{R}^3 : x_1^2 + x_2^2 + x_3^2 = 1\}$ is compact in $\mathbf{R}^3$, since $|x_i| \leq 1, \ \forall i = 1, 2, 3$ and S^2 is the set of all zeros of the continuous function, which is a polynomial:

$$f(x_1, x_2, x_3) = x_1^2 + x_2^2 + x_3^2 - 1$$

and hence S^2 is the inverse image $f^{-1}(0)$ of the closed set $\{0\}$ in the Euclidean line $\mathbf{R}$. Again, as S^2 is bounded, S^2 is compact in $\mathbf{R}^3$.

(ii) The real projective space $\mathbf{R}P^2$ is compact. Consider S^2 in $\mathbf{R}^3$ and the partition of S^2 into subsets having exactly two points which are antipodal points. Then $\mathbf{R}P^2$ is the corresponding identification space with identification map

$$p : S^2 \to \mathbf{R}P^2, x \mapsto [x] = \{x, -x\},$$

which is continuous. This shows that $\mathbf{R}P^2$ is the image of the compact space S^2 under the continuous map p and hence $\mathbf{R}P^2$ is compact.

Definition 1.2.32 A topological space X is said to be **Hausdorff** if every pair of distinct points of X have disjoint nbds.

Definition 1.2.33 A topological space X is said to be **normal** if for every pair of disjoint closed sets A and B, there is a continuous function

$$f : X \to [0, 1] \textit{ such that } f(x) = 0, \forall x \in A, \ f(x) = 1, \forall x \in B, \textit{ and } 0 \leq f(x) \leq 1, \ \forall x \in X,$$

i.e., if there exists a continuous function f such that

$$f : X \to \mathbf{R}. x \mapsto \begin{cases} 0, & \text{for all } x \in A \\ 1, & \text{for all } x \in B \end{cases}$$

and

$$0 \leq f(x) \leq 1 \text{ for all } x \in X.$$

1.2.7 Axioms of Countability and Regularity of a Topological Space

Definition 1.2.34 Let (X, τ) be a topological space. Then it is said to satisfy

(i) the **first axiom of countability**, if every point of X admits a countable open base in (X, τ);
(ii) the **second axiom of countability**, if (X, τ) has a countable open base.

Definition 1.2.35 Let (X, τ) be a topological space. Then it is said to be a **regular space** if for every nbd U of an arbitrary point $p \in X$, there exists an nbd V of the same point p such that $\overline{V} \subset U$.

1.2.8 Urysohn Lemma

Lemma 1.2.36 (Urysohn Lemma) *Let* $(\mathbf{R}, \sigma)$ *be the real line space. A topological space* (X, τ) *is normal if and only if every pair of disjoint closed sets* P, Q *in* (X, τ) *are separated by a continuous real-valued function*

$$f : (X, \tau) \to (\mathbf{R}, \sigma)$$

such that

$$f(x) = \begin{cases} 0, & \textit{for all } x \in P \\ 1, & \textit{for all } x \in Q \end{cases}$$

and

$$0 \leq f(x) \leq 1 \textit{ for all } x \in X.$$

1.2.9 Completely Regular Space

Definition 1.2.37 A topological space (X, τ) is said to be **completely regular** if

(i) its every one-pointing set is closed in (X, τ) and
(ii) for each point $p \in X$ and each closed set $A \subset X$ not containing the point p, there exists a continuous function

$$f : X \to \mathbf{I} = [0, 1], x \mapsto \begin{cases} 0, & \text{for } x = p \\ 1, & \text{for all } x \in A. \end{cases}$$

Example 1.2.38 Every normal space is completely regular by Urysohn Lemma 1.2.36.

Definition 1.2.39 Let $\mathcal{C}$ and $\mathcal{D}$ be two open coverings of a topological space (X, τ). Then $\mathcal{C}$ is said to be a **refinement of** $\mathcal{D}$ if every element of $\mathcal{C}$ is a subset of some element of $\mathcal{D}$.

Definition 1.2.40 Let (X, τ) be a topological space. A family $\mathcal{F}$ of open covering of X is said to be **locally finite** if every point of X has an nbd which intersects, nontrivially, only a finite number of members belonging to $\mathcal{F}$ *i.e.*, if for every point $x \in X$, there is an open set $U_x \in \tau$ containing x such that the set

$$\{U \in \mathcal{F} : U \cap U_x \neq \emptyset\} \text{ is finite.}$$

Definition 1.2.41 A Hausdorff space (X, τ) is said to be **paracompact** if every open covering $\mathcal{C}$ of X has an open, locally finite refinement, i.e., for every open covering $\mathcal{C}$ of X, there is a locally finite open covering $\mathcal{D}$ such that for every open set $V \in \mathcal{D}$, there is an open set $U \in \mathcal{C}$ with the property that $V \subset U$.

Theorem 1.2.42 *Let X be a normal space and $\mathcal{U} = \{U_i\}_{i \in \mathbf{A}}$ be a locally finite open covering of X. Then $\mathcal{U}$ has a shrinking in the sense that there is an open covering $\mathcal{V} = \{V_i\}_{i \in \mathbf{A}}$ such that*

$$\overline{V}_i \subset U_i, \ \forall i \in \mathbf{A},$$

where $\overline{V}_i$ denotes the closure of V_i.

Theorem 1.2.43 *Every paracompact Hausdorff space is normal.*

Definition 1.2.44 (*Partition of Unity*) Let X be a topological space with open covering $\mathcal{U}$. A partition of unity subordinate to $\mathcal{U}$ is a family $\{\psi_i\}_{i \in \mathbf{A}}$ of continuous maps

$$\psi_i : X \to [0, 1]$$

such that

(i) the set of $i \in \mathbf{A}$ for which $\psi_i(x) > 0$, $\forall x \in X$ is finite and
(ii) $\Sigma_{i \in \mathbf{A}} \psi_i(x) = 1$, $\forall x \in X$.

Theorem 1.2.45 *Let X be a topological space. Then it is paracompact iff its every open covering $\mathcal{U}$ has a partition of unity subordinate to $\mathcal{U}$.*

Theorem 1.2.46 *Every metric space is paracompact.*

Theorem 1.2.47 (Space-filling Curve Theorem) *There is a surjective continuous map $\psi : \mathbf{I} \rightarrow \mathbf{I}^2$.*

Remark 1.2.48 For the application of space-filling curve, see Chap. 3 of this volume.

Theorem 1.2.49 (Fubini) *Let X be a compact set in $\mathbf{R}^n$ such that every subset $X_t = X \cap (t \times \mathbf{R}^{n-1})$ has measure zero in the hyperplane $\mathbf{R}^{n-1}$. Then the Lebesgue measure $\mu(X) = 0$ on $\mathbf{R}^n$.*

Proof See Chap. 5, Basic Topology, Volume 1. ❑

1.3 Some Basic Results on Analysis

Let f be a smooth map defined on an open set U in $\mathbf{R}^n$ into $\mathbf{R}^m$. Given a point $x \in U$ and a vector $v \in \mathbf{R}^n$, the derivative of $f : U \rightarrow \mathbf{R}^m$ at the point x in the direction v, denoted by $df_x(v)$, is defined by

$$df_x(v) = lim_{\ t \rightarrow 0} \frac{f(x + tv) - f(x)}{t}.$$

Consider df_x as a map

$$df_x : \mathbf{R}^n \rightarrow \mathbf{R}^m, \ v \mapsto df_x(v).$$

Remark 1.3.1 df_x is also abbreviated as Df_x or $Df(x)$.

Definition 1.3.2 The map $f : U \rightarrow \mathbf{R}^m$ is called C^1 if

(i) $Df(x)$ exists for every $x \in U$ and
(ii) $Df : U \rightarrow \mathcal{L}(\mathbf{R}^m, \mathbf{R}^n), x \mapsto Df_x$ is continuous, where $\mathcal{L}(\mathbf{R}^n, \mathbf{R}^m)$ denotes the vector space of all linear transformations from $\mathbf{R}^n$ into $\mathbf{R}^m$.

Definition 1.3.3 Let $U \subset \mathbf{R}^n$ be an open set and $f : U \to \mathbf{R}^m$ be a C^1 map. By recursion, f is said to be a C^r for $2 \leq r < \infty$ if the map

$$Df : U \to \mathbf{R}^{mn} \text{ is } C^{r-1}.$$

It is said to be C^∞ if it is C^r for all r. On the other hand, for $f = (f_1, f_2, \dots, f_m) : U \to \mathbf{R}$, the map f is said to be (real) analytic or C^ω on U, if in some nbd of every point of U, it can be expanded in a convergent power series having a positive radius of convergent.

Derivative of a Linear Map and Jacobian Matrix

This subsection gives a method to differentiate a map $f = (f_1, f_2, \dots, f_m) : \mathbf{R}^n \to \mathbf{R}^m$, where

$$f_i : \mathbf{R}^n \to \mathbf{R},\ i = 1, 2, \dots, n$$

and

$$f(x) = (f_1(x), f_2(x), \dots, f_m(x)) \in \mathbf{R}^m,\ \forall x \in \mathbf{R}^n.$$

Theorem 1.3.4 *Let $U \subset \mathbf{R}^n$ be an open set and $f : U \to \mathbf{R}^m$ be a C^∞ map. Then the partial derivatives $\partial f_j / \partial x_i$ exist. The matrix representation of the linear map Df is given by $m \times n$; the matrix*

$$J(f) = \begin{pmatrix} \frac{\partial f_1}{\partial x_1} & \cdots & \frac{\partial f_1}{\partial x_n} \\ \cdots & \cdots & \cdots \\ \frac{\partial f_m}{\partial x_1} & \cdots & \frac{\partial f_m}{\partial x_n} \end{pmatrix}$$

of the first derivatives of the mapping f is called the ***Jacobian matrix of*** *f, denoted by $J(f)$, where every partial derivative is evaluated at $x = (x_1, x_2, \dots, x_n) \in U$. The Jacobian matrix is written as $\frac{\partial f}{\partial x}|_x$.*

Example 1.3.5 Let $f : \mathbf{R}^2 \to \mathbf{R}^3$, $(x, y) \to (x^3, x^2y, x^4y)$. Then $f = (f_1, f_2, f_3)$, where

$$f_1(x, y) = x^3,\ f_2(x, y) = x^2y,\ f_3(x, y) = x^4y.$$

Then

$$J(f) = \begin{pmatrix} \frac{\partial f_1}{\partial x} & \frac{\partial f_1}{\partial y} \\ \frac{\partial f_2}{\partial x} & \frac{\partial f_2}{\partial y} \\ \frac{\partial f_3}{\partial x} & \frac{\partial f_3}{\partial y} \end{pmatrix} = \begin{pmatrix} 3x^2 & 0 \\ 2xy & x^2 \\ 4x^3y & x^4 \end{pmatrix}.$$

Theorem 1.3.6 (Inverse Function Theorem) *Let $U \subset \mathbf{R}^n$ be an open set and $f : U \to \mathbf{R}^n$ be a differentiable map. If p is a point in U such that the Jacobian determinant $|J(f)|$ of f is nonzero at p, then there exists an nbd V of p and an nbd W of $f(p)$ such that*

(i) *f sends V homeomorphically on W and*
(ii) *the inverse of f from W onto V is also a differentiable map.*

Theorem 1.3.7 (Implicit Function Theorem) *Let $U \subset \mathbf{R}^n$ be an open set and $V \subset \mathbf{R}^m$ be an open set and $(x, y) \in \mathbf{R}^n \times \mathbf{R}^m \approx \mathbf{R}^{n+m}$. If $(x_0, y_0) \in U \times V \subset \mathbf{R}^{n+m}$ and $f : U \times V \to \mathbf{R}^m$ is a C^1 map such that $f(x_0, y_0) = 0$ and $det(\partial f/\partial y)$ is nonzero at the point (x_0, y_0), then there exists an open nbd $W \subset U$ of the point x_0 and a map $h : W \to V$ such that*

(i) $h(x_0) = y_0$,
(ii) $f(x, h(x)) = 0, \ \forall x \in W$ *and*
(iii) *the map h is unique.*

Moreover, if $(\partial f/\partial x)$ represents the matrix A and $(\partial f/\partial y)$ represents the matrix B, then h is a C^1 map with $(\partial h/\partial x) = -B^{-1}A$.

1.4 Category and Functor

This section conveys the introductory concepts of **category theory**, which unifies many basic concepts and results of mathematics in an accessible way. A category is a certain collection of mathematical objects (possibly with an additional structure) and morphisms which are like mappings agreeing with this structure. The objects may be a set, group, ring, vector space, module, a sequence of abelian groups, topological space, etc., and morphisms are collections of mapping preserving this structure. A natural transformation is a certain function from one functor to the other one satisfying some specific properties. These concepts together with their dual concepts form the foundation of category theory, which provides a convenient language to unify several mathematical results. This language born through the work of S. Eilenberg (1913–1998) and S. MacLane (1900–2005) during 1942–1945 is used throughout the present book. This section conveys the introductory concepts of category, functor and natural transformation in the language of category theory.

1.4.1 Introductory Concept of Category

Definition 1.4.1 **A category** $\mathcal{C}$ consists of

(i) a certain collection of objects $X, Y, Z, \ldots$ usually denoted by $ob(C)$;
(ii) for every ordered pair of objects $X, Y \in ob(C)$, a set of morphisms from X to Y, denoted by $[X, Y]$, is specified;
(iii) for every ordered triple of objects $X, Y, Z \in ob(C)$ and any pair of morphisms $f \in [X, Y], g \in [Y, Z]$, their composition, denoted by $g \circ f \in [X, Z]$, is defined by the following properties:
(iv) (associativity): if $f \in [X, Y],\ g \in [Y, Z],\ h \in [Z, W]$, then $h \circ (g \circ f) = (h \circ g) \circ f \in [X, W]$;
(v) (existence of identity morphism): for every object $Y \in ob(C)$, there is a morphism 1_Y such that for every $f \in [X, Y],\ g \in [Y, Z]$, the equality $1_Y \circ f = f$ and $g \circ 1_Y = g$ holds.

Remark 1.4.2 The identity morphism in a category is unique for each of its objects.

Example 1.4.3 **(i)** The collection of all sets and their mappings form category, denoted by $\mathcal{S}et$, where objects are all sets, morphisms are all possible mappings between them and the composition is the usual composition of functions.
(ii) The collection of all groups and their homomorphisms forms category, denoted by $\mathcal{G}rp$, where objects are all groups, morphisms are all possible homomorphisms between them and the composition is the usual composition of mappings.
(iii) The collection of all abelian groups and their homomorphisms forms category, denoted by $\mathcal{A}b$, where objects are all abelian groups, morphisms are all possible homomorphisms between them and the composition is the usual composition of mappings.
(iv) The collection of all rings and their homomorphisms form category, denoted by $\mathcal{R}ing$.
(v) The collection of all R-modules and their R-homomorphisms form category, denoted by $\mathcal{M}od_R$.
(vi) All morphisms may not be map: For example, let $(X, \leq)$ be a partial order set. There is an associated category $\mathcal{C}_X$ whose objects are all elements of X and the set $[x, y]$ of morphisms from x to y is either $\emptyset$ or the singleton set according to $x \not\leq y$ or $x \leq y$.
(vii) The collection of all vector spaces and their linear transformations for a category, denoted by $\mathcal{V}ect$.
(viii) The collection of all topological spaces and their continuous maps forms a category, denoted by $\mathcal{T}op$.

The concepts of bijective mappings of sets, isomorphism of groups, rings or vector spaces and so on can be unified through the concept of an equivalence in category theory.

Definition 1.4.4 A morphism $f \in [X, Y]$ in a category $\mathcal{C}$ is said to be an equivalence if there is a morphism $g \in [Y, X]$ in the category $\mathcal{C}$ such that

$$f \circ g = 1_Y, \ g \circ f = 1_X.$$

If f is an equivalence in $\mathcal{C}$, then g is also an equivalence in $\mathcal{C}$ and the objects X and Y are said to be equivalent.

Example 1.4.5 The equivalences and equivalent objects in the following categories are specified:

(i) In the category $\mathcal{S}et$, equivalences are bijective mappings and equivalent objects are precisely the sets having the same cardinality;
(ii) in the category $\mathcal{G}rp$, equivalences are group isomorphisms and equivalent objects are precisely isomorphic groups.
(iii) in the category $\mathcal{R}ing$, equivalences are ring isomorphisms and equivalent objects are precisely isomorphic rings.
(iv) in the category $\mathcal{V}ect_F$ over F, equivalences are vector space isomorphisms over F, and equivalent objects in the category of finite-dimensional vector spaces over F are precisely the vector spaces over F having the same dimension.
(v) in the category $\mathcal{T}op$, equivalences are homeomorphisms and equivalent objects are precisely homeomorphic topological spaces.

Remark 1.4.6 Important Examples of the categories from a topological viewpoint are available in **Basic Topology, Volume 3**.

1.4.2 Introductory Concept of Functors

A functor is a natural mapping from one category to the other in the sense that it preserves the identity morphism and composites of well-defined morphisms. It plays a key role in converting a problem of one category to the problem of other category to have a better chance for a solution.

Definition 1.4.7 Given two categories $\mathcal{C}_1$ and $\mathcal{C}_2$, **a covariant functor**

$$F : \mathcal{C}_1 \to \mathcal{C}_2, X \to F(X), f \mapsto F(f)$$

consists of

(i) an object function which assigns to every object $X \in \mathcal{C}_1$ an object $F(X) \in \mathcal{C}_2$ and
(ii) a morphism function which assigns to every morphism $f \in [X, Y]$ in the category $\mathcal{C}_1$, a morphism $F(f) \in [F(X), F(Y)]$ in the category $\mathcal{C}_2$ such that

(a) $F(1_X) = 1_{F(X)}$ for every identity morphism 1_X;
(b) for a pair of morphisms $f \in [X, Y], g \in [Y, Z]$ in the category $\mathcal{C}_1$, the equality $F(g \circ f) = F(g) \circ F(f)$ holds in the category $\mathcal{C}_2$.

A contravariant functor is defined dually.

Definition 1.4.8 Given two categories $\mathcal{C}_1$ and $\mathcal{C}_2$, **a contravariant functor**

$$F : \mathcal{C}_1 \to \mathcal{C}_2, X \to F(X), f \mapsto F(f)$$

consists of

(i) an object function which assigns to every object $X \in \mathcal{C}_1$ an object $F(X) \in \mathcal{C}_2$ and
(ii) a morphism function which assigns to every morphism $f \in [X, Y]$ in the category $\mathcal{C}_1$, a morphism $F(f) \in [F(Y), F(X)]$ in the category $\mathcal{C}_2$ such that

(a) $F(1_X) = 1_{F(X)}$;
(b) for morphisms $f \in [X, Y], g \in [Y, Z]$ in the category $\mathcal{C}_1$, the equality $F(g \circ f) = F(f) \circ F(g)$ holds in the category $\mathcal{C}_2$.

Example 1.4.9 **(i)** There is a covariant functor $F : \mathcal{G}rp \to \mathcal{S}et$ whose object function assigns to every group G its underlying set $|G|$ and to every group homomorphism $f : G \to K$ in the category $\mathcal{G}rp$ to its corresponding underlying set function $f : |G| \to |K|$ in the category $\mathcal{S}et$.
(ii) Given an object $A \in \mathcal{C}$, there is a covariant functor $h_C : \mathcal{C} \to \mathcal{C}$ whose object function assigns to every object $X \in \mathcal{C}$ the object $h_C(X) = [C, X]$ (the set of all morphisms from the object C to the object X in the category $\mathcal{C}$) and the morphism function assigns to every morphism $f \in [X, Y]$ in $\mathcal{C}$, the morphism $h_C(f) : h_C(X) \to h_C(Y), g \mapsto f \circ g$ in the category $\mathcal{C}$.
(iii) Given an object $A \in \mathcal{C}$, there is a contravariant functor $h^C : \mathcal{C} \to \mathcal{C}$ whose object function assigns to every object $X \in \mathcal{C}$ the object $h^C(X) = [X, C]$ (the set of all morphisms from the object X to the object C in the category $\mathcal{C}$) and the morphism function assigns to every morphism $f \in [X, Y]$ in $\mathcal{C}$, the morphism $h^C(f) : h^C(Y) \to h^C(X), g \mapsto g \circ f$ in the category $\mathcal{C}$.

Remark 1.4.10 Important Examples of the functors from a topological viewpoint are available in **Basic Topology, Volume 3**.

1.5 Euclidean Spaces and Some Standard Notations

In mathematical problems, subspaces of an n-dimensional Euclidean space arise frequently. Such spaces are used both in theory and application of topology. Some standard notations used throughout the book are given.

$\emptyset$	: empty set
$\mathbf{Z}$	: ring of integers (or set of integers)
$\mathbf{Z}_n$	: ring of integers modulo n
$\mathbf{R}$	: field of real numbers
$\mathbf{C}$	: field of complex numbers
$\mathbf{Q}$	: field of rational numbers
$\mathbf{H}$	: division ring of quaternions
$\mathbf{R}^n$	: Euclidean n-space, with $\|x\| = \sqrt{\sum_{i=1}^n x_i^2}$ and $\langle x, y\rangle = \sum_{i=1}^n x_i y_i$ for $x = (x_1, x_2, \ldots, x_n)$ and $y = (y_1, y_2, \ldots, y_n) \in \mathbf{R}^n$.
$\mathbf{C}^n$	: complex n-space
$\mathbf{I}$	: $[0, 1]$
$\dot{\mathbf{I}}$	: $\{0, 1\} \subset \mathbf{I}$
$\mathbf{I}^n$	: n-cube=$\{x \in \mathbf{R}^n : 0 \leq x_i \leq 1$ for $1 \leq i \leq n\}$ for $x = (x_1, x_2, \ldots, x_n)$
$\mathbf{D}^n$	: n-disk or n-ball=$\{x \in \mathbf{R}^n : \|x\| \leq 1\}$
S^n	: n-sphere=$\{x \in \mathbf{R}^{n+1} : \|x\| = 1\} = \partial\, \mathbf{D}^{n+1}$(the boundary of the $(n+1)$-disk D^{n+1})
$\mathbf{R}P^n$	: real projective space =quotient space of S^n with x and $-x$ identified for all $x \in S^n$
$\mathbf{C}P^n$	: complex projective space = space of all complex lines through the origin in the complex space $\mathbf{C}^{n+1}$
$\bigsqcup$	: disjoint union of sets or spaces
$\times, \Pi$	: product of sets, groups, modules, or spaces
$\cong$	: isomorphism
$\approx$	: homeomorphism
iff	: if and only if

$X \subset Y$or $Y \supset X$: set-theoretic containment (not necessarily proper).

References

Adhikari, M.R.: Basic Algebraic Topology and its Applications. Springer, India (2016)

Adhikari, M.R.: Basic Topology, Volume 3: Algebraic Topology and Topology of Fiber Bundles. Springer, India (2022)

Adhikari, M.R., Adhikari, A.: Groups. Rings and Modules with Applications. Universities Press, Hyderabad (2003)

Adhikari, M.R., Adhikari, A.: Textbook of Linear Algebra: An Introduction to Modern Algebra. Allied Publishers, New Delhi (2006)

Adhikari, M.R., Adhikari, A.: Basic Modern Algebra with Applications. Springer, New Delhi, New York, Heidelberg (2014)

Adhikari, A., Adhikari, M.R.: Basic Topology, Volume 1: Metric Spaces and General Topology. Springer, India (2022)

Alexandrov, P.S.: Introduction to Set Theory and General Topology. Moscow (1979)

Borisovich, Y.U., Bliznyakov, N., Izrailevich, Y.A., Fomenko, T.: Introduction to topology. Mir Publishers, Moscow (1985)
Bredon, G.E.: Topology and Geometry. Springer, New York (1983)
Dugundji, J.: Topology. Allyn & Bacon, Newton, MA (1966)
Hu, S.T.: Introduction to: General Topology. Holden-Day Inc, San Francisco (1969)
MacLane, S.: Categories for the Working Mathematician. Springer, New York (1971)
Simmons, G.: Introduction to Topology and Modern Analysis. McGraw Hill, New York (1963)
Williard, S.: General topology. Addision-Wesley, Boston (1970)

Chapter 2
Topological Groups and Topological Vector Spaces

This chapter studies certain topological-algebraic structures such as topological groups and topological vector spaces. The book **Basic Topology, Volume 1** of the present series of books studies the general properties of topological spaces and their continuous maps. But this chapter studies the topological spaces with other structures (algebraic) compatible with the given topological structures. For example, the circle group S^1 in the complex plane **C**, the 3-spheres S^3 (group of unit quaternions), the general linear group $GL(n, \mathbf{R})$, $GL(n, \mathbf{C})$, etc., admit a natural group structure under usual multiplication such that their usual topological and algebraic group structures are compatible in the sense that the corresponding group operations are continuous. It asserts that the concept of a topological group is precisely that concept in which the algebraic and topological structures are united and interrelated. This phenomenon leads to the concept of topological groups.

Lie groups named after Sophus Lie (1842–1899) arising in mathematics through the study of continuous transformations provide a rich supply of topological groups (see Chap. 4) which needs a discussion on topological groups. On the other hand, a topological vector space (linear space) is a vector space endowed with a topology such that the scalar multiplication and addition are both continuous. Topological vector spaces provide a rich supply of topological groups by Proposition 2.10.5. Most of the interesting concepts that arise in analysis are studied through vector spaces (real or complex), instead of single object such as a function, measure or operator. Topological vector spaces provide specified topological and algebraic structures for the study of functional analysis. It is interesting that the topology on topological vector space is localized in the sense that this topology can be determined by the family of open nbds of its origin 0.

Historically, the concept of topological groups arose through the study of groups of continuous transformations such as geometric transformations and these groups form an important class of topological manifolds (see Chap. 3). For example, if a Euclidean plane is endowed with a group of continuous transformations, then the

A. Adhikari and M. R. Adhikari, *Basic Topology 2*,
https://doi.org/10.1007/978-981-16-6577-6_2

group is transformed into a topological group. One of the basic concepts of the theory of topological groups is the concept of Lie groups (studied in Chap. 4).

This chapter presents introductory concepts of topological groups and topological vector spaces together with topological rings and fields having special structures. These special structures make their study interesting and display an interplay between modern algebra (including matrix algebra) and topology. For example, a topological group G is an algebraic group and there is a topology τ on G such that these two structures are linked together by the condition that the group operations (multiplication and inversion) are both continuous with respect to this topology τ (see Definition 2.1.1), which are said that group structure and topological structure on a topological group are compatible. Theorem 2.4.11 proves that every topological group can be viewed as a group of homeomorphisms of a topological space, which is analogous to Cayley's theorem in group theory. Topological rings, fields and vector spaces are defined in an analogous way. A link between a topological group and a topological vector space is given in Definition 2.10.4 (by groups of linear operators) and also in Proposition 2.10.5.

Various interesting applications of topological groups and topological vector spaces are also available in Sects. 2.12 and 2.12.9. The general theory of topological groups is well-connected with many families of topological groups such a topological groups of matrices and Lie groups, which form versatile important classes of topological groups. Classical matrix groups are studied in this chapter. On the other hand, Lie groups are studied in Chap. 4. They link topology with geometry, algebra and analysis. This chapter also studies topological group actions which are used to construct many geometrical objects (see Sect. 2.12.4). More study on topological groups is available in subsequent chapters.

For this chapter, the books Adhikari (2016, 2022); Adhikari and Adhikari (2003, 2006, 2014, 2022); Armstrong (1993), Bredon (1993), Chevalley (1946), Kelly and Namioka (1963); Morris and Wulbert (1967), Pontragin (1946); Rudin (1973), Simmons (1963), Sorani (1969), Zariski (1969) and some others are referred in Bibliography.

2.1 Topological Groups: Introductory Concepts

This section addresses the concept of topological groups and develops its theory by studying the properties of topological groups derived from the existence in these groups some topology. It provides important examples for the study of algebra, analysis, topology and geometry. From the logical view-point, this concept is a simple combination of abstract group and topological space interlinked by the continuity of group operations. This section also studies topological group actions which are used to construct many geometrical objects (see Sect. 2.12.4).

2.1.1 Basic Definitions and Examples

This subsection communicates some basic concepts related to topological groups with illustrative examples. Topological group theory relates algebra with geometry and analysis. The concept of topological groups was born through an exercise to connect two branches of mathematics such as abstract group theory and topology. This unification is an outcome of a great influence of Lie groups and direct consequence of various types of transformation groups. The basic concept of a topological group is that it is an abstract group endowed with a topology such that the multiplication and inverse operations are both continuous. This concept was accepted by mathematicians in the early 1930s.

Definition 2.1.1 A nonempty set G is said to be a **topological group**, if it satisfies the following axioms:

TG(1): G is algebraically a group;

TG(2): G is topologically a Hausdorff space, i.e., every pair of distinct points of G are strongly separated by disjoint open sets;

TG(3): The group multiplication

$$m : G \times G \to G, (x, y) \mapsto xy$$

is continuous, where the topology on $G \times G$ is endowed with product topology;

TG(4): The group inversion

$$v : G \to G, x \to x^{-1}$$

is continuous.

Remark 2.1.2 Some authors do not assume "Hausdorff property" for defining a topological group. Theorem 2.7.2 is proved without the assumption of Hausdorff axiom of a topological group.

Remark 2.1.3 The conditions **TG(3)** & **TG(4)** in Definition 2.1.1 are equivalent to the single condition

TG(5): the map

$$\psi : G \times G \to G, (x, y) \mapsto xy^{-1}$$

is continuous. In the language of nbds, it is formulated as follows:

(i) for any two points $x, y \in G$, every nbd W of xy, there exists a nbd U_x of x and a nbd U_y of y such that

$$U_x U_y \subset W, \text{ where } U_x U_y = \{ab : a \in U_x \text{ and } b \in U_y\};$$

(ii) given a point $x \in G$, for every nbd V of x^{-1}, there exists a nbd U of x such that $U^{-1} \subset V$.

Equivalently, given two points $x, y \in G$, for every nbd W of the element xy^{-1}, there exist one nbd U of x and another nbd V of y such that $UV^{-1} \subset W$.

Example 2.1.4 **(i)** Every (algebraic) group endowed with discrete topology is a topological group;

(ii) The real line space $\mathbf{R}$ endowed with the Euclidean topology induced by the Euclidean metric

$$d : \mathbf{R} \times \mathbf{R} \to \mathbf{R}, \ (x, y) \mapsto |x - y|$$

is a topological group under usual addition of real numbers, because it is algebraically a group and the usual operations of addition and inversion

$$f : \mathbf{R} \times \mathbf{R} \to \mathbf{R}, \ (x, y) \mapsto x + y$$

and

$$v : \mathbf{R} \to \mathbf{R}, \ x \mapsto -x$$

are both continuous;

(iii) $\mathbf{R}^n$ is a topological group under usual pointwise addition of real numbers and with topology induced by Euclidean metric on $\mathbf{R}^n$;

(iv) The punctured real line space $X = \mathbf{R} - \{0\}$ is a topological group under usual multiplication of real numbers and the topology on X induced by the usual Euclidean topology of $\mathbf{R}$;

(v) $\mathbf{C}^n$ is a topological group under usual addition of complex numbers and the usual product topology on $\mathbf{C}^n$;

(vi) Consider the unit circle $S^1 = \{z \in \mathbf{C} : |z| = 1\}$ in the complex plane $\mathbf{C}$ with the subspace topology on S^1 inherited from the usual topology on $\mathbf{C}$. Then it is a topological group with the group multiplication

$$m : S^1 \times S^1 \to S^1, \ (e^{i\theta}, e^{i\beta}) \mapsto e^{i(\theta+\beta)}$$

and inversion

$$v : S^1 \to S^1, \ e^{i\theta} \mapsto e^{-i\theta}.$$

Remark 2.1.5 Continuity conditions **TG(3)** & **TG(4)** in Definition 2.1.1 are essential to relate the group structure on G with the topological structure on G, because G may not be a topological group for an arbitrary topological structure on G (see Example 2.1.6).

Example 2.1.6 Every (algebraic) group endowed with nontrivial topology is not a topological group. For example, consider the set $G = \{x_1, x_2, x_3\}$ consisting of three distinct elements x_1, x_2 and x_3. Then G is algebraically a group under the multiplication defined by the table

$\cdot$	x_1	x_2	x_3
x_1	x_1	x_2	x_3
x_2	x_2	x_3	x_1
x_3	x_3	x_1	x_2

Define a topology τ on G by taking the collection of open sets

$$\tau = \{\emptyset, U_1 = \{x_1\}, U_2 = \{x_2\}, U_3 = \{x_1, x_2\}, G\}.$$

Consider the group inversion map $\upsilon : G \to G, x \mapsto x^{-1}$. Then

$$\upsilon(x_1) = x_1, \upsilon(x_2) = x_3, \upsilon(x_3) = x_2.$$

This shows that $\upsilon^{-1}(U_2) = \{x_3\}$, which is not an open set in the topology τ and hence the inversion map υ cannot be continuous. This asserts that G is algebraically a group but it is not a topological group with respect to the given group structure and the above topology τ on G.

For better understanding, as this group G is isomorphic to the group $(\mathbf{Z}_3, +)$, consider G as the additive group $\mathbf{Z}_3 = \{0, 1, 2\}$ and the topology

$$\tau = \{\emptyset, U_1 = \{0\}, U_2 = \{1\}, U_3 = \{0, 1\}, \mathbf{Z}_3\}.$$

Then $(\mathbf{Z}_3, +)$ is an algebraic group but not a topological group under the topology τ. But it is a topological group under discrete (or indiscrete) topology.

Remark 2.1.7 Example 2.1.6 shows that making an algebraic group G to be an interesting topological group, depends on both algebraic and topological structures (restricted) on G, though any algebraic group is a topological group under the discrete and indiscrete topologies.

Example 2.1.8 The additive algebraic group $(\mathbf{Z}, +)$ is a topological group under discrete topology but it not so for nontrivial topology on $\mathbf{Z}$. For example, $(\mathbf{Z}, +)$ is not a topological group under a nontrivial topology on $\mathbf{Z}$ defined in Proposition 2.12.43 (discussed later on).

Proposition 2.1.9 *If G is a topological group, then the following statements are equivalent on the underlying topological space of G*

(i) *G is a T_0 space;*
(ii) *G is a T_1 space;*
(iii) *G is a Hausdorff space.*

Proof It follows from Definition 2.3.1. ❑

Definition 2.1.10 Let G be a topological group and H be a nonempty subset of G. Then H is said to be a **subgroup** of the topological group G, if

(i) H is a subgroup of the abstract group G and
(ii) H is a closed subset of the underlying topological space of the topological group G, i.e., $\overline{H} = H$.

Remark 2.1.11 For the study of topological subgroups and normal subgroups; see Sect. 2.3.

2.1.2 Homomorphism, Isomorphism and Automorphism for Topological Groups

This section generalizes the standard concepts related to homomorphisms of abstract groups. In the theory of topological groups, two topological groups are said to be same if they have the same topological structures (up to homeomorphism) and the same algebraic structure (up to isomorphism).

Definition 2.1.12 Let G and H be two topological groups. A map $f : G \to H$ is said to be

(i) a **homomorphism**, if f is a homomorphism of abstract groups from G into H and f is also a continuous map from the topological space G into the topological space H;
(ii) an **isomorphism**, if f is an isomorphism of abstract groups from G onto H and f is also a homeomorphism from the topological space G onto the topological space H;
(iii) an **automorphism**, if f is an isomorphism of the topological group G onto itself;
(iv) a **monomorphism** if f is a homomorphism of topological groups and the map f is also injective;
(v) an **epimorphism**, if f is a homomorphism of topological groups and the map f is also surjective.

Remark 2.1.13 A topological group homomorphism between two topological groups is a continuous map of topological spaces such that it is also a homomorphism of abstract groups. A topological group isomorphism between two topological groups is a homeomorphism between topological spaces and it is also an isomorphism between abstract groups. Two topological groups are said to be **equivalent** if there exists a topological group isomorphism between them.

Example 2.1.14 The additive topological group $(\mathbf{R}, +)$ with usual topology and the multiplicative topological group $(\mathbf{R}^+, \cdot)$ of positive reals with topology inherited from the usual topology of $\mathbf{R}$, are isomorphic topological groups by a topological group isomorphism

$$f : (\mathbf{R}, +) \to (\mathbf{R}^+, \cdot),\ t \mapsto e^t.$$

So, we shall not distinguish the topological groups $(\mathbf{R}, +)$ and $(\mathbf{R}^+, \cdot)$, which are considered to be the same as the topological groups.

Example 2.1.15 Two isomorphic topological groups are algebraically isomorphic but its converse is not true, i.e., two topological groups may be isomorphic as abstract groups but they may not be isomorphic as topological groups.

(i) Consider the algebraic groups $G = \{x_1, x_2, x_3\}$ and $\mathbf{Z}_3 = \{0, 1, 2\}$ given in Example 2.1.6. The group (G, λ) endowed with indiscrete topology λ is a topological group and the group $(\mathbf{Z}_3, \mu)$ endowed with discrete topology μ is also a topological group. These two topological groups (G, λ) and $(\mathbf{Z}_3, \mu)$ are not topologically isomorphic but they are algebraically isomorphic, since

$$f : G \to \mathbf{Z}_3,\ x_1,\ x_2,\ x_3 \mapsto 0,\ 1,\ 2$$

is an algebraic isomorphism but it is not continuous.

(ii) The groups $(\mathbf{R}, +)$ and $(\mathbf{R}^2, +)$ are isomorphic, since both of them are vector space over the field $\mathbf{Q}$ having the same dimension. On the other hand, they are not homeomorphic under usual topology, since deletion of a single point from $\mathbf{R}$ makes $\mathbf{R}$ disconnected, but by deletion of a single point from $\mathbf{R}^2$ keeps $\mathbf{R}^2$ connected.

Example 2.1.16 Two non-isomorphic algebraic groups may be homeomorphic by making them topological groups. This asserts that preservation of topological structure under a homeomorphism does not guarantee preservation of algebraic structure by an isomorphism. For example, consider the set $M(3, \mathbf{R})$ of all 3×3 real matrices. It can be identified with the Euclidean space $\mathbf{R}^{3^2}$. Consider the topological groups

$(\mathbf{R}^3, +)$ and $G = \{\begin{pmatrix} 1 & a & b \\ 0 & 1 & c \\ 0 & 0 & 1 \end{pmatrix} : a, b, c \in \mathbf{R}\}$ with usual matrix multiplication.

The group G is nonabelian. On the other hand, the group $\mathbf{R}^3$ is abelian. Hence, these two algebraic groups cannot be isomorphic. But considering them as subspaces of $\mathbf{R}^{3^2}$ with Euclidean topology, their underlying spaces are homeomorphic.

2.2 Local Isomorphism and Local Characterization of Topological Groups

The discussion on topological groups in Sect. 2.1 shows that there is a close connection between algebraic and topological operations on a topological group G. In many situations, the study of a topological group G in a nbd of its identity element provides valuable information on the whole topological group G. This section studies local properties of topological groups in the sense that those properties which are true for all locally isomorphic groups.

For the study of local properties of a topological group G with identity e, we need to study only the behavior of G in an arbitrary small nbd U of e. So, a natural question arises: is it possible to study the nbd U independently without reference to the entire group G? The classical theory of Lie groups studied in Chap. 4 gives its answer. But the Proposition 2.6.9 prescribes a method for constructing a topological group locally isomorphic to a given topological group. On the other hand, Proposition 2.2.16 gives a sufficient condition on a family of subsets $\mathbf{U}$ of an abstract group G under which G becomes a topological group, where the topology τ is determined by $\mathbf{U}$.

2.2.1 Neighborhood System of the Identity Element of Topological Group

The abstract concept of a group G defines a topology on G in Proposition 2.2.16, by providing a sufficient condition on a family of subsets $\mathbf{U}$ of the abstract group G under which G becomes a topological group, where the topology τ is determined by $\mathbf{U}$. To study a topology on G, it is sufficient in many occasions to describe a complete system (family) of nbds of its identity element. For example, Theorem 2.5.4 proves that a topological group G with its identity element e is locally compact if there exists a compact nbd of e.

Definition 2.2.1 A topological group G is said to be **discrete** if it contains no limit point in the sense that every point of G has a nbd containing only a single point.

Remark 2.2.2 An abstract group G, endowing with its underlying space $|G|$, the discrete topological space is a discrete topological group. Proposition 2.2.3 characterizes discrete topological groups.

Proposition 2.2.3 *A topological group G with its identity element e is discrete if e is an isolated element of G.*

Proof Since e is an isolated element of G, it has a nbd which contains no point other than e, i.e., it is not a limit point of any nbd of e. Hence, the proposition follows from Definition 2.2.1. ❑

2.2.2 Local Properties and Local Isomorphism

This subsection studies special properties (called local properties) of topological groups that are determined by the behavior of the topological groups in the nbd of its identity element and coveys the concept of local isomorphism to define locally isomorphic topological groups.

Definition 2.2.4 introduces the concept of a local isomorphism for topological groups, which plays an important role in the study of topological group theory.

Definition 2.2.4 Let G be a topological group with identity element e_G and H be a topological group with identity element e_H. A **local isomorphism**

$$f : G \to H$$

is a homeomorphism from a nbd U of e_G in G onto a nbd V of e_H in H such that

(i) if $x, y \in U$ and $xy \in U$, then $f(xy) = f(x)f(y)$ and
(ii) if $h, t \in V$ and $ht \in V$, then $f^{-1}(ht) = f^{-1}(h)f^{-1}(t)$.

The topological groups G and H are said to be **locally isomorphic** if there exists a local isomorphism
$$f : G \to H.$$

Example 2.2.5 If two topological groups are isomorphic, then they are also locally isomorphic. But its converse is not necessarily true. For example, the additive group $(\mathbf{R}, +)$ and the multiplicative group $(S^1, \cdot)$ are locally isomorphic topological groups, but they are not isomorphic topological groups.

Proposition 2.2.6 *Let G be a topological group with identity element e_G and H be a topological group with identity element e_H. If*

$$f : G \to H$$

is a local isomorphism of a nbd U of e_G in G onto a nbd V of e_H, then

(i) $f(e_G) = e_H$;
(ii) *if $x, x^{-1} \in U$, then $f(x^{-1}) = f(x)^{-1}$.*

Proof **(i)** Since $e_G e_G = e_G$, it follows $f(e_G) = f(e_G)f(e_G)$, and hence $f(e_G) = e_H$;
(ii) Since $xx^{-1} = e_G \in U$, it follows that $e_H = f(e_G) = f(x)f(x^{-1})$. This implies that $f(x^{-1}) = f(x)^{-1}$.

❑

Proposition 2.2.7 *Local isomorphism relation between topological groups is an equivalence relation on the set of all topological groups.*

Proof It follows from Definition 2.2.4. ❑

2.2.3 *Local Characterization of Topological Groups*

This subsection studies local characterization of a topological group G. Its motivation comes from the fact that corresponding to every element $a \in G$, there exist two associated homeomorphisms (see Proposition 2.2.9)

$$L_a : G \to G, x \mapsto ax$$

and

$$R_a : G \to G, x \mapsto xa.$$

Definition 2.2.8 Given a topological group G and an element $a \in G$

(i) the map $L_a : G \to G,\ x \mapsto ax$ is called a **left translation** (map) by a and
(ii) the map $R_a : G \to G,\ x \mapsto xa$ is called a **right translation** (map) by a.

Proposition 2.2.9 *Let G be a topological group with identity element e. Then for every element $a \in G$, the left and right translation maps*

(i)

$$L_a : G \to G,\ x \mapsto ax;$$

(ii)

$$R_a : G \to G,\ x \mapsto xa,$$

are both homeomorphisms.

Proof Let G be a topological group and $a \in G$ be an arbitrary element.

(i) Consider the left translation map L_a. Since G is a topological group, the group operation is continuous and hence the map

$$L_a : G \to G, x \mapsto ax,\ \forall x \in G$$

is continuous.
Similarly, $L_{a^{-1}}$ is continuous. So, L_a and $L_{a^{-1}}$ are two continuous maps such that

$$L_a \circ L_{a^{-1}} = I_G = L_{a^{-1}} \circ L_a.$$

This asserts that L_a is a homeomorphism with $L_{a^{-1}}$ as its inverse.
(ii) For R_a, it is proved in a similar way.

❑

Corollary 2.2.10 *Let G be a topological group. If A is an open subset of G and B is any subset of G, then BA is also an open subset of G.*

Proof By hypothesis, A is an open set in G and by Proposition 2.2.9, for every $b \in B$, translation L_b is a homeomorphism. Since $BA = \bigcup_{b \in B} bA$, it follows that bA being the homeomorphic image of the open set A under L_b is open and hence BA is union of open sets in G. This asserts that BA is an open set in G. ❑

Corollary 2.2.11 *Let G be a topological group. If A is an open subset of G and B is any subset of G, then AB is also an open subset of G.*

Proof Proceed as in Corollary 2.2.10. ❑

Our next aim is to define a topology τ on an abstract group G, by means of nbds to make (G, τ) a topological group (see Proposition 2.2.16).

Definition 2.2.12 Let (X, τ) be a topological space. A collection $\mathcal{B}$ of open sets $\{U\}$ in X is said to be an **open base** for the topology τ if every open set in X can be expressed as a union of some members of $\mathcal{B}$. A basis $\mathcal{B}$ of (X, τ) is said to be a **complete system (complete family) of nbds** of the topological space X, if every open set of $\mathcal{B}$ is a nbd of every point contained in this open set.

Definition 2.2.13 gives an equivalent formulation of open base for a topology.

Definition 2.2.13 Let (X, τ) be a topological space. A collection of open sets $\mathcal{B}_x$, each member of which contains a point x in X, is called an **open base (or a local base)** for τ at the point x, if for every open set U containing x, there exists an open set $V_x \in \mathcal{B}_x$, such that $x \in V_x \subset U$.

Remark 2.2.14 Definition 2.2.12 asserts that to determine a topological space, it is sufficient to specify one of its open base. On the other hand, the nbds of a topological space (X, τ) is not completely determined by the closure operation in X, but they depend on the choice of the basis $\mathbf{B}$. So, speaking of nbds in our subsequent study, we assume that some specific basis has been selected. Moreover, Proposition 2.2.9 says that if the complete family open sets $\{U\}$ in X of nbds of the identity element e of a topological group G is known, then the complete family of nbds of any point can be determined.

Remark 2.2.15 Let G be a topological group with identity element e. The family $\mathbf{U} = \{U\}$ of nbds e in G, cannot be taken arbitrarily for forming a complete family of nbds of any point. They satisfy certain conditions called the **conditions (A)**, such as

(i) the intersection of any two sets belonging to $\mathbf{U}$ is in $\mathbf{U}$;
(ii) the intersection of all sets belonging to $\mathbf{U}$ is the set $\{e\}$;
(iii) any set containing a set belonging to $\mathbf{U}$ is in $\mathbf{U}$;
(iv) If $U \in \mathbf{U}$, there exists a set belonging to $V \in \mathbf{U}$ such that $VV \subset U$. (It follows from the continuity of multiplication in the topological group G at the point (e, e));
(v) the family of sets $\mathbf{V}$ such that $V^{-1} \in \mathbf{U}$ for every member $V \in \mathbf{V}$, then $\mathbf{V} = \mathbf{U}$. (It follows from the continuity of inversion map in the topological group G);
(vi) for any $g_0 \in G$, the family of sets $\{g_0 U g_0^{-1} : U \in \mathbf{U}\}$ coincides with $\mathbf{U}$. Because the set of all nbds of g_0 coincides, the collection of the sets of the form $\{g_0 U : U \in \mathbf{U}\}$ and hence every set of the form $\{U g_0 : U \in \mathbf{U}\}$ can also be represented in the form $\{g_0 U' : U' \in \mathbf{U}\}$ and conversely.

Proposition 2.2.16 gives a sufficient condition on a family of subsets $\mathbf{U}$ of an abstract group G under which G becomes a topological group, where the topology τ is determined by $\mathbf{U}$.

Proposition 2.2.16 *Let G be an abstract group with identity element e and $\mathbf{U}$ be a family of subsets satisfying the conditions (A) given in Remark 2.2.15. Then there exists a topology τ on G such that*

(i) *(G, τ) forms a topological group and*
(ii) *the given family $\mathbf{U}$ of subsets of G forms the complete nbd system of e under the topology τ.*

Proof Consider the family $\mathbf{V}$ of subsets V of G satisfying the condition if $g \in V$, then the set V contains some set of the form $gU : U \in \mathbf{U}$. Then any union of sets belonging to $\mathbf{V}$ is also in $\mathbf{V}$. Moreover, $\emptyset \in \mathbf{V}$ and $G \in \mathbf{V}$. Then it follows that the family $\mathbf{U}$ of subsets of G satisfies all the conditions of Remark 2.2.15. Hence, it determines a topology τ on G such that $\mathbf{U}$ forms the family of open sets with the family of nbds of a point $g \in G$ is the family of sets $\{gU : U \in \mathbf{U}\}$ in the topology τ. To show that (G, τ) forms a topological group, consider the map

$$\psi : G \times G \to G, (x, y) \mapsto xy^{-1}.$$

Let $(x_0, y_0) \in G \times G$ and $x_0 y_0^{-1} U$ be a nbd of the point $x_0 y_0^{-1}$. Then $U \in \mathbf{U}$. Let $U_1 \in \mathbf{U}$ be a set such that $U_1 U_1 \subset U$. Then the set $x_0 U_1^{-1}$ can be represented in the form $U_2 x_0^{-1} : U_2 \in \mathbf{U}$ by the Remark 2.2.15. This implies that

$$(x_0 U_2)(U_1 y_0)^{-1} = x_0 U_2 y_0^{-1} U_1 \subset x_0 y_0^{-1} U_1 U_1 \subset x_0 y_0^{-1} V.$$

This proves ψ is continuous and hence it is proved that (G, τ) is a topological group. ❑

Proposition 2.2.17 *Given a topological group G and elements $a, k \in G$, the homeomorphisms*

(i) $L_a : G \to G,\ x \mapsto ax$;
(ii) $R_a : G \to G,\ x \mapsto xa$;
(iii) $L_k : G \to G,\ x \mapsto kx$; *and*
(iv) $R_k : G \to G,\ x \mapsto xk$

satisfy the relations

$$R_a \circ R_k = R_{ka}$$

and

$$L_a \circ L_k = L_{ak}.$$

Proof It follows from the definitions of L_a and R_a for any point $a \in G$. ❑

Proposition 2.2.18 *Let G be a topological group. Then for every element $a \in G$, the inner automorphism*

$$\phi_a : G \to G,\ x \mapsto axa^{-1}$$

is a homeomorphism.

Proof The maps $L_a : G \to G,\ x \mapsto ax$ and $R_{a^{-1}} : G \to G\ x \mapsto xa^{-1}$ are both homeomorphisms. Hence, $\phi_a = R_{a^{-1}} \circ L_a : G \to G, x \mapsto axa^{-1}$ is a homeomorphism as it is the composite of two homeomorphisms. ❑

Proposition 2.2.19 *Let G be a topological group. Then the inversion map*

$$\upsilon : G \to G,\ x \mapsto x^{-1}$$

is a homeomorphism.

Proof As G is a topological group, υ is a continuous map such that

$$(\upsilon \circ \upsilon)(x) = \upsilon(\upsilon(x)) = \upsilon(x^{-1}) = x,\ \forall x \in G \implies \upsilon \circ \upsilon = 1_G \text{ (identity map on } G)$$

shows that υ is a bijection with self inverse. Consequently, the continuity of υ and its inverse $\upsilon^{-1}(= \upsilon)$ imply that υ is a homeomorphism. ❑

Definition 2.2.20 A topological space G is said to be **homogeneous** if for every pair of points $x, y \in G$, there exists a homeomorphism

$$\psi : G \to G \text{ such that } \psi(x) = y,$$

i.e., if ψ transforms x to y for any two elements $x, y \in G$.

Example 2.2.21 Every topological group is homogeneous as a topological space. Let G be a topological group. Since the left and right translations L_x and R_y are both homeomorphisms, for arbitrary pair of points $x, y \in G$, it follows that every topological group is homogeneous as a topological space. More precisely, for any two points $x, y \in G$, the homeomorphism

$$\psi = L_{yx^{-1}} : G \to G,\ x \mapsto (yx^{-1})x$$

is such that $\psi(x) = y$. Similar results hold for right translation. This implies that these translations have a certain **homogeneity property** as a topological space.

Remark 2.2.22 Let G be a topological group with the identity element e. Since left and right translations L_x and R_y are both homeomorphisms of G onto itself, for any two given points $x, y \in G$, the homeomorphism $L_{yx^{-1}} : G \to G$ maps x to y. This shows that for every $x \in G$, it is possible to move any $y \in G$ anywhere in G without changing its topological structure. This asserts that it is sufficient to consider the nbd system only at the identity point e.

Theorem 2.2.23 (or Corollary 2.2.24) asserts that if we know an open base $\mathcal{U}$ at the identity element e of a topological group G, then we can completely find open bases for the given topology on G at any point $g \in G$ by applying the concepts of left or right translations to the sets belonging in $\mathcal{U}$.

Theorem 2.2.23 *Let G be a topological group with identity element e and $\mathcal{U}$ be an open base at the point e. Then*

$$\mathcal{U}_l = \{gU : g \in G,\ U \in \mathcal{U}\}$$

and

$$\mathcal{U}_r = \{Ug : g \in G,\ U \in \mathcal{U}\}$$

are both open bases for the given topology on G.

Proof Let $g \in G$ and V be an open set in G, containing g. Since $L_{g^{-1}} : G \to G,\ x \mapsto g^{-1}x$ is a homeomorphism, it sends the open set V into the open set $g^{-1}V$. By hypothesis, $\mathcal{U}$ is an open base at the point e. Hence, there exists an open set $U \in \mathcal{U}$ such that $e \in U \subset g^{-1}V$. This shows that $g \in gU \subset V$. Hence, it follows that $\mathcal{U}_l$ constitutes an open base for the given topology on G. Similarly, it follows that $\mathcal{U}_r$ also constitutes an open base for the given topology on G. ❑

Corollary 2.2.24 *Let G be a topological group with identity e and $\mathcal{U}$ be an open base at the point e. Then the families of open sets*

$$g\mathcal{U} = \{gU : U \in \mathcal{U}\},\ \text{and}\ \mathcal{U}g = \{Ug : U \in \mathcal{U}\}$$

both constitute local bases at any point $g \in G$.

Proof It follows from the proof of Theorem 2.2.23. ❑

2.3 Topological Subgroups and Normal Subgroups

This section introduces the concepts of topological subgroups and topological normal subgroups of a topological group and studies them.

Definition 2.3.1 Let G be a topological group and H be a nonempty subset of G. Then H is said to be a **subgroup** of the topological group G, if

(i) H is a subgroup of the abstract group G; and
(ii) H is a closed subset of the underlying topological space of the topological group G, i.e., $\overline{H} = H$.

Remark 2.3.2 Definition 2.3.1 asserts that if H is a subgroup of a topological group G, then

(i) the topology of H is the subspace topology inherited from the topological group G and

(ii) H is not only a subgroup of the abstract group G, one additional topological condition is imposed on H which is its topological closureness property in the sense that $\overline{H} = H$.

Example 2.3.3 **(i)** $S^1 = \{z \in \mathbf{C} : |z| = 1\}$ is a topological subgroup of the topological group $\mathbf{C}^* = \mathbf{C} - \{0\}$ under usual multiplication of complex numbers. The topological group S^1 is called the **circle group** in the complex plane.

(ii) The set $\mathbf{Z}$ endowed with the discrete topology forms a topological subgroup of the topological group $(\mathbf{R}, +)$.

Remark 2.3.4 The concepts of isomorphism and subgroups for topological groups needs consideration of both the topological and algebraic structures. For example:

(i) a nonempty subset of a topological group is a subgroup if it is algebraically a subgroup as well as it has the subspace topology;

(ii) an isomorphism between two topological groups is a homeomorphism as well as it is a group isomorphism.

Proposition 2.3.5 *Let G be a topological group and H be a subgroup of G. Then H is itself a topological group.*

Proof To prove that H is a topological group, it is sufficient to show that the restriction of the continuous group operation of G on H is also continuous. Given two elements $x, y \in G$, let $xy^{-1} = z$. Then every nbd V of the point $z \in H$ in the topological space H is obtained as $V = H \cap W$ for some nbd W of the same point z in the topological space G. Since G is a topological group and $x, y \in G$, there exist a nbd U_x containing x and another nbd U_y containing y such that $U_x U_y^{-1} \subset W$. This asserts that $V_x = H \cap U_x$ and $V_y = H \cap U_y$ are also nbds of x and y under the subspace topology of H inherited from the topology of G. This shows that $V_x V_y^{-1} \subset W$ and also $V_x V_y^{-1} \subset H$, and hence H is a topological group. ❑

Definition 2.3.6 Let G be a topological group and N be a subgroup (topological) of G. Then N is said to be a normal subgroup of the topological group G, if N is a **normal subgroup** of the abstract group G.

Remark 2.3.7 Definition 2.3.6 asserts that if N is a normal subgroup of a topological group G, then

(i) the topology of N is the subspace topology inherited from the topological group G and

(ii) N is not only a normal subgroup of the abstract group G, but also one additional topological condition is imposed on N which is its topological closureness property in the sense that $\overline{N} = N$.

Remark 2.3.8 Let G be a topological group. If H is a subgroup of G or N is a topological subgroup of G, then G is not only an abstract group, but also the underlying topological space of the topological group G, that imposes one additional condition, which is the condition of closureness of H and N.

Proposition 2.3.9 *Let G be a topological group*

(i) *If H is a subgroup of the abstract group G, then $\overline{H}$ is a subgroup of the topological group G.*
(ii) *If N is a normal subgroup of the abstract group G, then $\overline{N}$ is a normal subgroup of the topological group G.*

Proof **(i)** Let H be a subgroup of the abstract group G and $x, y \in \overline{H}$ be any two points. Then $xy^{-1} \in \overline{H}$. To show it, let V be an open nbd of the point xy^{-1}. Then there exists an open nbd U_x of x and an open nbd U_y of y such that $U_x U_y^{-1} \subset V$. Since $x, y \in \overline{H}$, there exist points $a, b \in H$ such that $a \in U_x$ and $b \in U_y$ such that $ab^{-1} \in H$ and also $ab^{-1} \in V$. Hence, it follows that $xy^{-1} \in \overline{H}$. This asserts that $\overline{H}$ is a subgroup of the abstract group G and it is also a subgroup of the topological group G.

(ii) Let N be a normal subgroup of the abstract group G, and $x \in \overline{N}$ and $g \in G$ be arbitrary points. If W is an arbitrary open nbd of the point $g^{-1}xg$, then there exist an open nbd U of x such that $g^{-1}Ug \subset W$. Since, by hypothesis, $x \in \overline{N}$, there exists a point $n \in N$ such that $n \in U$. Moreover, $g^{-1}ng \in N$ and $g^{-1}xg \in W$ assert that $g^{-1}xg \in \overline{V}$, because an arbitrary open nbd W of the point $g^{-1}xg$ intersects N. This proves that $\overline{N}$ is a normal subgroup of the topological group G.

❑

Remark 2.3.10 For an alternative proof of Proposition 2.3.9; see proof of Proposition 2.3.29.

2.3.1 *Discrete Normal Subgroups*

This subsection studies **discrete normal subgroups**, which play an important role in the study of topological groups. For example, they define simple topological group. Let G be any topological group with its identity element e. It has two trivial normal subgroups such as the $\{e\}$ and the group G itself.

Definition 2.3.11 A normal subgroup of a topological group G with its identity element e is said to be discrete if e is an isolated element of G.

Definition 2.3.12 A topological group G is said to be **simple** if every normal subgroup of G is either discrete or coincides with G.

Example 2.3.13 Consider the topological group $(\mathbf{R}, +)$ under usual addition of real numbers and endowed with usual topology (see Example 2.1.4). Let $A \neq \mathbf{R}$ be a subgroup of the topological group $\mathbf{R}$. Then there exists a least positive integer k in A and A consists of all multiples of k. This implies that A is a discrete subgroup of the topological group $\mathbf{R}$. Hence, it follows by Definition 2.3.12 that $\mathbf{R}$ is a simple topological group. But $\mathbf{R}$ is not a simple abstract group from the view-point of abstract algebra.

Remark 2.3.14 A deep study of subgroups of the topological group $(\mathbf{R}, +)$ with an emphasis on its dense subgroups is available in Sect. 2.12.5.

2.3.2 *Some Topological Properties of Groups, Subgroups and Normal Subgroups*

This subsection proves some properties of subgroups of topological groups.

Theorem 2.3.15 *Let G be a topological group with identity element e. Then the connected component of e is a closed normal subgroup of G.*

Proof Let X be the connected component of e. Then X is the largest connected subset containing e, of G and hence it follows that $\overline{X} = X$. It implies that X is closed. To show that it is a normal subgroup of G, take any point $a \in X$. Consider the translation

$$L_{a^{-1}} : G \to G,\ g \mapsto a^{-1}g,$$

which is a homeomorphism by Proposition 2.2.9. Since X is connected and La^{-1} is a homeomorphism, it follows that its image $La^{-1}(X) = a^{-1}X$ is also connected. As $a^{-1}X$ is a connected subset containing $a^{-1}a = e$, and X is the maximal connected subset containing e in G, it follows that $a^{-1}X \subset X$ and hence $X^{-1}X = X$. This asserts that X is a subgroup of G. To show that X is a normal subgroup of G, take an element $g \in G$. Then the set $gXg^{-1} = (R_{g^{-1}} \circ L_g)(X)$ contains e and is also connected since it is the homeomorphic image of the connected set X. By the above argument, it follows that $gXg^{-1} \subset X$, which asserts that X is a normal subgroup of G. ❑

Corollary 2.3.16 *Let G be a topological group with identity element e and C_e be the connected component of e. Then the other connected components of G are cosets of C_e in G.*

Proof For any connected component C_g of $g \neq e \in G$, it follows that

$$L_g(C_e) = gC_e \subset C_g, \textit{and } L_{g^{-1}}(C_g) = g^{-1}C_g \subset C_e \implies C_g \subset gC_e,$$

since both gC_g and $g^{-1}(C_g)$ are connected, and by hypothesis, C_g is the connected component of g. This asserts that $C_g = gC_e$. ❑

Definition 2.3.17 Let (X, τ) be a topological space. It is said to be **locally connected** if every nbd of any point $x \in X$ contains a connected nbd of the point x.

Proposition 2.3.18 *Let (X, τ) be a locally connected space. Then every connected component of an open set in X is also an open set.*

Proof Let A be a connected component of an open set U in X. If $a \in A$, then U is an open nbd of a and hence U contains a connected nbd V of a. By hypothesis, A is a connected component and hence $V \cap A \neq \emptyset$. This implies that $V \subset A$ and a is an interior point of A. This proves that A is an open set. ❑

Remark 2.3.19 Any nbd of the identity element of a connected topological group G is a set of generators of the group G. It is proved in Theorem 2.12.30 .

The open base or basis at a point of a topological space given by open sets in Definition 2.2.13 is redefined by open nbds in Definition 2.3.20.

Definition 2.3.20 Let (X, τ) be a topological space and $x \in X$ be an arbitrary point. By a **nbd basis or a local basis** of the topological space (X, τ) at the point x, it is meant a family $\mathcal{F}_x$ of nbds of x in X such that every nbd of x contains a member of the family $\mathcal{F}_x$. Each member of this family is called the **basic nbd** of x in (X, τ).

Definition 2.3.21 Let G be a topological group and $X, Y \subset G$ be two given subsets. Then XY is the set defined by

$$XY = \{xy : x \in X, y \in Y\}$$

and X^{-1} is the set defined by

$$X^{-1} = \{x^{-1} : x \in X\}.$$

The subset X is said to be **symmetric** if $X = X^{-1}$.

Remark 2.3.22 Let G be a topological group.

(i) If X is a symmetric subset of G, then $X \cap X^{-1}$ and XX^{-1} are both symmetric;
(ii) If $X, Y \subset G$ are two symmetric subsets, then $X \cap Y$ is also symmetric.

Definition 2.3.23 Let (X, τ) be a topological space and $\mathcal{N}_x$ be the set of all open nbds of x $(\in X)$. Then a subset $\mathcal{U}_x$ of $\mathcal{N}_x$ is said to be a **fundamental system of nbds** of x or a base for τ if for an $N_x \in \mathcal{N}_x$, there is some $U_x \in \mathcal{U}_x$ such that $U_x \subset N_x$.

Proposition 2.3.24 *Let G be a topological group with its identity element e. Then there is a fundamental system of symmetric nbds of e.*

Proof Let $\mathcal{U} = \{U_i\}$ be an arbitrary fundamental system of nbds of e. To prove the proposition, it is sufficient to show that there is a symmetric subset of $\mathcal{U}$, which satisfies the conditions of Definition 2.3.23. Suppose $U \in \mathcal{U}$. Then U is open in G and U^{-1} is also an open nbd of e in G by Proposition 2.2.19. Hence, U and U^{-1} are both open nbds of e such that $V = U \cap U^{-1}$ is symmetric and open. Then the family $\mathcal{V} = \{V\}$ of such V forms the required fundamental system of symmetric nbds of e. ❑

Proposition 2.3.25 *Let G be a topological group with identity element e. Given any point $g \in G$ and any nbd N of g in G, there is a symmetric nbd S_y of e in G such that $S_y g S_y^{-1} \subset N$.*

Proof It follows from the continuity of multiplication of the topological group G. Because for continuous group multiplication

$$m : G \times G \to G$$

and any nbd N of g, there exist nbds U and V of g in G such that

$$U \times V \subset m^{-1}(N).$$

This implies that the nbd N of g contains the nbd $X = UV$ of g in G. Hence, it follows that the set $S_y = X \cap X^{-1}$ forms a symmetric nbd of e in G satisfying the required property. ❑

Corollary 2.3.26 *Let G be a topological group with identity element e. Given any nbd N of e in G, there is a symmetric nbd S_y of e in G such that $S_y S_y = S_y^2 \subset N$.*

Proof It follows from Proposition 2.3.25 as a particular case. ❑

Corollary 2.3.27 *Let G be a topological group with identity element e. Given any nbd N of e in G and any positive integer n, there is a symmetric nbd S_y of e in G such that $S_y^n \subset N$, where S^n is the product of n copies of the symmetric nbd S_y of e in G.*

Remark 2.3.28 Proposition 2.3.25 is applied to prove a basic result on topological quotient groups embodied in Theorem 2.6.5 An alternative proof of Proposition 2.3.9 is given in proof of Proposition 2.3.29.

Proposition 2.3.29 *Let G be a topological group and H be a subgroup of G. Then*

(i) *$\overline{H}$ is also a subgroup of G;*
(ii) *if H is normal in G, then $\overline{H}$ is also a normal subgroup of G.*

Proof **(i)** Since for arbitrary elements $x, y \in \overline{H}$. the product element $xy \in \overline{H}$ and the inverse element $x^{-1} \in \overline{H}$, it follows that $\overline{H}$ is also a subgroup of G;

(ii) By hypothesis, H is a normal subgroup of G. Then $gHg^{-1} \subset H$, $\forall g \in G$. This asserts that $g\overline{H}g^{-1} \subset \overline{H}$, $\forall g \in G$. Similarly, for any $g^{-1} \in G$, the opposite inclusion also holds. Hence, it follows that $\overline{H}$ is a normal subgroup of G.

❑

Proposition 2.3.30 *Let G be a topological group with identity element e and V be a nbd of e in G. Then there exists a symmetric nbd U of e with the property*

$$UU^{-1} = UU \subset V.$$

Proof Since the multiplication $m : G \times G \to G$ is continuous by hypothesis, it follows that there exist nbds W and T of e such that

$$WT \subset U.$$

If $M = W \cap T$, then M^{-1} is also a nbd of e and hence $U = M \cap M^{-1}$. Consider the nbd

$$U = M \cap M^{-1}.$$

It follows that U is a symmetric nbd of e such that

$$UU^{-1} = UU \subset MM \subset WT \subset V.$$

❑

Proposition 2.3.31 *Let G and K be two topological groups and $f : G \to K$ be a homomorphism such that f is continuous at a point $g_0 \in G$. Then f is continuous at all points $g \in G$.*

Proof Let $g \in G$ be an arbitrary point and U be a nbd of $f(g)$ in K. Then by left translation the image $L_{g_0 g^{-1}}(U)$ is a nbd of $f(g)$. Hence, there exists a V of g_0 such that

$$f(V) \subset L_{g_0 g^{-1}}(U),$$

which implies that $L_{gg_0^{-1}}(V)$ is a nbd of g such that

$$f(L_{gg_0^{-1}}(V)) \subset U.$$

❑

2.4 Topological Group Action and Transformation Group

This section studies topological group actions on topological spaces to study many geometrical objects such as real and complex projective spaces, torus, lens spaces, etc., as orbit spaces obtained by specifying actions of topological groups (see

Sect. 2.12.4). The concept of topological group action is used to prove Theorem 2.4.11, which asserts that every topological group can be viewed as a group of homeomorphisms of a topological space, **which is analogous to Cayley's theorem in group theory.**

Definition 2.4.1 Let G be a topological group with identity element e, and X be a topological space. If $G \times X$ has the product topology, then G **is said to act on** X **from the left** if there is continuous map

$$\mu : G \times X \to X,$$

the image $\mu(g, x)$, denoted by $g(x)$ or gx such that

(i) $e(x) = x$ for all $x \in X$;
(ii) $(gk)(x) = g(k(x))$, for all $x \in X$, and $g, k \in G$.

The group G or the pair (G, X) is then called a **topological transformation group** of X relative to the group action $\mu : G \times X \to X$. It is sometimes denoted by the triple (G, X, μ). A topological space X endowed with a left G-action on X is said to be a **left** G**-space.**

A **right action and a right** G**-space** are defined in an analogous way.

Remark 2.4.2 There is a one-to-one correspondence between the left and right G-structures on X. So it is sufficient to study only one of them according to the situation.

Example 2.4.3 For the general linear group $G(n, \mathbf{R})$, the Euclidean n-space $\mathbf{R}^n$ is a left $G(n, \mathbf{R})$-space under usual multiplication of matrices.

Definition 2.4.4 Let X be a left G-space. Two given elements x, y in X are called G-equivalent, if $y = g(x)(i.e., y = gx)$ for some $g \in G$. The relation of being G-equivalent is an equivalence relation on X and the corresponding quotient space $X mod\ G$ endowed with quotient topology induced from X (i.e., the largest topology such that the projection map

$$p : X \to X mod\ G, x \mapsto G(x)$$

is continuous, where $G(x) = \{g(x) : g \in G\}$ is called the **orbit space of** $x \in G$.

For an element $x \in X$, $G(x)$ is called the orbit of x and the subgroup G_x of G defined by

$$G_x = \{g \in G : g(x) = x\}$$

is called the **stabilizer or isotropy group at** x of the corresponding group action.

Example 2.4.5 (*Geometrical*) The actions of a given topological group on the same topological space may be different. For example, let T be the torus in $\mathbf{R}^3$ obtained by rotating the circle $C : (x - 3)^2 + z^2 = 1$ about the z-axis and $\mathbf{Z}_2 =< h >$, generated by h. Consider the three actions

(i)

$$\psi_1 : \mathbf{Z}_2 \times T \to T, \ (h, (x, y, z)) \mapsto (x, -y, -z),$$

which geometrically represents rotation of T through an angle 180° about the x-axis;

(ii)

$$\psi_2 : \mathbf{Z}_2 \times T \to T, \ (h, (x, y, z)) \mapsto (-x, -y, z),$$

which geometrically represents rotation of T through an angle 180° about the z-axis;

(iii)

$$\psi_3 : \mathbf{Z}_2 \times T \to T, \ (h, (x, y, z)) \mapsto (-x, -y, -z)$$

which geometrically represents refection of T about the origin.
Each of ψ_1, ψ_2 and ψ_3 is a homeomorphism of T of order 2 with the sphere, torus and Klein bottle as the resulting orbit spaces respectively.

Proposition 2.4.6 *Let X be a left G-space and $g \in G$ be an arbitrary point. Then the map*

$$\psi_g : X \to X, \ x \mapsto g(x) \ (= gx)$$

is a homeomorphism.

Proof ψ_g is continuous for every $g \in G$, since the group action is continuous. So, ψ_g and $\psi_{g^{-1}}$ are two continuous maps such that $\psi_g \circ \psi_{g^{-1}} = I_X = \psi_{g^{-1}} \circ \psi_g$ and hence it follows that ψ_g is a homeomorphism. ❑

Remark 2.4.7 Let X be a G-left space and **homeo(X)** be the set of homeomorphisms $\psi_g : X \to X$ for all $g \in G$. Then **homeo(X)** $= \{\psi_g : g \in G\}$ is a group under usual composition of mappings. Proposition 2.4.8 shows that this group is closely related to the group action of G on X.

Proposition 2.4.8 *Let X be a left G-space. Then the map*

$$f : G \to \mathbf{homeo(X)}, \ g \mapsto \psi_g$$

is a group homomorphism.

Proof Since for each $g \in G$, by Proposition 2.4.6, ψ_g is a homeomorphism, the map f is well defined. Again, since for $g, k \in G$, $f(gk) = \psi_{gk} = f(g)f(k)$ asserts that f is a group homomorphism. ❑

Definition 2.4.9 Let X be a left G-space and $G(x) = \{g(x) : g \in G\}$ be orbit space of $x \in G$. Then the action of G on X or the topological transformation group G is said to be:

(i) **transitive**, if given any two $x, y \in X$, there exists an element $g \in G$, such that $g(x) = y$, i.e., if there is one orbit produced by this action;

(ii) **free** if isotropy group $G_x = \{g \in G : g(x) = x\} = \{e\}$ for every $x \in X$, which means that $g(x) = x$ for some $x \in X \implies g = e$;

(iii) **effective or trivial**, if $g(x) = x$ for every $x \in X$, then $g = e$, i.e., if the homomorphism $f : G \to \textbf{homeo(X)},\ g \mapsto \psi_g$ is a monomorphism.

Geometrically, a free action of a topological group G on a topological space X asserts that every point $g(\neq e) \in G$ moves each point of X. Let G be a group of homeomorphisms of X. Define a group action

$$\sigma : G \times X \to X,\ (g, x) \mapsto g(x).$$

Every topology such as the compact open topology on **Homeo(X)** induces a subspace topology on G, because G is a subset of **Homeo(X)**. It is a natural question: what topology of G is to be taken to make the action σ continuous? A positive answer is available, if X is a locally compact Hausdorff space and G is endowed with compact open topology, then the action σ is continuous.

Proposition 2.4.10 *Let X be a G-space. If the action μ of G on X is effective, then μ induces an embedding*

$$f : G \to \textbf{homeo(X)},\ g \mapsto \psi_g.$$

Proof It follows from the defining condition of f. ❑

Theorem 2.4.11 gives an an independent proof of Proposition 2.4.10

Theorem 2.4.11 *Let G be a topological group with the identity element e and X be a topological space. If the action μ of G on X is effective, then μ induces an embedding*

$$\psi : G \to \textbf{homeo(X)},\ g \mapsto \psi_g,$$

where ψ_g is the homeomorphism given in Proposition 2.4.6.

Proof Let $\mu : G \times X \to X$ be an effective action. Then for $g \neq e$, the homeomorphism ψ_g is not the identity homeomorphism. Consider the homomorphism

$$\psi : G \to \textbf{homeo(X)},\ g \mapsto \psi_g.$$

Since for any two distinct elements $g, k \in G$, the images $\psi_g(x) \neq \psi_k(x)$ for any $x \in X$, shows that $\psi(g) = \psi_g \neq \psi_k = \psi(k)$. Hence, it follows that

$$\psi : G \to \textbf{homeo(X)},\ g \mapsto \psi_g$$

is an embedding. ❑

Proposition 2.4.12 *Let $Xmod\ G$ be the orbit space obtained by an action of the topological group G on the topological space X equipped with the quotient topology, i.e., a subset U in $Xmod\ G$ is defined to be open if $p^{-1}(U)$ is open in X, where*

$$p : X \to Xmod\ G, x \mapsto G(x) = orb(x)$$

is the natural projection map. Then p is an open map.

Proof Given an open set U in X,

$$p^{-1}(p(U)) = \bigcup_{g \in G} \{ gU : g \in G\}$$

is a union of open sets gU and hence it is an open set in X. This asserts that $p(U)$ is an open set in the orbit space $Xmod\ G$ for every open set U in X. ❑

Remark 2.4.13 The above quotient maps

$$p : X \to Xmod\ G, x \mapsto G(x) = orb\ (x)$$

are identified with the covering maps in the study of covering space. Moreover, the concept of properly discontinuous action of a topological group given in Definition 2.4.14 is important in the study of covering spaces and fundamental groups (see **Basic Topology, Volume 3** of the present series of books).

Definition 2.4.14 Given a topological space X, a topological group G with its identity e is said to **act on** X **properly discontinuously** if for every point $x \in X$, there is a nbd U such that

$$U \cap gU \neq \emptyset \implies g = e$$

Example 2.4.15 The action of any finite topological group on a Hausdorff space is properly discontinuous.

2.5 Local Compactness and its Characterization

This section studies locally compact topological groups which form an important family of topological groups. Theorem 2.5.4 gives a characterization of locally compact topological groups with the help of compact nbds of its identity element.

2.5.1 Locally Compact Topological Groups

This subsection studies locally compactness property of subgroups of a topological group. Locally compact groups are characterized in Theorem 2.5.4.

Definition 2.5.1 A topological group (X, τ) is said to be **locally compact** if for each point $x \in X$, there is an open set $U \in \tau$ such that $x \in U$ and $\overline{U}$ is compact.

Example 2.5.2 Euclidean space $\mathbf{R}^n$ (under usual addition), the matrix group $M(n, \mathbf{R})$ (under usual addition) and general linear group $\mathrm{GL}(n, \mathbf{R})$ (under usual multiplication) endowed with the Euclidean topology form important classes of locally compact topological groups (see Corollary 2.8.16 and Proposition 2.8.23).

Proposition 2.5.3 *Let G be a topological group and A be a subgroup of G*

(i) *If G is compact, then A is also compact;*
(ii) *If G is locally compact, then A is also locally compact.*

Proof To prove it, we consider the underlying topological space of the topological group G

(i) If G is compact, then A is a closed subgroup of G. Hence, A is compact, since every closed subset of a compact space is compact;
(ii) If G is locally compact and $x \in A$, then there exists an open nbd U of x in G such that $\overline{U}$ is compact. Consider the open nbd $V = A \cap U$ in A. Since A is closed in G, then $\overline{V} \subset A$. This shows that $\overline{V}^* = \overline{V} \cap A = \overline{V}$ and also $\overline{V} \subset \overline{U}$. Since every closed subset of a compact set is closed, it follows that $\overline{V}$ is compact. Hence, V is an open nbd of x in A such that $\overline{V}$ is compact. This proves the local compactness of A.

❑

2.5.2 Characterization of Locally Compact Topological Groups

Theorem 2.5.4 gives a characterization of locally compact topological groups with the help of compact nbds of its identity element.

Theorem 2.5.4 *Let G be a topological group with its identity element e. Then G is locally compact iff there exists a compact nbd of e.*

Proof First suppose that G is locally compact. Then there exists a nbd U of e such that $\overline{U}$ is compact. Conversely, suppose that U is a compact nbd of e. Then by using Exercises 53 and 54 of Sect. 2.13 , there exists a nbd W of e such that

$$W^2 \subset U \text{ and } \overline{W} \subset W^2.$$

Because in a topological group G, for any nbd U of e, there exists a symmetric nbd W of e such that $W^n \subset U$ for any positive integer n. Again, for any nbd W of e and any subset K of G, $\overline{K} \subset KW \implies \overline{W} \subset W^2$ (taking $K = W$, in particular). Hence, $\overline{W} \subset U \subset \overline{U}$. This implies that W is a compact nbd of e. For any $g \in G$, clearly gW is a nbd of g. Since the right translation $R_g : G \to G$ is a homeomorphism, it follows that $g\overline{W}$ is a compact nbd of g and hence G is locally compact. ❑

Remark 2.5.5 More results on locally compact topological groups are given in Exercises 3, 13 and 14 of Sect. 2.13.

2.6 Quotient Structure of a Topological Group

This section studies quotient structure of a topological group by extending the standard results for abstract groups in algebra to topological groups. This idea is used to construct many important topological groups.

Definition 2.6.1 Let G be a topological group and A be a subset of G. The set $G/A = \{gA : g \in G\}$ of all left cosets of A can be endowed with a topology τ induced by the natural projection $p : G \to G/A$, which is the largest topology on G/A such that

$$p : G \to G/A, \; g \mapsto gA$$

is continuous. This asserts that p is the identification map. **This identification topologyτ is called the quotient topology on G/A and the identification space $(G/A, \tau)$ is called the corresponding quotient space of G.**

Definition 2.6.2 G be a topological group and N be a normal subgroup of G. The set G/N of cosets in the abstract group G form an abstract group under usual multiplication of cosets

$$\mu : G/N \times G/N \to G/N, \; (gN, h/N) \mapsto ghN.$$

The set G/N also admits the quotient topological structure by Definition 2.6.1 such that the group operations in the set G/N are continuous with respect to this topology. This makes the set G/N a topological group by Theorem 2.6.7, called the topological **quotient group or factor group** of the topological group G by the normal subgroup N.

The concept of group of components of a topological group as a particular type of a quotient group is introduced in Definition 2.6.3, followed by its general study in this section.

Definition 2.6.3 Let G be a topological group with identity element e and A be a connected component of e. Then it is a closed normal subgroup of G by Theorem 2.3.15. The **quotient group** G/A is a topological group, called the **group of components of G**. Its elements are the cosets of A in G and they are also the components of G by Corollary 2.3.16.

Proposition 2.6.4 gives the topological structure of quotient group of a locally connected topological group.

Proposition 2.6.4 *Let G be a locally connected topological group with identity element e and A be a component of e. The quotient group is discrete.*

Proof By hypothesis, G is locally connected. Hence, there exists a connected nbd U of e. Consider the natural projection map

$$p : G \to G/A, \; g \mapsto gA.$$

Since $U \subset A$, the image $p(U)$ is a point in G/A and it is an open nbd of the identity element of the topological group G/A. Hence, the proposition follows by using Proposition 2.2.3. ❑

Theorem 2.6.5 studies quotient spaces G/A, when A is a closed subgroup of G.

Theorem 2.6.5 *Let A be a closed subgroup of a topological group G. Then the quotient space G/A of cosets is a Hausdorff space under the quotient topology τ induced by the natural projection*

$$p : G \to G/A, \; g \mapsto gA$$

such that the map p is open and continuous.

Proof Let $V \subset G$ be an arbitrary open subset of G. Then $p^{-1}(p(V)) = VA = \bigcup\{Va : a \in A\}$ is an open set in G, since it is a union of open sets Va. Again the definition of quotient topology on G/A asserts that this is the largest topology on G/A such that p is continuous and $p(V)$ is an open set in the quotient space G/A. Consequently, the map p is open and continuous. To prove that quotient space G/A is Hausdorff, take two distinct points gA, kA in G/A. Then $gA \neq kA$ and hence $g^{-1}k$ is not an element of A. This shows that the element $g^{-1}k$ is in the open set $G - A$. Hence, it follows by Proposition 2.3.25 that there exists a symmetric open nbd N_e of e with the property that $(N_e g^{-1} k N_e) \cap A = \emptyset$, since $N_e g^{-1} k N_e \subset G - A$. It shows that $g^{-1}kN_e \cap N_e A = \emptyset$. This implies that $kN_e \cap gN_e A = \emptyset$. Consequently, $kN_e A \cap gN_e A = \emptyset$. This asserts that

$$p(g) = gA \subset gN_e A, \; p(k) = kA \subset kN_e A.$$

Since $gN_e A$ and $kN_e A$ are disjoint open sets in the quotient space $(G/A, \tau)$ containing gA and kA, respectively, i.e., gA and kA are strongly separated in G/A, it follows that the space $(G/A, \tau)$ is Hausdorff. ❑

Definition 2.6.6 Let G be a topological group and A be a normal subgroup of G. Let $p : G \to G/A, g \mapsto gA$ be the natural projection map. Define a topology on the algebraic group G/A as follows: a subset U is open in G/A iff $p^{-1}(U)$ is open in G. This topology is called the **quotient topology** on G/A.

Theorem 2.6.7 *Let G be a topological group with continuous group multiplication m and A be a closed normal subgroup of G. Then the quotient space $(G/A, \tau)$ is a topological group.*

Proof By hypothesis, A is closed normal subgroup of G. Then G/A is algebraically a group such that by Proposition 2.6.5, the quotient space $(G/A, \tau)$ is Hausdorff and the projection map

$$p : G \to G/A, \ g \mapsto gA$$

is open and continuous. We claim that the group

(i) multiplication $\mu : G/A \times G/A \to G/A, \ (gA, kA) \mapsto gkA$ is continuous and
(ii) inversion map $G/A \to G/A, \ gA \mapsto g^{-1}A$ is also continuous.

Proof I: Let $g, k \in G$ and U be an open nbd of the point $p(gk)$ in the space $(G/A, \tau)$. Since the group multiplication in the topological group G is continuous, there exist open nbds U_g and U_k in G of g, k, respectively, such that

$$gk \in U_g U_k \subset p^{-1}(U).$$

Since p is an open onto map, $p(U_g)$ and $p(U_k)$ are both open nbds in $(G/A, \tau)$ of the points $p(g)$ and $p(k)$, respectively. This asserts that the group multiplication in $(G/A, \tau)$ is continuous. Similarly, the group inversion in $(G/A, \tau)$ is also continuous. This proves that $(G/A, \tau)$ is a topological group.

Proof II: To prove (i), consider the commutative rectangle as shown in Fig. 2.1, where the horizontal arrows indicate

(i) continuous group multiplications

$$m : G \times G \to G$$

and

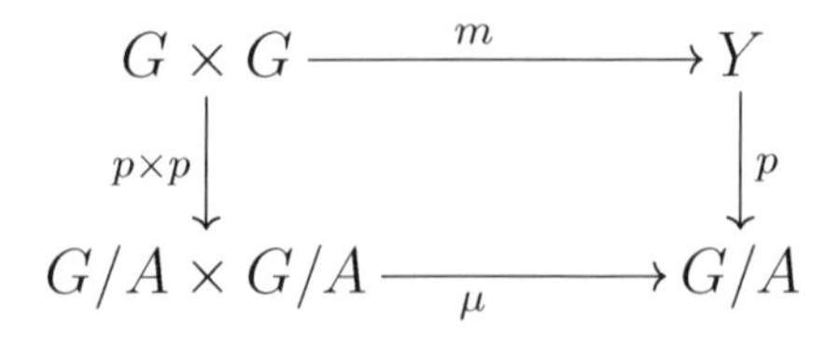

Fig. 2.1 Diagram representing multiplications m and μ

(ii) the usual multiplication $\mu : G/A \times G/A \to G/A,\ (gA, kA) \mapsto gkA.$

The two vertical arrows represent the maps p and $p \times p$. Then p is open and it is an identification map and hence $p \times p : G \times G \to G/A \times G/A$ is also an identification map. Since the map

$$\mu = p \circ m \circ (p \times p)^{-1} : G/A \times G/A \to G/A$$

is the product of three continuous maps, it is continuous. To show that the group multiplication

$$\mu : G/A \times G/A \to G/A,\ (gA, kA) \mapsto gkA$$

is continuous, it is sufficient, to prove that for any open set U in X/A, its inverse image $\mu^{-1}(U)$ in $G/A \times G/A$ is open. Equivalently, it is to be proved that the inverse image of $\mu^{-1}(U)$ under the map $p \times p$ is open in the product space $G \times G$, which is true, since the maps p and m are both continuous, showing that $m^{-1}(p^{-1}(U)$ is open in $G \times G$.
Similarly, (ii) follows by showing that the inversion map

$$\upsilon : G/A \to G/A,\ gA \mapsto g^{-1}A$$

is also continuous. ❑

Proposition 2.6.8 proves that the compactness and locally compactness properties of a quotient topological group are induced by the corresponding properties of its mother topological group.

Proposition 2.6.8 *Let G be a topological group.*

(i) *If G is compact, then its every quotient topological group is also compact.*
(ii) *If G is locally compact, then its every quotient topological group is also locally compact.*

Proof **(i)** Let G be a compact topological group and G/A be a quotient group. Consider the natural projection map

$$p : G \to G/A,\ g \mapsto gA.$$

Since p is a continuous onto map by Theorem 2.6.5 and compactness is a topological property, it follows that the topological space G/A is also compact.

(ii) Let G be a locally compact topological group with identity element e and G/A be a quotient group. Then there exists an open nbd U of e such that $\overline{U}$ is compact. Consider the natural projection map

$$p : G \to G/A,\ g \mapsto gA.$$

Then p s a continuous onto map by Theorem 2.6.5. Hence, it follows that $f(U)$ is an open set and $f(\overline{U})$ is a compact subset in the topological space G/A such that $A \subset f(U)$. By the regularity of the space G/A, there exists an open nbd V of the point A in the topological space G/A, such that $\overline{V} \subset f(\overline{U})$. Then $\overline{V}$ is also compact, since it is a closed subset of the compact set $f(\overline{U})$. This proves that the space G/A is also locally compact.

❑

Proposition 2.6.9 prescribes a method for constructing a topological group locally isomorphic to a given topological group. A generalization of this Proposition is given in Exercise 65 of Sect. 2.13.

Proposition 2.6.9 *Let G be a topological group and N be discrete normal subgroup of G. Then the topological groups G and $G/N = Q$ are locally isomorphic.*

Proof Consider the natural homomorphism

$$p : G \to G/N = Q, \; g \mapsto gN.$$

Let W be an open nbd of e_G in the topological space G such that no element of N other than e_G is in W. Let U be an open nbd in G such that

$$UU^{-1} \subset W \quad \text{and} \quad p(U) = V \subset Q.$$

Then p is an onto open and continuous map (see Chap. 1). Claim that p is a local isomorphism. To show this, let $x, y \in U$ be two elements such that $p(x) = p(y)$. Then $xy^{-1} \in N$ but $xy^{-1} \in W$ and hence $xy^{-1} = e_G$ by the choice of W. It shows that $x = y$. Consequently, it follows that f is a local isomorphism. ❑

Proposition 2.6.10 *Let G and H be two topological groups and $f : G \to H$ be an open homomorphism onto H. Then*

(i) *$ker\ f = K$ is a normal subgroup of the topological group G;*
(ii) *the topological groups H and G/K are isomorphic.*

Proof It follows from abstract group theory that $K = f^{-1}(e_H)$ is a normal subgroup of the abstract group G.

(i) Since f is continuous, K is a closed subset of the topological space X. This proves that K is a normal subgroup of the topological group G.
(ii) It follows from abstract group theory that the abstract groups H and G/K are isomorphic by an isomorphism

$$h : G/K \to H, gK \mapsto f(g).$$

The quotient space G/K is a topological group by Theorem 2.6.7. Moreover, h and h^{-1} are both continuous. Hence, it follows that h is a homeomorphism, since h is an isomorphism of abstract groups.

❑

Definition 2.6.11 Let G be a topological group and A be a closed normal subgroup of G. Then $(G/A, \tau)$ is a topological group by Theorem 2.6.7. This group is called the **topological quotient group or factor group** of G with respect to the normal subgroup A and quotient topology.

Example 2.6.12 $\mathbf{Z}$ endowed with discrete topology is a normal subgroup of the real line space $\mathbf{R}$ under usual addition. Then the factor group $\mathbf{R}/\mathbf{Z}$ with identification topology is a topological group called the topological quotient group of $\mathbf{R}$ with respect to $\mathbf{Z}$.

Proposition 2.6.13 *The topological quotient group* $\mathbf{R}/\mathbf{Z}$ *is isomorphic to the topological circle group* S^1.

Proof Consider the map

$$\psi : \mathbf{R} \to S^1, x \mapsto e^{2\pi i x}.$$

Then ψ is a group homomorphism with $ker\ \psi$ is the subgroup $\mathbf{Z}$ of $\mathbf{R}$. Moreover, it maps open sets in $\mathbf{R}$ to the open sets in S^1 and hence it is an identification map, since ψ is onto. Since any two points $x_1, x_2 \in \mathbf{R}$ are identified iff $x_1 - x_2$ in an integer, the map ψ induces a map

$$\tilde{\psi} : \mathbf{R}/\mathbf{Z} \to S^1 : x + \mathbf{Z} \mapsto \psi(x)$$

which is a homeomorphism such that it is also a group isomorphism by isomorphism theorem of groups. ❑

Geometrically, two points $x, y \in \mathbf{R}$ are identified by ψ in Proposition 2.6.13 iff $x - y$ is an integer.

Example 2.6.14 Let A be the subgroup of the topological group $\mathbf{R}^n$ (under usual addition) consisting of points having integral coordinates. Then A is a closed discrete subgroup of $\mathbf{R}^n$. The quotient space $\mathbf{R}^n/A$ is called the ***n*-dimensional torus**, denoted by $\mathbf{T}^n$. In particular, for $n = 2$, the two torus $\mathbf{T}^2$ is represented as $\mathbf{R}^2/\mathbf{Z}^2$ which is the Euclidean plane $\mathbf{R}^2$ modulo the equivalence relation

$$(x, y) \sim (u, v)$$

iff $(x - u)$ and $(y - v)$ are both integers. $\mathbf{T}^2$ is homeomorphic to the product space $S^1 \times S^1$.

2.7 Product and Direct Product of Topological Groups

This section studies the product and direct product of topological groups endowed with product topology.

2.7.1 Product of Topological Groups

This subsection addresses the product of topological groups endowed with product topology.

Definition 2.7.1 Let $\mathcal{G} = \{G_i\}_{i\in\mathbf{A}}$ be a family of topological groups . Then their direct product $G = \Pi_{i\in\mathbf{A}} G_i$ with coordinate-wise multiplication and product topology is a topological group called the **product of the topological groups** $\{G_i\}$. Then each projection map

$$p_k : G \to G_k,\ x = (x_1, x_2, \dots, x_k, \dots,) \mapsto x_k,\ \forall\, k \in \mathbf{A}$$

is continuous.

Theorem 2.7.2 does not assume the Hausdorff property of topological groups.

Theorem 2.7.2 *Let $\mathcal{G} = \{G_i\}_{i\in\mathbf{A}}$ be a family of topological groups with $e_i \in G_i$ identity elements and $G = \Pi_{i\in\mathbf{A}} G_i$ be their topological product group with product topology. Then*

- **(i)** *G is Hausdorff iff every G_i is Hausdorff ;*
- **(ii)** *G is compact iff every G_i is compact;*
- **(iii)** *if all G_i are locally compact, but all G_i excepting only a finite number of G_i are compact, then G is locally compact.*

Proof $G = \Pi_{i\in\mathbf{A}} G_i$ be the topological product group with identity element $e = (e_1, e_2, \dots)$, where $e_i \in G_i$ is the identity element.

(i) Suppose that every G_i space is Hausdorff and $x \in G$ is a point such that $x \neq e$. Then for this, $x = (x_1, x_2, \dots,)$. There is some $k \in \mathbf{A}$ such that $x_k \neq e_k$. Consequently, there exists some open set U_k in G_k containing e_k such that $x_k \notin U_k$. Consider the open set in G

$$p_k^{-1}(U_k) = G_1 \times G_2 \times \cdots \times G_{k-1} \times U_k \times G_{k+1} \times \cdots$$

containing the point e but it does not contain the point x. Hence, G is a T_0-space. As G is a topological group, it follows from Proposition 2.1.9 that G is Hausdorff. Conversely, suppose that G is Hausdorff. Then for any two distinct elements $x, y \in G$, there exist open sets U and V such that $x \in U$ and $y \in V$ and $U \cap V = \emptyset$. Consider the topological group G_k for any $k \in \mathbf{A}$ and the distinct points $x_k, y_k \in G_k$. Let $a = (z_1, z_2, \dots, z_{k-1}, x_k, z_{k+1}, \dots)$ and $b = (z_1, z_2, \dots, z_{k-1}, y_k, z_{k+1}, \dots)$ be two distinct points in G. Since by hypothesis, G is Hausdorff, there exist open sets V and W in G such that $a \in V$ and $b \in W$ and $V \cap W = \emptyset$. This implies that $p_k(V) \cap p_k(W) = \emptyset$. Hence, it follows that G_k is a Hausdorff space.

(ii) Suppose that G_i is compact for every $i \in \mathbf{A}$. Then by Tychonoff theorem saying that the topological product of any family of compact spaces is also compact, it follows that the product group G is also compact. Conversely, assume that G is compact. Since compactness is preserved under every continuous map, it follows that $p_i(G) = G_i$ is also compact, since p is continuous and onto.

(iii) Suppose that every group G_i is locally compact and there are only finitely many $G_i \in \mathcal{G}$ which fail to be compact. Without loss of generality, assume that G_i is not compact only for $1 \leq i \leq n$. Hence, it follows by (ii) that the product group $\Pi_{i>n} G_i$ is compact and thus G is locally compact.

❑

2.7.2 *Intersection of Subgroups of a Topological Group*

This subsection generalizes to topological groups the concept of intersection of subgroups of abstract groups.

Proposition 2.7.3 *Let G be a topological group and $\mathcal{S}$ be a family of subgroups of G and $M = \bigcap_{S \in \mathcal{S}} S$.*

(i) *M is a subgroup of the topological group G;*

(ii) *If every $S \in \mathcal{S}$ is a normal subgroup of the topological group G, then M is also a normal subgroup of G.*

Proof **(i)** It follows from abstract group theory that $M = \bigcap_{S \in \mathcal{S}} S$ is subgroup of the abstract group G. Since every member $S \in \mathcal{S}$ is a closed set in the topological space G, their intersection M is also a closed set in G. This implies that M is a subgroup of G.

(ii) Proceed as in (i).

❑

Proposition 2.7.4 *Let G be a topological group and H be a nonempty subset of G. Then there exists a unique*

(i) *minimal subgroup of the topological group G that contains the set H;*
(ii) *minimal normal subgroup of the topological group G that contains the set H.*

Proof **(i)** Let $\mathcal{S}$ be the family of all subgroups S of G that contain the set H and $M = \bigcap_{S \in \mathcal{S}} S$ be their intersection. Then M is a subgroup of the topological group G by Proposition 2.7.3. Since $M = \bigcap_{S \in \mathcal{S}} S$, it follows that M is the minimal subgroup containing H which is unique.

(ii) Let $\mathcal{S}$ be the family of all normal subgroups S of G that contain the set H and $M = \bigcap_{S \in \mathcal{S}} S$ be their intersection. Then M is a normal subgroup of the topological group G by Proposition 2.7.3. Since $M = \bigcap_{S \in \mathcal{S}} S$, it follows that M is the minimal normal subgroup containing H which is unique.

❑

Proposition 2.7.5 *Let G be a topological group, H be a subgroup of G and N be a normal subgroup of G. Then $M = H \cap N$ is a normal subgroup of the topological group H.*

Proof Since H is a subgroup and N is a normal subgroup of the abstract G it follows from abstract group theory that $M = H \cap N$ is a normal subgroup of the abstract group H. Since M is a closed subset of the topological space G, it is also closed in the topological space H. Hence, it follows that $M = H \cap N$ is a normal subgroup of the topological group H. ❑

2.7.3 Product of Subgroups of a Topological Group

This subsection generalizes to topological groups the concept of product of subgroups of an arbitrary abstract group.

Proposition 2.7.6 *Let G be a topological group, H be a subgroup of G and N be a normal subgroup of G. Suppose that $HN = \{hn : h \in H,\ n \in N\}$ is a closed subset of the topological space G.*

(i) *Then $HN = NH$ is a subgroup of the topological group G;*
(ii) *If in particular, H is a normal subgroup of the topological group G, then HN is also a normal subgroup of the topological group G;*
(iii) *If either of the groups H or N is compact and the topological space G satisfies the second axiom of countability, then HN is a closed subset of the topological group G.*

Proof Since H is a subgroup of the abstract group G and N is a normal subgroup of G, it follows from abstract group theory that HN is a subgroup of the abstract group. Moreover, if H is also a normal subgroup of the abstract group G, then HN is also a normal subgroup of the abstract group G.

(i) By hypothesis, HN is a closed subset of the topological space G. Hence it follows that $HN = NH$ is a subgroup of the topological group G;

(ii) By hypothesis, HN is a closed subset of the topological space G. Suppose that H is a normal subgroup of the topological group G, then it follows that HN is also a normal subgroup of the topological group G;

(iii) Suppose H is compact and G satisfies the second axiom of countability. Let $\{x_n\}$ be a sequence of points in HN such that $lim_{n\to\infty}x_n = x$. Then each x_n can be represented as $x_n = h_n y_n : h_n \in H,\ y_n \in N$ for $n = 1, 2, \ldots$. By compactness of H, it follows that we can choose from the sequence x_n a subsequence x_{n_i} which converges to a point $h \in H$. Hence, it follows from convergence of the sequences $\{x_{n_i}\}$ and $\{h_{n_i}\}$ that the sequence $\{y_{n_i}\}$ converges to the point $h^{-1}x \in N$, since N is closed in the topological space G. This implies that $x = h(h^{-1}x) \in HN$. This proves that HN is a closed subset of the topological space G.
For the other alternative, if N is compact, the proof is similar.

❑

Corollary 2.7.7 *Let G be a topological group and the topological product $P = N_1 N_2 \cdots N_m$ of m normal subgroups of the group G be closed in G. Then P is also a normal subgroup of G.*

Proof It follows from the theory of abstract groups that P is a normal subgroup of the abstract group G. Since by hypothesis, P is closed in the topological space G, the corollary follows. ❑

Remark 2.7.8 Part (iii) of the Proposition 2.7.6 asserts that the closure condition of HN in the part (i) of the same proposition is always satisfied by a compact topological group G possessing the second axiom of countability property. For more result on product of subgroups of a topological group; see Exercise 19 of Sect. 2.13.

Theorem 2.7.9 *Let G be a locally compact topological group satisfying the second axiom of countability, H be a subgroup of G and N be a normal subgroup of G. If the product space $P = HN$ is a closed subset of the topological space G and if $M = H \cap N$, then the quotient groups H/M and P/N are isomorphic as topological groups.*

Proof Consider the map

$$f : H \to HN/N,\ h \mapsto hN.$$

Then f is an epimorphism of abstract groups with $ker\ f = H \cap N$. Then it follows from the theory of abstract groups that the abstract groups $H/ker\ f$ is isomorphic to the abstract group $Im(f) = HN/N$ by an isomorphism (algebraic)

$$\tilde{f} : H/kerf \to HN/N,\ h\, ker\ f \mapsto f(h).$$

Hence, it follows under the given conditions that $\tilde{f}$ is the required isomorphism of topological groups from the quotient group $H/H \cap N$ to the quotient group HN/N. This completes the proof of the theorem. ❑

2.7.4 Direct Product of Topological Groups

This subsection studies direct product of topological groups, which is a generalization to topological groups of the concept of direct product of subgroups of an arbitrary abstract group.

Definition 2.7.10 Let G be a topological group with identity element e and H, N be two normal subgroups of the topological group G. Then G is said to be decomposed into the **direct product** of its subgroups H and N if

$$G = HN \ and \ H \cap N = \{e\}.$$

Definition 2.7.11 Let H and K be two topological groups and $G = H \times K = \{(h, k) : h \in H, k \in K\}$ Then G is an abstract group under the composition defined by

$$(h, k)(h', k') = (hh', kk').$$

Again, G is a topological space, which is the product space of the topological spaces H and K. The topological group G thus obtained is called the **direct product** of the topological groups H and K.

2.7.5 Isomorphism Theorems

This subsection proves two isomorphism theorems which follow as a consequence of Theorem 2.7.9.

Theorem 2.7.12 *Let G be a locally compact topological group satisfying the second axiom of countability. If G is decomposed into the direct product of its subgroups H and N of G, then H is isomorphic to the quotient group G/N.*

Proof Let G be decomposed into the direct product of its subgroups H and N. Then $G = HN \ and \ H \cap N = \{e\}$ by Definition 2.7.10. Hence, the theorem follows from Theorem 2.7.9. ❑

Theorem 2.7.13 *Let G be a locally compact topological group satisfying the second axiom of countability and be decomposed into the direct product of its subgroups H and N. Suppose that*

(i) *the topological group H' is isomorphic to the topological group H and*

(ii) *the topological group N' is isomorphic to the topological group N.*

If G' is the direct product of the topological groups H' and N', then the topological group G is also isomorphic to G'.

Proof By hypothesis, the topological group H is isomorphic to the topological group H' and the topological group N is isomorphic to the topological group N'. Then there exist isomorphisms of topological groups

$$f : H' \to H \quad \text{and} \quad g : N' \to N.$$

Define a map

$$\psi : G' \to G, \ (x, y) \mapsto f(x)g(x).$$

The map ψ is well defined, because by hypothesis, G' is the direct product of groups H' and N' and the right-hand multiplication is the usual multiplication in G. Then ψ is an isomorphism of abstract groups. To show that ψ is continuous, take an element $z = xy$ such that $x \in H$ and $y \in N$. Let W_z be a nbd of z in G. Then there exist a nbd U_x of x in the topological space H and another nbd U_y of y in the topological space N such that

$$U_x U_y \subset U_z.$$

If $U = U_x \cap H$ and $V = U_y \cap N$, then U is a nbd of x in H and V is a nbd of y in N. Suppose $x' = f^{-1}(x)$ and $y' = f^{-1}(y)$. Then there exist a nbd U' of x' such that $f(U') \subset U$ and another nbd V' of y' such that $f(V') \subset V$. If $W = \{(x, y) : x \in U', \ y \in V'\}$, then W is a nbd of the point (x', y') such that $\psi(W') \subset W_z$. This implies that ψ is continuous. Thus, ψ is algebraically an isomorphism and is topologically continuous. This implies that ψ is an open. Hence it follows by the given conditions that $\psi : G' \to G$ is an isomorphism of topological groups. ❑

2.8 Classical Topological Groups of Matrices

This section studies some classical topological groups such as $\mathrm{M}(n, \mathbf{R})$, $\mathrm{GL}(n, \mathbf{R})$, $\mathrm{SL}(n, \mathbf{R})$, $\mathrm{O}(n, \mathbf{R})$ and $\mathrm{SO}(n, \mathbf{R})$ with topology induced from the Euclidean space $\mathbf{R}^{n^2}$ and their complex analogues with topology induced from the Euclidean space $\mathbf{R}^{2n^2}$ together with quaternionic analogues of orthogonal and unitary groups. More study on some classical topological groups of matrices from the view of Lie groups is available in Chap. 4.

Definition 2.8.1 Let $\mathrm{M}(n, \mathbf{R})$ be the set of all square matrices of order n over the field $\mathbf{R}$. Then it is an abelian group under usual addition of matrices called the **matrix group** over $\mathbf{R}$.

Definition 2.8.2 (*General linear group*) Let $\mathrm{GL}(n, \mathbf{R})$ be the set of all $n \times n$ nonsingular matrices over $\mathbf{R}$. It forms a group under usual multiplication of matrices called general linear group over $\mathbf{R}$.

Definition 2.8.3 (*Special linear group*) Let $\mathrm{SL}(n, \mathbf{R})$ be defined by $\mathrm{SL}(n, \mathbf{R}) = \{A \in \mathrm{GL}(n, \mathbf{R}) : \det A = 1\}$. It is a subgroup of $\mathrm{GL}(n, \mathbf{R})$ called special linear group.

Definition 2.8.4 (*Orthogonal group*) Let $\mathrm{O}(n, \mathbf{R})$ be defined by $\mathrm{O}(n, \mathbf{R}) = \{A \in \mathrm{GL}(n, \mathbf{R}) : AA^t = I\}$ is a subgroup of $\mathrm{GL}(n, \mathbf{R})$. It is a group called orthogonal group.

Definition 2.8.5 (*Special orthogonal group*) Let $\mathrm{SO}(n, \mathbf{R})$ be defined by $\mathrm{SO}(n, \mathbf{R}) = \{A \in \mathrm{O}(n, \mathbf{R}) : \det A = 1\}$. It forms a group called special orthogonal group.

Complex analogues of $\mathrm{M}(n, \mathbf{R})$, $\mathrm{GL}(n, \mathbf{R})$, $\mathrm{O}(n, \mathbf{R})$ and $\mathrm{SL}(n, \mathbf{R})$ are $\mathrm{M}(n, \mathbf{C})$, $\mathrm{GL}(n, \mathbf{C})$, $\mathrm{U}(n, \mathbf{C})$ and $\mathrm{SL}(n, \mathbf{C})$, respectively, where $\mathrm{M}(n, \mathbf{C})$ is identified with the Euclidean space $\mathbf{R}^{2n^2}$.

Definition 2.8.6 Let $\mathrm{GL}(n, \mathbf{C})$ be the set of all $n \times n$ nonsingular matrices over $\mathbf{C}$. It forms a group under usual multiplication of matrices called **general linear group** over $\mathbf{C}$.

Definition 2.8.7 Let $\mathrm{SL}(n, \mathbf{C})$ be defined by $\mathrm{SL}(n, \mathbf{C}) = \{A \in \mathrm{GL}(n, \mathbf{C}) : \det A = 1\}$. It is a subgroup of $\mathrm{GL}(n, \mathbf{C})$, called **special unitary or special linear group**.

Definition 2.8.8 Let $\mathrm{U}(n, \mathbf{C})$ be defined by $\mathrm{U}(n, \mathbf{C}) = \{A \in \mathrm{GL}(n, \mathbf{C}) : AA^* = I\}$ is a subgroup of $\mathrm{GL}(n, \mathbf{C})$, called **unitary group**, where A^* is the complex conjugate transpose of A.

Example 2.8.9 $\mathrm{U}(1, \mathbf{C}) = \{z \in \mathbf{C} : |z| = 1\}$ and hence $\mathrm{U}(1, \mathbf{C})$ and the circle group $S^1 (\cong \mathbf{R}/\mathbf{Z})$ are isomorphic topological groups, which are abelian.

Definition 2.8.10 Let $\mathbf{H}$ be the division ring of Hamilton's quaternionic numbers $q = a + bi + cj + dk$, such that $a, b, c, d \in \mathbf{R}$ with its conjugation $\overline{q} = a - bi - cj - dk$ and $\overline{q}q = a^2 + b^2 + c^2 + d^2 = |q|^2$. This defines the norm $|q|$ of q as

$$|q| = [a^2 + b^2 + c^2 + d^2]^{\frac{1}{2}}$$

The set $M(n, \mathbf{H})$ of all $n \times n$ matrices over $\mathbf{H}$ is identified with the Euclidean space $\mathbf{R}^{4n^2}$. Let $\mathrm{GL}(n, \mathbf{H})$ be the set of all $n \times n$ nonsingular matrices over $\mathbf{H}$. It forms a group under usual multiplication of matrices called general linear group over $\mathbf{H}$. The group $\mathrm{Sp}(n, \mathbf{H})$ defined by $\mathrm{Sp}(n, \mathbf{H}) = \{A \in \mathrm{GL}(n, \mathbf{H}) : AA^* = I\}$ is a subgroup of $\mathrm{GL}(n, \mathbf{H})$ called **symplectic group**, where A^* is the quaternionic conjugate transpose of A, conjugation is in the sense of reversal of all three imaginary components. The quaternionic analogue of orthogonal and unitary groups is the symplectic group but there is no analogue of special orthogonal group or special unitary group in the symplectic case.

Example 2.8.11 $\mathrm{Sp}(1, \mathbf{H}) = \{q \in \mathbf{H} : |q| = 1\} = \{q = a + bi + cj + dk : a, b, c, d \in \mathbf{R}\ with\ a^2 + b^2 + c^2 + d^2 = 1\}$ and hence $\mathrm{Sp}(1, \mathbf{H})$ is homeomorphic to the 3-sphere S^3. Clearly, $\mathrm{Sp}(1, \mathbf{H})$ and $\mathrm{U}(2, \mathbf{C}) = \{A \in \mathrm{GL}(2, \mathbf{C}) : AA^* = I\}$ are isomorphic topological groups.

2.8.1 Topological Groups of Matrices

This subsection studies $GL(n, \mathbf{R})$, $GL(n, \mathbf{C})$ and symplectic groups $Sp(n, \mathbf{H})$ by using topological concepts. For more study; see Sects. 2.12.7 and 2.13.

Theorem 2.8.12 *The general real linear group $GL(n, \mathbf{R})$ of all invertible $n \times n$ matrices over* $\mathbf{R}$

(i) *is a topological group;*
(ii) *the group $GL(n, \mathbf{R})$ is neither compact nor connected.*

Proof **(i)** Let $\mathrm{M}(n, \mathbf{R})$ be the set of all $n \times n$ matrices over $\mathbf{R}$ and $A = (a_{ij}) \in \mathrm{M}(n, \mathbf{R})$ be an arbitrary matrix. If $\mathbf{R}$ has the Euclidean topology and $\mathbf{R}^{n^2}$ has the product topology, then the map

$$f : \mathrm{M}(n, \mathbf{R}) \to \mathbf{R}^{n^2}, (a_{ij}) \mapsto (a_{11}, a_{12}, \dots, a_{1n}, a_{21}, a_{22}, \dots, a_{2n}, \dots, a_{n1}, a_{n2}, \dots, a_{nn})$$

identifies $\mathrm{M}(n, \mathbf{R})$ with the Euclidean space $\mathbf{R}^{n^2}$.
A topology called the Euclidean topology is defined on $\mathrm{M}(n, \mathbf{R})$ by declaring a subset $U \subset \mathrm{M}(n, \mathbf{R})$ to be open iff $f(U)$ is open in $\mathbf{R}^{n^2}$. With this topology, $\mathrm{M}(n, \mathbf{R})$ is homeomorphic to the topological space $\mathbf{R}^{n^2}$. Since the space $\mathbf{R}^{n^2}$ is Hausdorff, noncompact and locally compact, so is the space $\mathrm{M}(n, \mathbf{R})$. For each pair of integers i, j satisfying $1 \leq i, j \leq n$, there is a continuous projection map

$$p_{ij} : \mathrm{M}(n, \mathbf{R}) \to \mathbf{R},\ (a_{i,j}) \mapsto a_{i,j},$$

which sends a matrix $A = (a_{i,j})$ to its ij-th entry.

Consider the general real linear group $\mathrm{GL}(n, \mathbf{R})$ as a subspace of $\mathrm{M}(n, \mathbf{R})$. To show that the usual matrix multiplication

$$m : \mathrm{GL}(n, \mathbf{R}) \times \mathrm{GL}(n, \mathbf{R}) \to \mathrm{GL}(n, \mathbf{R})$$

is continuous with respect to above topology, let $A{=}(a_{ij}),\ B{=}(b_{ij}) \in \mathrm{GL}(n, \mathbf{R})$.

Then the ij-th entry in the product $m(A, B)$ is $\sum_{k=1}^{n} a_{ik} b_{kj}$.
Hence, the matrix multiplication m is continuous if and only if the composite maps

$$\mathrm{GL}(n, \mathbf{R}) \times \mathrm{GL}(n, \mathbf{R}) \xrightarrow{m} \mathrm{GL}(n, \mathbf{R}) \xrightarrow{p_{ij}} \mathbf{R}$$

are continuous. But $p_{ij}(m(A, B)) = \sum_{k=1}^{n} a_{ik} b_{kj}$. Since the usual multiplication and addition functions in the real number space $\mathbf{R}$ are both continuous, it follows that

$$m : \mathrm{GL}(n, \mathbf{R}) \times \mathrm{GL}(n, \mathbf{R}) \to \mathrm{GL}(n, \mathbf{R})$$

is continuous. Consequently, $GL(n, \mathbf{R})$ topologized as a subspace of the topological space $\mathrm{M}(n, \mathbf{R})$, is such that the usual matrix multiplication

$$GL(n, \mathbf{R}) \times GL(n, \mathbf{R}) \to GL(n, \mathbf{R}), (A, B) \mapsto AB$$

is continuous. We next show that in this subspace topology, the inverse map

$$v : GL(n, \mathbf{R}) \to GL(n, \mathbf{R}),\ A \mapsto A^{-1}$$

is continuous. Moreover, the determinant

$$det : M(n, \mathbf{R}) \to \mathbf{R}, A \mapsto detA$$

is continuous, since it is just a polynomial in the sense that $detA$ is an usual sum of usual products of real entries in A with proper signs and hence det is a polynomial function in n^2 variables called **det polynomial**. This implies that the determinant function restricted on its subspace $GL(n, \mathbf{R})$ is also continuous. Since $detA \neq 0$, $\forall A \in GL(n, \mathbf{R})$ the composite maps $p_{jk} \circ v$ are continuous for every pair of integers j, k with $1 \leq j, k \leq n$. This asserts that the inversion map

$$v : GL(n, \mathbf{R}) \to GL(n, \mathbf{R}),\ A \mapsto A^{-1}$$

is continuous. Consequently, $GL(n, \mathbf{R})$ is a topological group.

(ii) We now show that the group $GL(n, \mathbf{R})$ is neither compact nor connected. Clearly, $GL(n, \mathbf{R})$ is the inverse image of nonzero real numbers under the determinant function

$$det : M(n, \mathbf{R}) \to \mathbf{R}, A \mapsto detA.$$

Consider the continuous surjective determinant function

$$det : GL(n, \mathbf{R}) \to \mathbf{R} - \{0\}, A \mapsto detA.$$

Since the continuous image of $GL(n, \mathbf{R})$ under the determinant function is the subspace $\mathbf{R} - \{0\}$ of the real line space $\mathbf{R}$ and the subspace $\mathbf{R} - \{0\}$ is neither compact nor connected, it follows that $GL(n, \mathbf{R})$ is neither compact nor connected, because compactness and connectedness are both topological properties. ❑

Remark 2.8.13 Proposition 2.8.17 asserts that $GL^+(n, \mathbf{R}) = \{A \in GL(n, \mathbf{R}) : det\ A > 0\}$ and $GL^-(n, \mathbf{R}) = \{A \in GL(n, \mathbf{R}) : det\ A < 0\}$ are the connected components of $GL(n, \mathbf{R})$.

Corollary 2.8.14 *Every subgroup of the topological group* $GL(n, \mathbf{R})$ *is also a topological group.*

Corollary 2.8.15 $GL(n, \mathbf{R})$ *is an open subset of* $M(n, \mathbf{R})$.

Proof Since the determinant function

$$det : M(n, \mathbf{R}) \to \mathbf{R},\ A \mapsto det A$$

is continuous, it follows that

$$GL(n, \mathbf{R}) = M(n, \mathbf{R}) - (det)^{-1}\{0\}$$

is open, because, the set $\{0\}$ is closed in $\mathbf{R}$ implies $(det)^{-1}\{0\}$ is closed in $M(n, \mathbf{R})$. ❑

Corollary 2.8.16 M$(n, \mathbf{R})$ *endowed with the Euclidean topology is Hausdorff, noncompact and locally compact.*

Proof The space M$(n, \mathbf{R})$ endowed with the Euclidean topology is homeomorphic to the space $\mathbf{R}^{n^2}$. Since the space $\mathbf{R}^{n^2}$ is Hausdorff, noncompact and locally compact, so is the space M$(n, \mathbf{R})$. ❑

Proposition 2.8.17 *Let* $GL^+(n, \mathbf{R}) = \{A \in GL(n, \mathbf{R}) : det A > 0\}$ *and* $GL^-(n, \mathbf{R}) = \{A \in GL(n, \mathbf{R}) : det\ A < 0\}$. *They are*

(i) *disjoint open sets in* $GL(n, \mathbf{R})$;
(ii) *connected components of* $GL(n, \mathbf{R})$;
(iii) *homeomorphic.*

Proof The proof follows from the continuity of the determinant function

$$det : GL(n, \mathbf{R}) \to \mathbf{R} - \{0\},\ \ A \mapsto det\ A$$

❑

Remark 2.8.18 For more study of these classical topological groups of matrices over $\mathbf{R}$, $\mathbf{C}$ *and* $\mathbf{H}$; see Sects. 2.12.7 and 2.13.

2.8.2 Rotation Group SO(2,R)

This subsection studies 2-**dimensional rotation group,** which is the special orthogonal group SO(2,**R**) with topology induced by Euclidean metric on $\mathbf{R}^4$, which is homeomorphic to the unit circle S^1 with the usual topology inherited from **C** (identified with $\mathbf{R}^2$).

Proposition 2.8.19 **(i)** *$SO(2, \mathbf{R})$ is a closed subset of $M(2, \mathbf{R})$, i.e., in the space $\mathbf{R}^{2\times 2}$. It is also compact;*

(ii) *$SO(2, \mathbf{R})$ is homeomorphic to the unit circle S^1;*

(iii) *$SO(2, \mathbf{R})$ is a compact normal subgroup of the orthogonal group $O(2, \mathbf{R})$.*

Proof **(i)** A rotation of the Euclidean plane $\mathbf{R}^2$ about the origin is a linear operator and its matrix representation M with respect to the standard basis is an orthogonal matrix having its determinant 1. This asserts that a matrix M represents a rotation of $\mathbf{R}^2$ iff $M \in SO(2, \mathbf{R}) = \{M \in GL(2, \mathbf{R}) : M^t M = I, \, det M = 1\}$. Since the composition of two rotations of $\mathbf{R}^2$ about the origin is also a rotation, the group $SO(2, \mathbf{R})$ is called the 2-**dimensional rotation group** . It is the solution set in $\mathbf{R}^{2\times 2}$ determined by the following system of equations:
For any matrix $M = (x_{ij}) \in SO(2, \mathbf{R})$, the relations $M^t M = I_2$ and $det\ M = 1$ assert that

$$x_{11}^2 + x_{21}^2 = 1,$$

$$x_{11}x_{12} + x_{21}x_{22} = 0,$$

$$x_{12}^2 + x_{22}^2 = 1,$$

and

$$x_{11}x_{22} - x_{12}x_{21} = 1,$$

Consider the polynomial functions:

(a) $f_1 : \mathbf{R}^4 \to \mathbf{R}, \ (x_{11}, x_{12}, x_{21}, x_{22}) \mapsto x_{11}^2 + x_{21}^2 - 1;$
(b) $f_2 : \mathbf{R}^4 \to \mathbf{R}, \ (x_{11}, x_{12}, x_{21}, x_{22}) \mapsto x_{11}x_{12} + x_{21}x_{22};$
(c) $f_3 : \mathbf{R}^4 \to \mathbf{R}, \ (x_{11}, x_{12}, x_{21}, x_{22}) \mapsto x_{12}^2 + x_{22}^2 - 1;$
(d) $f_4 : \mathbf{R}^4 \to \mathbf{R}, \ (x_{11}, x_{12}, x_{21}, x_{22}) \mapsto x_{11}x_{22} - x_{12}x_{21} - 1.$

Since each f_i is continuous and $\{0\}$ is closed in $\mathbf{R}$, it follows that each of the solution sets $X_i = f_i^{-1}(\{0\})$ is closed in $\mathbf{R}^4$, for $i = 1, 2, 3, 4$ and hence $SO(2, \mathbf{R}) = X_1 \cap X_2 \cap X_3 \cap X_4$ is closed and bounded in the Euclidean space $\mathbf{R}^{2\times 2}$. Consequently, $SO(2, \mathbf{R})$ is also compact by Heine-Borel theorem. For an alternative proof see Corollary 2.8.21.

(ii) Since by the defining properties of $SO(2, \mathbf{R})$, each matrix $M \in S(O, \mathbf{R})$ has its unit column vectors which are also orthogonal. It asserts that an arbitrary matrix $M \in SO(2, \mathbf{R})$ is of the form $\begin{pmatrix} \cos\theta & -\sin\theta \\ \sin\theta & \cos\theta \end{pmatrix}$. Geometrically, M represents a rotation of $\mathbf{R}^2$ about the origin through an angle θ. So, to prove that the topological group $SO(2, \mathbf{R})$ is isomorphic to the unit circle S^1, it is sufficient to consider the map

$$f : SO(2, \mathbf{R}) \to S^1, \quad \begin{pmatrix} \cos\theta & -\sin\theta \\ \sin\theta & \cos\theta \end{pmatrix} \mapsto e^{i\theta}.$$

f is a group isomorphism and also a homeomorphism. It asserts that the groups $SO(2, \mathbf{R})$ and S^1 are isomorphic as topological groups by Definition 2.1.12.

(iii) $SO(2, \mathbf{R})$ is a subgroup of the orthogonal group $O(2, \mathbf{R})$ having only the other cosset $K = \{M \in O(2, \mathbf{R}) : det M = -1\}$ in the group $O(2, \mathbf{R})$. Hence the index $[\,O(2, \mathbf{R}) : SO(2, \mathbf{R})] = 2$ implies that $SO(2, \mathbf{R})$ is algebraically a normal subgroup of the orthogonal group $O(2, \mathbf{R})$. It is topologically compact, since f defined in (ii) is a homeomorphism and S^1 is compact.

❑

Corollary 2.8.20 *The orthogonal group $O(2, \mathbf{R})$ is a closed subset of $M(2, \mathbf{R})$ (i.e., in $\mathbf{R}^{2\times 2}$). It is also compact.*

Proof Consider X_1, X_2 *and* X_3 defined in Proposition 2.8.19, As $O(2, \mathbf{R}) = X_1 \cap X_2 \cap X_3$ is also closed. Since $O(2, \mathbf{R})$ is closed and bounded, it follows that it is compact by Heine-Borel theorem. ❑

Corollary 2.8.21 *$SO(2, \mathbf{R})$ is compact in $M(2, \mathbf{R})$.*

Proof Since S^1 is compact, the corollary follows from Proposition 2.8.19 by using the homeomorphism f defined therein. ❑

Example 2.8.22 Corollary 2.8.16 asserts that $\mathrm{M}(n, \mathbf{R})$ endowed with the Euclidean topology is a Hausdorff, noncompact and locally compact topological group.

Proposition 2.8.23 *$GL(n, \mathbf{C})$ endowed with the Euclidean topology is a Hausdorff, noncompact and locally compact topological group.*

Proof Proceed as in Corollary 2.8.16 ❑

2.9 Topological Rings and Topological Semirings

This section introduces the concepts of topological rings and topological semirings. A topological ring (field) R is algebraically a ring (field) R and there is a topology τ on R such that the ring (field) operations are continuous with respect to the topology τ, i.e., the algebraic and topological structures on R are compatible. These concepts link algebraic structure with topological structure.

Definition 2.9.1 Given a ring $(R, +, \cdot)$, a topology τ on the set R is said to a **ring topology** if

(i) the additive group $(R, +)$ equipped with the topology τ is a topological group;
(ii) the multiplication map

$$m : R \times R \to R, \ (x, y) \mapsto xy$$

is continuous, where $R \times R$ is endowed with the product topology.

Definition 2.9.2 A topological space (R, τ) is said to be a **topological ring**, if R is a ring (algebraically) and the topology τ is a ring topology on R. Equivalently, with respect to the topology τ, the ring operations in R are continuous, i.e., the mappings

(i) $R \times R \to R, \ (x, y) \mapsto x - y$ and
(ii) $R \times R \to R, \ (x, y) \mapsto xy$ are both continuous, where $R \times R$ is endowed with the product topology.

Example 2.9.3 Under usual addition and multiplication of real numbers, $(\mathbf{R}, +, \cdot)$ is a topological ring.

A semiring considered as a common generalization of an associative ring and a distributive lattice plays an important role in computer science and hence structural results of semirings from algebraic and topological view-point are important (see Adhikari and Adhikari, 2014).

Definition 2.9.4 A **topological semiring** is an algebraic system $(R, +, \cdot)$ endowed with a topology τ on R such that

(i) $(R, +)$ is a semigroup with continuous addition function $R \times R, \ (x, y) \mapsto x + y$;
(ii) $(R, \cdot)$ is a semigroup with continuous multiplication function $R \times R, \ (x, y) \mapsto xy$.
(iii) addition and multiplication are connected by distributive laws (both sides).

Example 2.9.5 Let $\mathbf{R}^*$ be the set of all nonnegative integers. Then under usual addition and multiplication the semiring $(\mathbf{R}^*, +, \cdot)$ is a topological semiring.

Example 2.9.6 Let $(S, +, \cdot)$ be an algebraic system, where $S = \{0, 1\}$ and

$$1 + 0 = 0 + 1 = 1, \ 1 + 1 = 1, \ 0 + 0 = 0, \ 1 \cdot 0 = 0 \cdot 1 = 0, \ 1 \cdot 1 = 1, \ 0 \cdot 0 = 0.$$

Then the semiring $(S, +, \cdot)$ endowed with Sierpinski topology τ defined in Chap. 3 of **Basic Topology, Volume 1** is a topological semiring. If $\{0\}$ is the closed set, then the open sets of Sierpinski topology τ are precisely, $\{\emptyset, \{1\}, \{0, 1\}\}$ and its closed sets are precisely, $\{\emptyset, \{0\}, \{0, 1\}\}$.

Example 2.9.7 Let (X, τ) be a topological space and $\mathcal{C}$ be the collection of all closed subsets of (X, τ). Then $\mathcal{C}$ forms a bounded distributive lattice and hence $\mathcal{C}$ admits a semiring structure $(\mathcal{C}, +, \ \cdot)$ by taking set-theoretic intersection as multiplication and set-theoretic union as addition. Similar result is true for the collection of all open subsets of (X, τ).

Definition 2.9.8 Given a division ring R, a topology τ on R is said to be a **division ring topology (or skew-field topology)** on R if

(i) τ is a ring topology on R;
(ii) the inversion map

$$v : (R - \{0\}) \to (R - \{0\}) : x \mapsto x^{-1}$$

is continuous, where $(R - \{0\})$ has the topology relative to τ.

Definition 2.9.9 A topological space (R, τ) is said to be a **topological division ring** if R is a division ring or skew-field (algebraically) and the topology τ is a division ring topology on R.

Definition 2.9.10 A topological space (F, τ) is said to be a **topological field** if F is a field (algebraically) such that the algebraic operations operating on F are continuous in the topological space. This topology is also called **field topology** on F.

Remark 2.9.11 A ring R is said to be a topological ring if R is a topological space and the algebraic ring operations on R are continuous. A division ring R is said to be a topological division ring if it is a topological ring satisfying the condition: for any nonzero element $r \in R$ and for any nbd V of r^{-1}, there is a nbd U of r such that $U^{-1} \subset V$. A commutative topological division ring is said to be a topological field.

Example 2.9.12 The fields **R** and **C** are important examples of topological fields. The division ring **H** of real quaternions is also an important example of topological division ring.

Definition 2.9.13 Two topological fields (F, τ) and (K, σ) are said to be **isomorphic** if there exists a homeomorphism

$$f : (F, \tau) \to (K, \sigma),$$

which preserves the operations of addition and multiplication.

Remark 2.9.14 Exercise 60 of Sect. 2.13 gives a deep result saying that a second countable locally compact connected topological division ring is isomorphic to one of the topological fields **R** or **C** or to the topological division ring **H** of real quaternions.

2.10 Topological Vector Space and Topological Algebra

This section introduces the concept of topological vector spaces. A topological vector (linear) space is algebraically a vector space V and there is a topology τ on V such that the vector space operations (such as group operations and scalar multiplication) are continuous with respect to the topology τ. Topological vector spaces provide a rich supply of topological groups by Proposition 2.10.5. Most of the interesting concepts that are studied in analysis deal with vector spaces (real or complex), instead of single object such as a function, measure or operator. Moreover, for the study of functional analysis, specified topological and algebraic structures provided by the topological vector spaces are needed.

It is interesting that the topology on topological vector space is localized in the sense that this topology can be determined by the family of open nbds of its zero element 0. This asserts that a linear transformation is continuous if it is continuous at 0, and hence it is so if it is uniformly continuous. Throughout the rest of this chapter, the field **F** represents **R** or **C**, endowed with usual topology (i.e., Euclidean topology). Different applications and more properties of topological vector spaces are available in Sect. 2.12.

2.10.1 Basic Concepts of Topological Vector Spaces

This subsection begins with a study of vector spaces endowed with a topology to introduce the concept of topological vector spaces and also studies basic concepts related to topological vector spaces.

Definition 2.10.1 A **topological vector space or linear space** (over **F**) is algebraically an abstract vector space (linear space) X and topologically it is a Hausdorff space with a topology τ such that

(i) the addition

$$+ : X \times X \to X, (x, y) \mapsto x + y$$

and

(ii) scalar multiplication

$$m : F \times X \to X, \ \ (\alpha, x) \mapsto \alpha \cdot x = \alpha x$$

are both continuous with respect to the topology τ on X and the corresponding product topologies on $X \times X$ and $\mathbf{F} \times X$.

This topology τ is called a **vector space topology** and the pair (X, τ) is called a **topological vector space**. The additive zero element of X is denoted by 0, called the **origin** of the vector space X.

Remark 2.10.2 The continuity of addition map $(x, y) \mapsto x + y$ in Definition 2.10.1 means that if U is a nbd of $x + y$ in X, then there exist nbds U_x of x and U_y of y in X such that

$$U_x + U_y \subset U, \quad \text{where } U_x + U_y = \{a + b : a \in U_x, \ b \in U_y\}.$$

Similarly, the continuity of scalar multiplication map $(\alpha, x) \mapsto \alpha x$ means that if U is a nbd of αx in X, then for some real number $t > 0$ and some nbd V of x in X, there is some $\beta \in \mathbf{F} = \mathbf{R}$ *or* $\mathbf{C}$ such that

$$\beta V \subset U, \ whenever, |\beta - \alpha| < t.$$

Remark 2.10.3 Some authors do not include the Hausdorff axiom for defining a topological vector space, because of Remark 2.10.24.

Definition 2.10.4 (*Groups of linear operators*) Let X be a topological vector space and G be a topological group with identity e. Then G is said to **act as a group of linear operator** on X, if

(i) given an $g \in G$, there is a continuous linear operator $T_g : X \to X$ such that

$$T_e = 1_X(Identity\ map), \ and\ T_{gh} = T_g \circ T_h, \ \forall\, g, h \in G;$$

(ii) the map $H : G \times X \to X, \ (g, x) \mapsto T_g(x)$ is continuous.

Proposition 2.10.5 *Every topological vector space is a topological group.*

Proof Let X be a topological vector space. Then by definition , it is Hausdorff and the scalar multiplication multiplication $m : \mathbf{F} \times X \to X$ is continuous. It shows that the function

$$f : X \to X, x \mapsto m(-1, x) = (-1)x = -x$$

is continuous for every $x \in X$. This asserts that the function

$$g : X \times X \to X, (x, y) \mapsto x + f(y) = x - y,$$

is also continuous for every pair of elements $x, y \in X$. This proves that X is also a topological group. ❑

Example 2.10.6 **(i)** The Euclidean n-space $\mathbf{R}^n$ is a topological vector space over $\mathbf{R}$ with topology induced by Euclidean metric.

(ii) $\mathbf{C}^n$ is a topological vector space over $\mathbf{C}$ with topology induced by the metric obtained by the norm

$$||z|| = |z_1| + |z_2| + \cdots + |z_n|$$

or the norm

$$||z|| = max\{|z_1|, |z_2|, \ldots, |z_n|\}$$

for $z = (z_1, z_2, \ldots, z_n) \in \mathbf{C}^n$.

(iii) Let X be a vector space over the field $\mathbf{R}$ or $\mathbf{C}$. Then X endowed with the trivial (indiscrete) topology is a topological vector space.

(iv) On the other hand, every nontrivial vector space over the field **F** (**R** or **C**) endowed with discrete topology is not a topological vector space. Suppose to the contrary that every nontrivial vector space over the field **F** endowed with discrete topology is a topological vector space. Consider the sequence $\{\beta_n = \frac{1}{n^2}\}$ in the topological field **F**. Since the scalar multiplication in a vector space is continuous, it follows for any $x(\neq 0) \in X$, $\beta_n x \to 0$ and hence for every nbd U of $0 \in X$, there is a positive integer k such that

$$\beta_n x \in U, \ \ \forall n \geq k.$$

For the particular open set $U = \{0\}$ under the discrete topology in X, it shows that $\beta_n x = 0$. This asserts that $x = 0$, which contradicts our assumption.

(v) Let X be a normed linear space and τ be the topology obtained by the metric induced by the norm of X. Then (X, τ) is a topological vector space.

(vi) Given a topological space X, the linear space $C(X, \mathbf{R})$ of all bounded continuous real-valued functions on X with topology τ on $C(X, \mathbf{R})$, induced by the metric

$$d : C(X, \mathbf{R}) \times C(X, \mathbf{R}) \to C(X, \mathbf{R}), (f, g) \to \sup_{x \in X} |f(x) - g(x)|$$

forms a topological vector space, called the **topological vector space of all bounded real-valued continuous functions** on the space X.

(vii) The real Hilbert space H is a topological vector space.

Definition 2.10.7 Let X and Y be two topological vector spaces over the same field **F** and $T : X \to Y$ be a linear transformation of vector spaces. Then T is said to be a **topological**

(i) **linear homomorphism** if T is continuous and open;
(ii) **linear monomorphism** if T is an injective topological linear homomorphism;
(iii) **linear epimorphism** if T is a surjective topological linear homomorphism;
(iv) **linear isomorphism** if T is both an injective and surjective topological linear homomorphism;
(v) **linear automorphism** if T is a topological linear isomorphism of X onto itself.

Definition 2.10.8 Given a topological vector space X over **F** (**R** or **C**), an arbitrary point $b \in X$ and any nonzero scalar $\alpha \in \mathbf{F}$, the maps

$$T_b : X \to X, \ x \mapsto b + x$$

and

$$M_\alpha : \mathbf{F} \times X \to X, (\alpha, x) \mapsto \alpha x$$

are respectively called a**translation operator** and a **multiplication operator**.

Proposition 2.10.9 *Given a point $b \in X$ and a nonzero scalar $\alpha \in$ **F**,*

(i) *the translation operator* $T_b : X \to X, x \mapsto b + x$ *and*
(ii) *the multiplication operator* $M_\alpha : \mathbf{F} \times X \to X, x \mapsto \alpha x$

on the topological vector space X *over* **F** *are both homeomorphisms of* X.

Proof As X is algebraically a vector space, it follows that the maps T_b, and M_α are both bijective with T_{-b}, and $M_{1/\alpha}$ their respective inverses. Again, the continuity of the vector space operations asserts that the four maps T_b, M_α, T_{-b} and $M_{1/\alpha}$ are continuous and hence each of them is a homeomorphism. ❑

Corollary 2.10.10 *Let* X *be a topological vector space over* **F**. *Then*

(i) *given an arbitrary point* $b \in X$, *the translation map*

$$T_b : X \to X, \ x \mapsto b + x,$$

is a homeomorphism;
(ii) *given an arbitrary nonzero scalar* $\alpha \in \mathbf{F}$, *the dilation map*

$$d_\alpha : X \to X, \ x \mapsto \alpha x,$$

is an automorphism of X. *Finally, as* d_α *is an isomorphism from* X *onto itself and hence it is an automorphism of* X.

Proof It follows from Proposition 2.10.9. More precisely, continuity of T_b and d_α follows from Definition 2.10.1. Their bijectivity follows from vector space structure. Hence, it follows that they are homeomorphic with inverse maps $x \mapsto x - b$ and $x \mapsto \frac{1}{\alpha}x$, respectively.

❑

Corollary 2.10.11 *Let* U *be an open set in a topological vector space* X. *Then* $x + U$ *(called the* ***translate of*** U*) and* αU *are both open sets in* X *for each* $x \in X$ *and for each nonzero* $\alpha \in \mathbf{F}$.

Proof It follows from Corollary 2.10.10. ❑

Example 2.10.12 Proposition 2.10.9 asserts that the topology of a topological vector space X is a translation invariant in the sense that all translations in X are homeomorphisms. But a translation invariant under a topology τ on an abstract vector space can not guarantee that (X, τ) is a topological vector space. For example, d be a metric (induced by a norm) on vector space X. Then d is a translation invariant in the sense that

$$d(x + z, y + z) = d(x, y), \ \forall x, y, z \in X.$$

Under the topology τ_d on X induced by d the addition is continuous and translation is a topological invariant. On the other hand, if d is taken as discrete metric on x, its induced topology τ_d is also discrete but this topology is not always compatible with scalar multiplication (see Example 2.10.6).

Definition 2.10.13 Let X be a topological vector space with vector topology τ. A family $\mathcal{B}$ of nbds of 0 is said to be a **local base at** 0 **for the topology** τ if each nbd of 0 contains a member of the family $\mathcal{B}$.

Remark 2.10.14 In the context of topological vector space, local base means a local base at 0. Let X be a topological vector space with vector topology τ. It follows from Proposition 2.10.9 that the open sets in τ are precisely those subsets which are unions of translates of members of the family of $\mathcal{B}$ and hence τ is completely determined by any local base.

Definition 2.10.15 Let (X, τ) be a topological vector space over $\mathbf{F} = (\mathbf{R}\ or\ \mathbf{C})$ and A be a topological vector subspace of X with subspace topology induced from X. Then A is also a topological vector space. The space A is called a **topological vector subspace** of X if

(i) A is topological vector space and
(ii) A is also a topological subspace of X.

Example 2.10.16 Let X be a real topological vector space and $Y \subset X$ be a vector subspace endowed with the relative topology induced from X. Then Y is a topological vector subspace of X.

Definition 2.10.17 Let X be a real topological vector space and $Y \subset X$ be nonempty. Then Y is said to be **convex** if for any two vectors $u, v \in Y$,

$$ru + (1-r)v \in Y \ for\ every\ r \in \mathbf{R} \text{ satisfying } 0 \leq r \leq 1.$$

Proposition 2.10.18 *Let X be a real topological vector space and $Y \subset X$ be a convex subset of X containing the vector 0. Then $rY \subset sY$ for any two nonnegative real numbers r, s such that $r < s$.*

Proof Let $v \in rY$. Then $v = ru$ for some $u \in Y$. Let $c = r/s$ for some $s > r$. Then c is a real number such that $0 < c < 1$. Again, since the Y is convex in X by hypothesis, $cu + (1-c)0 \in Y$. This asserts that $v = ru = (sc)u = s(cu) \in sY$. ❑

Remark 2.10.19 Different types of topological vector spaces are introduced in Definition 2.10.20 for our future study.

Definition 2.10.20 Let X be a topological vector space over $\mathbf{F}$ with topology τ and origin 0.

(i) X is said to be **locally convex** if every nbd U of the origin 0 contains a convex nbd of 0;
(ii) A subset $Y \subset X$ is said to be **bounded** if given a nbd U of the origin 0 in X, there is a positive integer n such that $Y \subset nU$;
(iii) A subset $Y \subset X$ is said to be **totally bounded** if given a nbd U of the origin 0 in X, there is a finite set B such that $Y \subset B + U$;

(iv) X is said to be **locally bounded** if there exists a bounded nbd U of the origin 0 in X.

(v) X is said to be **locally compact** if 0 has a nbd U such that its closure $\overline{U}$ is compact.

(vi) X is said to be **metrizable** if its topology τ coincides with the topology τ_d induced by some metric d on X.

(vii) X is said to be an **F-space** if topology τ coincides with the topology τ_d induced by a complete invariant metric d on X.

(viii) X is said to be **Fréchet** if X is locally convex **F**-space.

(ix) X is said to be **normable** if there exists a norm on X such the topology τ of X coincides with the topology τ_d, where d is metric induced by norm on X.

(x) X is said to have **Heine- Borel property** if its every closed and bounded subset is compact. Topology τ coincides with the topology τ_d, where d is metric induced by norm on X.

(xi) X is said to be **metrizable** if its topology τ coincides with the topology τ_d induced by some metric d on X.

Theorem 2.10.21 *Let X be a locally bounded topological real vector space with continuous scalar multiplication μ. Then it has a countable local base at its origin 0.*

Proof As by hypothesis, X is a locally bounded topological real vector space, there exists a bounded nbd U_0 of the origin 0 in X. Given a nonzero scalar $r \in \mathbf{R}$, the map

$$\psi_r : X \to X,\ x \mapsto \mu(r, x) = rx$$

is continuous, since the scalar multiplication μ is continuous. If $s = r^{-1}$, then

$$\psi_r \circ \psi_s = 1_X = \psi_s \circ \psi_r$$

asserts that ψ_r is a homeomorphism. Again, as $\psi_r(0) = 0$, it follows that

$$rU_0 = \psi_r(U_0)$$

is nbd of 0 for every nonzero real number r. Consider the sequence

$$\mathcal{N} = \{1/n\ U_0 : n \in \mathbf{N}\}$$

of nbds of the origin 0. Let V be a nbd of the origin 0 in X. As U_0 is bounded, there is a positive integer n such that $U_0 \subset nV$. It asserts that

$$1/n\ U_0 \subset 1/n\ (nV) = V.$$

Hence, it follows by Definition 2.10.13 that $\mathcal{N}$ forms a countable local base at its origin 0. ❑

Proposition 2.10.22 *Let X be a topological vector space and U be a nbd of 0 in X. Then there is a nbd V of 0 such that*

$$V = -V(\text{ i.e., } V \text{ is symmetric }) \text{ and } V + V \subset U.$$

Proof By hypothesis, X is topological vector space and U is a nbd of 0 in X. Let

$$f : X \times X \to X, (x, y) \to x + y$$

be the continuous vector addition. Since $0 + 0 = 0$, it follows that there exist nbds U_1 and U_1 of 0 in X such that $U_1 + U_2 \subset U$. If we take

$$V = U_1 \cap U_2 \cap (-U_1) \cap (-U_2),$$

then V satisfies the required property. ❑

Theorem 2.10.23 proves a separation property of topological vector spaces.

Theorem 2.10.23 *Let X be a topological vector space with topology τ and A, B be two disjoint subsets of X such that A is compact and B is closed in X. Then there is a nbd U of 0 such that*

$$(A + U) \cap (B + U) = \emptyset. \tag{2.1}$$

Proof Since $A + U$ is the union of translates of $x + U : x \in A$, it is an open set that contains A.

(i) If $A = \emptyset$, then the theorem is trivial.

(ii) So, we assume that $A \neq \emptyset$. Let $x \in A$ be an arbitrary point. Then $x \notin B$ by assumption that $A \cap B = \emptyset$. Since B is closed, $x \notin B$ and the topology τ is invariant under translation, it follows that from Proposition 2.10.22 that 0 has a symmetric nbd U_x such that

$$(x + U_x + U_x) \cap (B + U_x) = \emptyset \tag{2.2}$$

By hypothesis, A is compact. Hence we can choose a finite number of points $x_1, x_2, \ldots, x_n \in A$ such that

$$A \subset (x_1 + U_{x_1}) \cup (x_2 + U_{x_2}) \cup \cdots \cup (x_n + U_{x_n}).$$

Taking, $U = U_{x_1} \cap U_{x_2} \cap \cdots \cap U_{x_n}$ it follows that

$$A + U \subset \bigcup_{k=1}^{n} (x_k + U_{x_k} + U) \subset \bigcup_{k=1}^{n} (x_k + U_{x_k} + U_{x_k})$$

Since by (2.2) , no term in $\{x_k + U_{x_k} + U_{x_k}\}$ intersects $B + U$, the theorem is proved.

❑

Remark 2.10.24 Theorem 2.10.23 proves a separation property of topological vector spaces. The term **separation property** is justified, because, $A + U$ being a union of translates $(x + U : x \in A)$ of U is an open set containing A. Similarly. $B + U$ is also an open set such that these two open sets are disjoint by Theorem 2.10.23. This asserts in particular that every pair of distinct points have distinct nbds, showing that Hausdorff separation axiom holds in X.

Theorem 2.10.25 *Let X be a topological vector space over* $\mathbf{F}$. *Then*

(i) *for any subset $Y \subset X$, its closure $\overline{Y} = \bigcap(Y + U)$, where U runs over the family of all nbds of* 0;
(ii) *for any two subsets Y and Z of X*

$$\overline{Y} + \overline{Z} \subset \overline{Y + Z};$$

(iii) *the closure of a subspace is also a subspace in the sense that for any subspace A of X, its closure $\overline{A}$ is also a subspace of X;*
(iv) *the closure of a convex set is also a convex subset in the sense that for any convex subset B of X, its closure $\overline{B}$ is also a convex subset of X;*
(v) *the closure of a bounded set is also bounded in the sense that for any bounded subset C of X, its closure $\overline{C}$ is also a bounded subset of X.*

Proof **(i)** $x \in \overline{Y}$ iff $(x + U) \cap Y \neq \emptyset$ for each nbd U of 0 in X. This holds iff $x \in (Y - U)$ for each such nbd U of 0 in X. The proof follows, since, $-U$ is a nbd of 0 iff U is so.
(ii) Let $y \in \overline{Y}, z \in \overline{Z}$ and V be any nbd of $y + z$ in X. Then there exist a nbd V_1 of y and a nbd V_2 of z in X such that $V_1 + V_2 \subset V$. Since by assumption, $y \in \overline{Y}, z \in \overline{Z}$, it follows that there exist elements $a \in Y \cap V_1$ and $b \in Z \cap V_2$. This shows that $a + b \in (Y + Z) \cap V$, since $y \in \overline{Y}, z \in \overline{Z}$. It implies that $(Y + Z) \cap V \neq \emptyset$, This asserts that $y + z \in \overline{Y + Z}$, which proves (ii).
(iii) For any two nonzero scalars, $\alpha, \beta \in \mathbf{F}, \alpha\overline{A} = \overline{\alpha A}$ and $\beta\overline{A} = \overline{\beta A}$ and for $\alpha = \beta = 0$, these relations are trivial. Consequently,

$$\alpha\overline{A} + \beta\overline{A} = \overline{\alpha A} + \overline{\beta A} \subset \overline{\alpha A + \beta A} \subset \overline{A} \quad \text{since, by hypothesis, A is a subspace of X.}$$

This proves that $\overline{A}$ is also a subspace of X.
(iv) Proceed as in (iii) to prove that convex sets in X have convex closures.
(v) Given any nbd U of 0, there is a nbd V of 0 such that $\overline{V} \subset U$ by Exercise 29 of Sect. 2.13. By hypothesis, C is bounded and hence $C \subset rV$ for sufficiently large $r > 0$. Then $\overline{C} \subset r\overline{V} \subset rU$ for this choice of r. This proves that $\overline{C}$ is also bounded in X.

❑

Definition 2.10.26 Let X be a topological vector space over $\mathbf{F}$. A subset $A \subset X$ is said to be **balanced** if

$$\alpha A \subset A, \ \forall \alpha \in \mathbf{F} \text{ with } |\alpha| \leq 1.$$

Theorem 2.10.27 *Let X be a topological vector space over $\mathbf{R}$ and U be a nbd of 0.*

(i) *If $0 < t_1 < t_2 < \cdots$ is an increasing sequence $\{t_n\}$ of positive real numbers such that $t_n \to \infty$ as $n \to \infty$, then $X = \bigcup_{n=1}^{\infty} t_n U$.*
(ii) *If A is a compact subset of X, then A is bounded.*
(iii) *If $\{t_n\}$ is a decreasing sequence of positive real numbers such that $t_n \to 0$ as $n \to \infty$ and U is bounded, then the family*

$$\{t_n U : n = 1, 2, \ldots, \}$$

forms a local base for X.

Proof **(i)** Given an element $x \in X$, consider the map

$$\psi : \mathbf{R} \to X, t \mapsto tx.$$

Since ψ is continuous, the set

$$\{t \in \mathbf{R} : tx \in U\} \subset \mathbf{R}$$

is open and contains 0. This shows that it also contains $1/t_n$ for n, sufficiently, large. This asserts that

$$(1/t_n)x \in U \quad \text{or } x \in t_n U$$

for sufficiently, large n, which proves (i).

(ii) Let V be a balanced nbd of 0 such that $V \subset U$. Then by using (i), it follows that

$$A \subset \bigcup_{n=1}^{\infty} nV.$$

Since, by hypothesis, A is compact and V is balanced, there exists positive integers $n_1, n_2, \ldots, n_t$ such that

$$A \subset n_1 V \cup n_2 V \cup \cdots \cup n_t V = n_t V.$$

For any $s > n_t$, it follows that $A \subset sV \subset sU$, which proves (ii).

(iii) Let $\{t_n\}$ be a decreasing sequence of positive real numbers such that $t_n \to 0$ as $n \to \infty$ and U be bounded nbd of 0 in X. Then there exists some $t > 0$ such that

$$U \subset sV, \forall s > t.$$

If n is chosen sufficiently large such that $tt_n < 1$, then it follows that

$$U \subset (1/t_n)V.$$

This implies that V contains all excepting only a finite many of the sets $t_n U$. This proves that the family

$$\{t_n U : n = 1, 2, \ldots, \}$$

forms a local base for X.

□

2.10.2 *Topological Algebra*

This subsection introduces the concept of topological algebra. Example 2.10.29 provides a rich supply of topological algebras which are $\mathcal{C}(X)$ endowed with compact open topology. A characterization of completely regular Hausdorff spaces X by the closed ideals of the topological algebra $\mathcal{C}(X)$ is given in Exercise 35 of Sect. 2.13. For more study of the topological algebra $\mathcal{C}(X)$, the paper Morris and Wulbert (1967) is referred.

Definition 2.10.28 A topological algebra $\mathcal{A}$ is an algebra over **R** or **C** endowed with a topology τ such that with respect to the topology τ

(i) $\mathcal{A}$ is a topological vector space and
(ii) multiplication in $\mathcal{A}$ is jointly continuous.

Example 2.10.29 Given a topological space X, let $\mathcal{C}(X)$ denote the algebra of all continuous complex (or real)-valued functions with pointwise addition. Then $\mathcal{C}(X)$ endowed with compact open topology forms a topological algebra.

2.10.3 *Representation of a Topological Group*

This subsection conveys the algebraic concept of representation θ of an algebraic group G on a vector space V in Definition 2.10.30. On the other hand, Definition 2.10.32 introduces the concept of representation of degree n of a topological group G which is used for **representation of a compact subgroups** of $GL(n, \mathbf{C})$ on a finite dimensional vector space in Exercise 44 of Sect. 2.13.

Definition 2.10.30 introduces the concept of representation of an algebraic group G on a finite dimensional vector space.

Definition 2.10.30 Let V be an n-dimensional vector space over $\mathbf{F}$ and $GL(V)$ be the group of all invertible linear operators $T : V \to V$, with multiplication as the usual composition of maps. Then every choice of a basis $\mathcal{B}$ of V, defines an isomorphism

$$\psi_{\mathcal{B}} : GL(V) \to GL(n, F),\ T \mapsto M_T\ (matrix\ representation\ of\ T\ with\ respect\ to\ \mathcal{B}).$$

Given a group G, a representation of G on V is a homomorphism

$$\theta : G \to GL(V)$$

and dimension of V is defined to be the dimension of θ. The homomorphism θ is said to be a **representation of** G on the n-dimensional vector space V.

Remark 2.10.31 Matrix representation of a group G is considered as representation θ of G over the vector space $n \times n$ matrices over $\mathbf{F}^n$ of column vectors. The totality of all nonsingular $n \times n$ matrices over $\mathbf{F}^n$ matrices form a topological group $GL(n, \mathbf{F})$ under usual multiplication of matrices (see Theorem 2.8.12).

We are interested to study compact topological groups. Definition 2.10.32 introduces the concept of representation of degree n of a topological group G.

Definition 2.10.32 Let G be a topological group and $GL(n, \mathbf{F})$ be the topological group of nonsingular matrices. A homomorphism

$$\theta : G \to GL(n, \mathbf{F})$$

is called a **representation** of degree n of the topological group G.

2.10.4 Topological Quotient Vector Space

This subsection studies quotient vector topology and topological quotient vector spaces. In this section the field $\mathbf{F}$ represents $\mathbf{R}$ or $\mathbf{C}$, endowed with usual topology.

Definition 2.10.33 Let (X, τ) be a topological vector space over $\mathbf{F}$ and A be a topological vector subspace of X with subspace topology induced from X. Then A is also a topological vector space. The space A is called a topological vector subspace of X. Then X/A is said to be a **topological quotient vector space** if

(a) if X/A is the quotient vector space and
(b) the quotient space X/A has the topology $\tau_{X/A}$ such that a subset U in X/A is open iff $p^{-1}(U)$ is open in (X, τ), where

$$p : X \to X/A, x \mapsto x + A$$

is the natural projection map.

This topology $\tau_{X/A}$ on X/A is a quotient topology called a **quotient vector topology** and p is called a quotient map.

Remark 2.10.34 More precisely, the topological space $(X/A, \tau_{X/A})$ is said to be a **topological quotient vector space or linear topological quotient space** if the usual addition and scalar multiplication:

(i)

$$+ : X/A \times X/A \to X/A,\ (x + A, y + A) \mapsto (x + y) + A$$

and

(ii)

$$m : F \times X/A \to X/A,\ (\alpha, x + A) \mapsto \alpha x + A,$$

are both continuous under the product topologies on $X/A \times X/A$ and $\mathbf{F} \times X/A$.

Unless stated otherwise, it is assumed that the quotient space X/A has the quotient topology $\tau_{X/A}$ with quotient map

$$p : X \to X/A,\ x \mapsto x + A.$$

Theorem 2.10.35 *Let X be a topological vector space over* $\mathbf{F}$ *with topology τ and A be a topological vector subspace of X. Let $(X/A, \tau_{X/A})$ be a topological quotient vector space with*

$$p : X \to X/A, x \mapsto x + A$$

its quotient map.

(i) *p is linear, continuous, and open;*
(ii) *Let Y be a topological vector space over* $\mathbf{F}$*. Then a linear map $f : X/A \to Y$ is continuous (open) iff the composite map $f \circ p : X \to Y$ is continuous (open).*

Proof (i) The quotient map $p : X \to X/A, x \mapsto x + A$ is continuous by definition of quotient topology $\tau_{X/A}$ on X/A and p is linear by its definition. To show that p is open, take an arbitrary open set U in (X, τ). Claim that $p(U)$ is also open in $(X/A, \tau_{X/A})$. Consider the set $p^{-1}(p(U)) = U + A$, which is open in (X, τ) since the sum of the open set U and any set is open obtained by translation homeomorphisms, it follows that p is open.

(ii) Let V be a subset in Y. Then $f^{-1}(V)$ is open in X/A iff $p^{-1}(f^{-1}(V)) = (f \circ p)^{-1}(V)$ is open in X, which holds iff $f \circ p$ is continuous. Since p is both open and continuous by (i), proceeding in a similar way, it follows that f is open iff $f \circ p$ is open. ❑

Remark 2.10.36 The Hausdorff property of the topological quotient vector space X/A holds iff A is closed in X.

2.10.5 Topological Modules

This subsection defines topological modules by extending the concept of topological vector spaces.

Definition 2.10.37 Let R be a topological ring and M be an R-module endowed with a topology τ. Then M is said to be a **topological** R**-module** if the mappings

(i) $M \times M \to M, \ (x, y) \mapsto x - y$
(ii) $R \times M \to M, \ (r, x) \mapsto rx$

are both continuous.

Proposition 2.10.38 *Let M be a topological R-module.*

(i) *$(M, +)$ is a topological group;*
(ii) *The multiplication of elements of R by elements of M is continuous;*
(iii) *Every topological ring R is also a topological R-module.*

Proof It follows from Definition 2.10.37 ❑

Remark 2.10.39 A topological ring or a topological module may not be Hausdorff according to their definitions. If the zero 0 of a topological manifold M is a closed set , then M is a Hausdorff space.

2.11 Linear Transformations and Linear Functionals

This section studies continuity of linear transformations and continuous linear functionals on topological vector spaces. An analogous study from the view-point of normed linear spaces is available in Chap. 2 of **Basic Topology, Volume 1** of the present book series.

2.11.1 Continuity of Linear Transformations

This subsection conveys the concept of continuity of linear transformations on topological vector spaces. This concept is localized at a point in the sense of Definition 2.11.2.

Definition 2.11.1 Let (X, τ) and (Y, σ) be two topological vector spaces. Then a linear transformation $T : X \to Y$ is said to be **continuous at a point** $x \in X$ if for every nbd U of $T(x)$ in Y, there is a nbd V of x such that $T(V) \subset U$.

The concept of continuity of a linear transformation can be localized as given in Definition 2.11.2.

Definition 2.11.2 Let (X, τ) and (Y, σ) be two topological vector spaces with local bases $\mathcal{V}$ and $\mathcal{U}$ at the origin 0 respectively. Then a linear transformation $T : X \to Y$ is said to be continuous at a point $p \in X$, if for each nbd $U \in \mathcal{U}$ of the point $T(p)$, there exists a nbd V of p in $\mathcal{V}$ such that

$$T(V + p) \subset U + T(p)$$

i.e.,

$$if\ and\ only\ if\ \ T(V) \subset U.$$

Remark 2.11.3 In view of Definition 2.11.2, to examine the continuity of a linear transformation $T : X \to Y$ between topological vector spaces at a point $x \in X$, it is sufficient to do so at the origin 0 only. Hence, it asserts that if T is continuous at 0, it is also continuous at every point $x \in X$.

Definition 2.11.4 Let X and Y be two topological vector spaces and $T : X \to Y$ be a linear transformation. Then T is said to be **uniformly continuous** if for each nbd V of $0 \in Y$, there is a nbd U of $0 \in X$ such that $x - y \in U$ implies $T(x) - T(y) \in V$.

Proposition 2.11.5 *Let (X, τ) and (Y, σ) be two topological vector spaces and $T : X \to Y$ be a continuous linear transformation. Then T is uniformly continuous.*

Proof Let U be an open nbd in (Y, σ) of $0 \in Y$. Then there is a nbd V in (X, τ) of 0 such that for every pair of elements $x, x' \in X$ with $x - x' \in V$, the element $T(x) - T(x') = T(x - x')$ is in U. It implies that two elements become close to each other, if their difference is near to 0, which gives a uniform concept of nearness. It asserts that T is uniformly continuous. ❑

2.11.2 Continuity of Linear Functionals

This subsection concentrates on the study of the concept of real or complex linear functionals (which are special types of linear transformations) defined on topological vector spaces. For more properties of linear functionals and their continuity; see Sect. 2.12.8, Exercises 37 and 50 of Sect. 2.13.

Definition 2.11.6 Let X be a topological vector space over $\mathbf{R}$. Then a function $f : X \to \mathbf{R}$ is said to be a **linear functional** if

$$f(rx + sy) = rf(x) + sf(y),\ \forall x, y \in X,\ and\ \forall r, s \in \mathbf{R}.$$

Definition 2.11.7 Let X be a topological vector space over $\mathbf{R}$ or $\mathbf{C}$. Consider $\mathbf{R}$ as topological vector space over itself. Then a functional $T : X \to \mathbf{R}$ is said to be a continuous real linear functional if T is a **continuous linear transformation** of vector spaces.

Similarly, a continuous complex linear functional is defined.

Example 2.11.8 Given a point $x = (x_1, x_2, \ldots, x_n) \in \mathbf{R}^n$ and $r \in \mathbf{R}$, the function

$$f : \mathbf{R}^n \to \mathbf{R},\ x \mapsto \Sigma_{i=1}^n r x_i,$$

is a continuous linear functional.

Definition 2.11.9 Let X be a topological vector space over $\mathbf{R}$. The **dual space** of X, denoted by X^* is the vector space consisting of all continuous linear functionals on X, denoted by $\mathcal{C}(X, \mathbf{R})$, where addition and scalar multiplication on X^* are defined by

$$(f + g)(x) = f(x) + g(x),\ and\ (rf)(x) = rf(x),\ \ \forall f, g \in X^*,\ x \in X,\ r \in \mathbf{R}.$$

For a topological vector space X over $\mathbf{C}$, its dual space X^* is defined in a similar way and is denoted by $\mathcal{C}(X, \mathbf{C})$.

Remark 2.11.10 For the particular case, when $Y = \mathbf{R}$ in $\mathcal{C}(X, Y)$, then the normed linear space $\mathcal{C}(X, Y) = \mathcal{C}(X, \mathbf{R})$ is the dual space X^* of X. On the other hand, if X is a topological vector space over $\mathbf{C}$, then a complex linear functional on X is in X^* iff its real part is continuous and every real linear map $X \to \mathbf{R}$ is the real part of a unique linear functional $T \in X^*$.

Example 2.11.11 Given a topological vector space X over $\mathbf{R}$ or $\mathbf{C}$, the elements of the set $\mathcal{C}(X, \mathbf{R})$ or $\mathcal{C}(X, \mathbf{C})$ are continuous linear functionals or simply functionals and this set of functionals is abbreviated as X^*.

Example 2.11.12 Let $\mathbf{I} = [0, 1]$ be the subspace of the real number space $\mathbf{R}$ and $\mathcal{C}(\mathbf{I}, \mathbf{R})$ be the space of all continuous functions $f : \mathbf{I} \to \mathbf{R}$. Given a function $g \in \mathcal{C}(I, \mathbf{R})$, define a function

$$\psi : \mathcal{C}(\mathbf{I}, \mathbf{R}) \to \mathbf{R},\ f \mapsto \int_0^1 f(t)g(t)dt,$$

where the integral is the Riemann integral. Then ψ is a linear map. Since,

$$|\psi(f)| = |\int_0^1 f(t)g(t)\,dt| \leq \int_0^1 |f(t)g(t)|\,dt \leq ||f|| \int_0^1 |g(t)|\,dt,\ \forall f \in C(I, \mathbf{R}),$$

it follows that ψ is bounded and continuous. Moreover

$$||\psi|| = \int_0^1 |g(t)|\,dt.$$

Hence, it follows that $\psi : \mathcal{C}(\mathbf{I}, \mathbf{R}) \to \mathbf{R},\ \ f \mapsto \int_0^1 f(t)g(t)dt$ is a continuous linear functional.

2.12 Applications

This section conveys some interesting applications of topological vector spaces in the areas of geometry, matrices, group theory, analysis and in some other areas. Given a linear subspace A of a linear topological space X, it is assumed that the quotient space X/A has the quotient topology $\tau_{X/A}$ with quotient map $p : X \to X/A,\ x \mapsto x + A$, unless stated otherwise. In this section, $\mathbf{F} = \mathbf{R}$ **or** $\mathbf{C}$ is endowed with usual topology.

2.12.1 Linear Isomorphism Theorem

This subsection proves linear isomorphism theorem analogous to first isomorphism theorem of groups.

Proposition 2.12.1 *Let X and Y be two topological vector spaces over* $\mathbf{F}$ *and $f : X \to Y$ be a continuous linear map. If Y is Hausdorff and $K = ker\ f = \{x \in X : f(x) = 0_Y \in Y\}$ is the kernel of f, then K is closed in X.*

Proof By hypothesis, Y is Hausdorff and $0_Y \in Y$. This implies that $\{0_Y\}$ is closed in Y. Again, since f is continuous, it follows that $K = f^{-1}(0_Y)$ is closed in X. ❑

Theorem 2.12.2 gives a convenient form of a **linear isomorphism theorem** which is analogous for group theory.

Theorem 2.12.2 *Let X and Y be two linear spaces over* $\mathbf{F}$ *and $f : X \to Y$ be a linear map with $K = ker\ f$. Then*

(i) *the induced map*

$$\tilde{f} : X/K \to Im\ f \subset Y,\ x + K \mapsto f(x)$$

is continuous (open) iff f is continuous (open) and

(ii) *f is continuous and open iff $\tilde{f}$ is a topological isomorphism.*

Proof Clearly, the natural projection map

$$p : X \to X/K,\ x \mapsto x + K$$

is continuous and open and $\tilde{f} \circ p = f$.

(i) Let $U \subset Im\ f$ be open and f be continuous. Then $f^{-1}(U)$ is open in X and hence $p(f^{-1}(U))$ is open in the quotient space X/K. This shows that $\tilde{f}^{-1}$ is open in the quotient space X/K and hence it proves that the map $\tilde{f}$ is continuous. Next suppose that $\tilde{f}$ is continuous. Since p is continuous and $\tilde{f} \circ p = f$, it follows that f is continuous, since f is the composite of two continuous maps p and $\tilde{f}$.

(ii) Let $V \subset Im\ f$ be a subset. Then $\tilde{f}^{-1}(V)$ is open in the quotient space X/K iff $p^{-1}(\tilde{f}^{-1}(V)) = (\tilde{f} \circ p)^{-1}(V)$ is open in X and hence $\tilde{f}$ is continuous iff $\tilde{f} \circ p = f$ is continuous. It is proved in an analogous way that $\tilde{f}$ is open iff $\tilde{f} \circ p$ is open. Since, a topological isomorphism is a linear isomorphism which is also a homeomorphism, hence the second part follows from the above discussion.

❑

2.12.2 Finite Dimensional Topological Vector Spaces

This subsection studies finite dimensional topological vector spaces over the field $\mathbf{F} = \mathbf{R}$ **or** $\mathbf{C}$ endowed with metric topology. It is well known from linear algebra that if X is a vector space over the field $\mathbf{F}$ is of dimension n, with a basis $\{e_1, e_2, \ldots, e_n\}$, then given any vector $x \in X$, there exist unique scalars $x_1, x_2, \ldots, x_n$ in $\mathbf{F}$ such that

$$x = x_1e_1 + x_2e_2 + \cdots + x_ne_n$$

In can be expressed as a mapping

$$f : \mathbf{F}^n \to X,\ (x_1, x_2, \ldots, x_n) \mapsto x_1e_1 + x_2e_2 + \cdots + x_ne_n$$

which is (algebraically) a linear isomorphism. This asserts that if X is a finite dimensional linear space over $\mathbf{F}$ then X is (algebraically) linear isomorphic to $\mathbf{F}^{\,dim\ X}$. This results leads to the results embodied in Theorem 2.12.10

Proposition 2.12.3 *Let X be a topological vector space over* $\mathbf{F}$. *Given any vector $v \in X$, the map*

$$\psi_v : \mathbf{F} \to X,\ \alpha \mapsto \alpha v$$

is continuous.

Proof Let X be a topological vector space over $\mathbf{F}$ with continuous scalar multiplication

$$\mu : \mathbf{F} \times X \to X,\ (\alpha, v) \mapsto \alpha v.$$

Then given $v \in X$,

$$\psi_v : \mathbf{F} \to X,\ \alpha \mapsto \alpha v = \mu(\phi_v(\alpha)),$$

where the mapping

$$\phi_v : \mathbf{F} \to \mathbf{F} \times X,\ \alpha \mapsto (\alpha, v)$$

is continuous. This shows that $\psi_v = \mu \circ \phi_v$ and hence ψ_v is continuous, since it the composite of two continuous maps. ❑

The boundedness of a topological vector space given in Definition 2.10.20 can be reformulated in Definition 2.12.4.

Definition 2.12.4 Let X be a topological vector space over $\mathbf{F}$ with origin 0. A subset $A \subset X$ is said to be **bounded** if corresponding to every nbd U of 0 in X, there is a real number $r > 0$ such that

$$A \subset t\,U, \;\; \forall t > r.$$

Theorem 2.12.5 *Let X be a topological vector space over $\mathbf{F}$. If f is a linear functional on X such that $f(x) \neq 0$ for some $x \in X$, then the following statements are equivalent:*

(i) *f is continuous;*
(ii) *$ker\ f = \{x \in X : f(x) = 0\}$ is closed in X;*
(iii) *$ker\ f$ is not dense in X;*
(iv) *f is bounded in some nbd of 0 in X.*

Proof By hypothesis, $f : X \to \mathbf{R}$ is a linear transformation such that $f(x) \neq 0$ for some $x \in X$.

1. $(i) \implies (ii)$: Let f be continuous. Then $K = ker\ f = f^{-1}(\{0\})$ is closed in X, since $\{0\}$ is a closed set in the scalar field $\mathbf{F}$.
2. $(ii) \implies (iii)$: Let $ker\ f$ be closed in X. Since by hypothesis, $f(x) \neq 0$ for some $x \in X$, it follows that $ker\ f \neq X$. It implies that $ker\ f$ is not dense in X.
3. $(iii) \implies (iv)$: Let $ker\ f$ be not dense in X. Then the complement of $ker\ f$ in X has nonempty interior. Since every nbd of 0 contains a balanced nbd of 0, it follows that there is a balanced nbd U of 0 such that

$$(x + U) \cap ker\ f = \emptyset \text{ for some } x \in X. \tag{2.3}$$

Then $f(U)$ is a balanced subset of the scalar field $\mathbf{F}$. This implies that either $f(U)$ is bounded in $\mathbf{F}$ or $f(U) = \emptyset$. The first case proves (iv) and the later case is not possible by (2.3).

$(iv) \implies (i)$: Let f be bounded in some nbd U of 0 in X. There there exists a real number $R_0 > 0$ such

$$|f(x)| < R_0, \; \forall x \in U.$$

Since, for an $\epsilon > 0$, if $V = (\epsilon/R_0)U$, then

$$|f(x)| < \epsilon, \; \forall x \in V,$$

it follows that f is continuous at the origin 0. It proves that f is continuous.

❑

Theorem 2.12.6 *Let X be a topological vector space over* $\mathbf{C}$ *and A be a subspace of X with dim A =n. Then every linear isomorphism $f : \mathbf{C}^n \to A$ is a homeomorphism, where $\mathbf{C}^n$ is endowed with the Euclidean topology.*

Proof Let $f : \mathbf{C}^n \to A$ be a linear isomorphism for all $n \geq 1$. Suppose T_n is the statement that every linear isomorphism $f : \mathbf{C}^n \to A$ is a homeomorphism for all $n \geq 1$. We prove the validity of the statement T_n by induction on n. For $n = 1$, consider $f : \mathbf{C} \to A$. Then by hypothesis, $f : \mathbf{C} \to A$ is a linear isomorphism. Suppose $f(1) = v \in A$. Then $f(\alpha) = \alpha v$. The continuity of f follows from the continuity of vector space operations of the topological vector space A. Its inverse map $f^{-1} : A \to \mathbf{C}$ exists by hypothesis and it is linear functional on A having its null space $\{0\}$ a closed set and hence f^{-1} is continuous by Theorem 2.12.5. This proves that f is homeomorphism and hence the statement T_n is true for $n = 1$.

Next, suppose that $n > 1$ and T_{n-1} is true. Let $\{e_1, e_2, \ldots, e_n\}$ be the standard basis of $\mathbf{C}^n$ and $f : C^n \to A$ be a linear isomorphism. If $f(e_i) = v_i$ for $i = 1, 2, \ldots, n$, then the map

$$f : \mathbf{C}^n \to A, (\alpha_1, \alpha_2, \ldots, \alpha_n) \mapsto \alpha_1 v_1 + \alpha_2 v_2 + \cdots + \alpha_n v_n$$

is continuous by the continuity of vector space operations of the topological vector space A. Clearly, $\{v_1, v_2, \ldots, v_n\}$ forms a basis of A by the linear isomorphism of f. This asserts that there are linear functionals $f_1, f_2, \ldots, f_n$ on A such that every vector $v \in A$ has a unique expression of the form

$$v = f_1(v)v_1 + f_2(v)v_2 + \cdots + f_n(v)v_n.$$

Since every $f_i : A \to \mathbf{C}$ has the null space of dimension $n - 1$ in A, which is a closed set in A, by our assumption of validity of T_{n-1}, it follows by Theorem 2.12.5 that each f_i is continuous for $i = 1, 2, \ldots, n$. This shows that f is continuous. Its inverse map

$$f^{-1} : \mathbf{C}^n \to A, \ v \mapsto (f_1(v), f_2(v), \ldots, f_n(v))$$

is also continuous and hence T_n is a homeomorphism. This shows that if T_{n-1} is true, then T_n is also true and hence the theorem is proved by the principle of mathematical induction. ❑

Corollary 2.12.7 *Let X be topological vector space over* $\mathbf{C}$ *and A be a subspace of X with $dim A = n$. Then A is closed in X.*

Proof It follows from Theorem 2.12.6. ❑

Theorem 2.12.8 *The dimension of every locally compact topological vector space is finite.*

Proof Let X be a locally compact topological vector space over $\mathbf{R}$ and $0 \in X$ has a nbd U such that its closure $\overline{U}$ is compact. Then $\overline{U}$ is bounded by Theorem 2.10.27

and the family of sets $\{\frac{1}{2^n}U : n = 1, 2, \ldots, \}$ forms a local base for X. Since $\overline{U}$ is compact, there exist $v_1, v_2, \ldots, v_r \in X$ such that

$$\overline{U} \subset (v_1 + \frac{1}{2}U) \bigcup \cdots \bigcup (v_r + \frac{1}{2}U).$$

If A be the vector subspace of X spanned by the vectors $v_1, v_2, \ldots, v_r \in X$, then $dim\ A \leq r$. It follows that A is a closed subspace of X by Corollary 2.12.7. Since $\alpha A = A,\ \forall \alpha \neq 0 (\in \mathbf{R})$ and $U \subset A + \frac{1}{2}U$, it shows that

$$\frac{1}{2}U \subset A + \frac{1}{2^2}U.$$

Hence, it follows that

$$U \subset A + \frac{1}{2}U \subset A + A + \frac{1}{2^2}U = A + \frac{1}{2^2}U.$$

Proceeding in this way, we find that

$$U \subset \bigcap_{n=1}^{\infty} (A + \frac{1}{2^n}U).$$

Since $\{\frac{1}{2^n}U\}$ is a local base, it follows by Theorem 2.10.25 that

$$U \subset \overline{A} = A \implies U \subset A \implies tU \subset A. \forall t = 1, 2, \ldots \implies A = X$$

by Theorem 2.10.25 . This asserts that the dimension of X is $\leq r$. ❑

Corollary 2.12.9 *Every locally bounded topological vector space having the Heine-Borel property has finite dimension.*

Proof By hypothesis, 0 has a bounded nbd U in X. Hence, it follows by Theorem 2.10.25 that $\overline{U}$ is also bounded. It implies by Heine-Borel property that $\overline{U}$ is compact. This shows that X is locally compact. Hence, the corollary follows from Theorem 2.12.8. ❑

Theorem 2.12.10 *Let X be a finite dimensional topological vector space of dimension n. Then*

(i) *X is topologically isomorphic to the linear space $\mathbf{F}^n$;*
(ii) *every linear functional f on X is continuous;*
(iii) *in more general, every linear map f from X to any other topological linear space Y is continuous;*

Proof Let X be topological vector space of dimension n over $\mathbf{F}$ with vector addition $\mathcal{A}$ and scalar multiplication μ. If $\{e_1, e_2, \ldots, e_n\}$ is a basis of X, consider the map

$$f : \mathbf{F}^n \to X,\ (x_1, x_2, \ldots, x_n) \mapsto x_1e_1 + x_2e_2 + \cdots + x_ne_n.$$

Clearly, f is (algebraically) a linear isomorphism. We claim that f is also a homeomorphism.

f **is continuous** : For $n = 1$, the map $f : \mathbf{F} \to X, x \mapsto x_1e_1$ is continuous by Proposition 2.12.3. For $n > 1$ consider the map

$$f : \mathbf{F}^n \to X,\ (x_1, x_2, \ldots, x_n) \mapsto x_1e_1 + x_2e_2 + \cdots + x_ne_n = \mathcal{A}(f_{e_1}(x_1), f_{e_2}(x_2), \ldots, f_{e_n}(x_1))$$

$$= \mathcal{A}((f_{e_1} \times f_{e_2} \times \cdots \times f_{e_n})(x_1, x_2, \ldots, x_n)),\ \forall x = (x_1, x_2, \ldots, x_n) \in \mathbf{F}^n,$$

where

$$f_{e_k}(x_k) = x_ke_k : k = 1, 2, 3, \ldots.$$

This shows that

$$f = \mathcal{A} \circ (f_{e_1} \times f_{e_2} \times \cdots \times f_{e_n})$$

is continuous, because it is the composite of continuous maps.

f^{-1} **is also continuous** : Use induction on dimension n. For $n = 1$, consider the map

$$f^{-1} : X \to \mathbf{F}, x = \alpha e_1 \mapsto \alpha.$$

It is clearly, continuous, because, $K = ker\ f^{-1} = \{0\}$ is closed in X and hence f^{-1} is continuous by Theorem 2.12.5. Since, $f : \mathbf{F}^n \to X$ is a linear isomorphism such that f and f^{-1} are both continuous. Hence, it follows that f is a homeomorphism. The remaining part of the theorem is left as an exercise. ❑

Theorem 2.12.11 (Tychonoff theorem) *For any $n \in \mathbf{N}$, the only topology that endows $\mathbf{F}^n$ a Hausdorff topological vector space (TVS) is the Euclidean topology.*

Proof Consider the Hausdorff topological vector space $\mathbf{F}^n$ endowed with Euclidean topology τ. Let σ be any topology on $\mathbf{F}^n$ such that $(\mathbf{F}^n, \sigma)$ is a Hausdorff topological vector space. Then by Theorem 2.12.10, the identity map

$$1_d : (\mathbf{F}^n, \tau) \to (\mathbf{F}^n, \sigma)$$

is an isomorphism of topological vector spaces. Then $\tau = \sigma$. Because, for any open set $U \in \sigma$, the open set $1_d^{-1}(U) = U \in \tau$. This implies that $\sigma \subset \tau$. Similarly, $\tau \subset \sigma$. This implies that $\sigma = \tau$. This proves the theorem.

❑

Corollary 2.12.12 (Another form of Tychonoff theorem) *There exists a unique topology τ on a finite dimensional vector space X such that (X, τ) is a Hausdorff topological vector space.*

Proof X be a finite dimensional topological vector space over **F** of dimension n. Then it follows by Theorem 2.12.11 that there exists a unique topology τ on X such that (X, τ) is a Hausdorff topological vector space. ❑

2.12.3 Completeness of Topological Vector Spaces

This subsection generalizes the notion of Cauchy sequence and studies completeness of topological vector spaces.

Recall that if (X, d) is a metric space, then a sequence $\{x_n\}$ is a Cauchy sequence if corresponding to every $\epsilon > 0$, there is an integer n_0 such that

$$d(x_m, x_n) < \epsilon, \text{ whenever } m > n_0 \text{ and } n > n_0.$$

If X is a topological vector space with topology τ and if $\mathcal{B}$ is a local base for τ, then a Cauchy sequence is redefined.

Definition 2.12.13 A sequence $\{x_n\}$ in X is said to be a Cauchy sequence if corresponding to every $U \in \mathcal{B}$, there is an integer n_0 such that

$$x_m - x_n \in U, \ \forall m, n > n_0.$$

Clearly, two local bases for τ determine the same class of Cauchy sequences.

Definition 2.12.14 Let X be a topological vector space. A net $\{(X_i, \geq) : i \in \mathbf{A}\}$ in X is said to be a **Cauchy net** if for every nbd U of 0, there is some $k \in \mathbf{A}$ with both i, j following k in the order $\geq$ such that $x_i - x_j \in U$. In other words, if X is a Cauchy net, then $x_i - x_j$ converges to zero. A cauchy sequence in X is sequence which is a Cauchy net. If a subset $A \subset X$ has the property that every Cauchy net in A converges to a point in A, then A is said to be **complete.** A is said to be sequentially complete if every Cauchy sequence in A converges to a point in A.

Definition 2.12.15 Let X be a topological vector space. A family $\mathcal{F}$ of subsets of X is said to contain **small sets** if for every nbd U of 0 in X, there is a member $A \in \mathcal{F}$ and an element $x \in X$ such that $A \subset x + U$, where $x + U$ is called a **translate** of the open set U.

Theorem 2.12.16 (A characterization of completeness) *Let X be a topological vector space and A be a subset X. Then A is complete in X iff every family $\mathcal{F}$ of closed subsets of X satisfying the conditions*

(i) *$\mathcal{F}$ has the finite intersection property (FIP) and*
(ii) *$\mathcal{F}$ contains small sets,*

has a nonempty intersection in the sense that

$$\bigcap_{F \in \mathcal{F}} F \neq \emptyset.$$

Proof Let X be a topological vector space and $A \subset X$. Suppose A is complete in X and $\mathcal{F}$ is a family of closed subsets of X satisfying the conditions (i) and (ii).

We claim that

$$\bigcap_{F \in \mathcal{F}} F \neq \emptyset.$$

Let $\mathcal{B}$ be the family of those sets which are the intersection of only a finite number of members belonging to $\mathcal{F}$ and $\mathcal{B}$ be directed by set-theoretic inclusion $\subset$. Construct a net $\{(x_B, \subset) : B \in \mathcal{B}\}$ by choosing an element from each $B \in \mathcal{B}$. By hypothesis (ii), $\mathcal{B}$ also contains small sets, since $\mathcal{F}$ contains small sets. Hence it follows that the above net is a Cauchy net that converges to a point $y \in A$, since, by hypothesis, A is complete. For any member $M \in \mathcal{B}$, the net $\{(x_B, \subset) : B \in \mathcal{B}\}$ is in M, by set inclusion relation and M is closed and $y \in B$, $\forall\, B \in \mathcal{B}$. This asserts that

$$\bigcap_{B \in \mathcal{B}} B \neq \emptyset,$$

which implies that

$$\bigcap_{F \in \mathcal{F}} F \neq \emptyset.$$

This proves the necessity of the condition. To prove the sufficiency of the condition, let $\{(x_i, \geq) : i \in \mathbf{A}\}$ be a Cauchy net. Construct a family $\mathcal{B}$ of sets containing small sets as follows: for every $i \in \mathbf{A}$, let $B_i = \{x_i : j \geq i\}$ and $\mathcal{B} = \{B_i : i \in A\}$. Since by hypothesis, $\{(x_j, \geq) : i \in \mathbf{A}\}$ is a Cauchy net, the family $\mathcal{B}$ contains small sets. Again, since, $(A, \geq)$ is a directed set, the intersection of its finitely many members belonging to $\mathcal{B}$ is nonempty ($\neq \emptyset$). Let $\mathcal{C}$ be the family of closures of the sets belonging to $\mathcal{B}$. Then $\mathcal{C}$ is a family of closed sets having the properties proved for $\mathcal{B}$ as above. Suppose that the intersection of this family is $\neq \emptyset$. Then there exists a point y such that $y \in \overline{B_i}$ for every i. Hence, it follows that the net converges to y. ❑

Theorem 2.12.17 characterizes compactness property of subsets of topological vector spaces X in terms of totally boundedness and completeness properties of X together.

Theorem 2.12.17 (Hausdorff theorem on total boundedness) *A subset A of a topological vector space X is compact iff A is totally bounded and complete.*

Proof By hypothesis, X is a topological vector space. Let A be a totally bounded and complete subset of X. To prove that A is compact, let $\mathcal{C}$ be a family of closed subsets of A such that it has the FIP. Then $\mathcal{C}$ is contained in a maximal family of this type. Without loss of generality, we assume that $\mathcal{C}$ is such a maximal family. We claim that the intersection

$$\bigcap_{C \in \mathcal{C}} C \neq \emptyset.$$

Since by hypothesis, A is complete, it is sufficient for this proof is to show that $\mathcal{C}$ contains small sets and hence it is sufficient to prove that if $\{A_i : i = 1, 2, \ldots, n\}$ forms a finite covering of A by closed subsets of A, then $A_i \in \mathcal{C}$ for some i, because, it is possible to cover A by a finite number of translates of a small open set. Since the family $\mathcal{C}$ is maximal, a closed subset $A_i(\subset A) \notin \mathcal{C}$, otherwise, its adjunction to $\mathcal{C}$ contradicts FIP. This shows that there exists a subfamily $\mathcal{A}_i \subset \mathcal{C}$ having empty intersection with A_i. This implies that there exists a finite subfamily such that $\bigcup\{\mathcal{A}_i : i = 1, 2, \ldots, n\}$ of $\mathcal{C}$ with empty intersection with every A_i and hence its intersection is empty, which is a contradiction. The converse part follows immediately, because every compact subset of a topological vector space is totally bounded and complete. ❑

Remark 2.12.18 For more results on completeness property of a topological vector space; see Exercises of Sect. 2.13

2.12.4 Torus, Lens Space, Real and Complex Projective Spaces as Orbit Spaces

This subsection identifies some important well-known topological spaces such a **torus, real and complex projective spaces** as orbit spaces obtained by actions of topological groups. This gives interesting applications of topological group actions.

Example 2.12.19 (n-torus T^n) Let $\mathbf{R}$ be the real line space and $\mathbf{R} \times \mathbf{R}$ be the Euclidean space with product topology. Consider the translation $h = \psi_1 : \mathbf{R} \to \mathbf{R}, x \mapsto x + 1$. Again, for each integer n,

$$\psi_n : \mathbf{R} \to \mathbf{R}, x \mapsto x + n$$

is also a translation and hence it is a homeomorphism. Then the cyclic group $< h >$ generated by h is isomorphic to $\mathbf{Z}$. The group $\mathbf{Z}$ is endowed with discrete topology acts as a group of homeomorphisms. The action ψ_n is free and the resulting orbit space $\mathbf{R}\ mod\ \mathbf{Z} = \mathbf{R}/\mathbf{Z}$ is homeomorphic to the circle group S^1. More precisely, in forming the orbit space $\mathbf{R}\ mod\ \mathbf{Z}$, two points $x, y \in \mathbf{R}$ are identified iff they differ by an integer. Again, the exponential map

$$p : \mathbf{R} \to S^1,\ x \mapsto e^{2\pi i x}$$

is an identification map and two points on $\mathbf{R}$ are identified iff they differ by an integer. This induces a homeomorphism

$$\psi_p : \mathbf{R}/\mathbf{Z} \to S^1,\ x + \mathbf{Z} \mapsto p(x).$$

For every ordered pair $(n, m) \in \mathbf{Z} \times \mathbf{Z}$, consider the action $\psi_{n.m}$ in a similar way:

$$\psi_{n.m} : (\mathbf{Z} \times \mathbf{Z}) \times (\mathbf{R} \times \mathbf{R}) \to \mathbf{R} \times \mathbf{R},\ (m, n)(x, y) \mapsto (x + m, y + n).$$

The action $\psi_{n.m}$ sends the point $(x, y) \in \mathbf{R}^2$. This action is free with the corresponding orbit space $\mathbf{R}^2 \mod \mathbf{Z}^2$ is the 2-torus $T^2 = S^1 \times S^1$. It consists of equivalence classes determined by the equivalence relation on $\mathbf{R}^2 : (x\ , y) \sim (s,\ t)$ iff $x - s$ and $y - t$ are both integers. Similarly, the n- torus $T^n = S^1 \times S^1 \times \cdots \times S^1 (n$ factors) is obtained as an orbit space.

Example 2.12.20 (Real projective space $\mathbf{R}P^n$) Let $f : S^n \to S^n, x \mapsto -x, \forall n \geq 1$ be the antipodal map. Then f is a homeomorphism, called a central symmetry. By definition of group action, identity element gives rise to the identity homeomorphism. Hence, the multiplicative group $G = \{f, f \circ f = 1_d\}$ is a group of homeomorphisms and is generated by the antipodal map. This group contains two elements and is isomorphic to the additive group $\mathbf{Z}_2$. The group $\mathbf{Z}_2$ endowed with discrete topology becomes a topological group and acts on S^n by identifying pairs of antipodal points and hence the orbit space S^n *mod* $\mathbf{Z}_2$, called the n-dimensional real projective space is denoted by $\mathbf{R}P^n$. This construction implies that the n dimensional real projective space $\mathbf{R}P^n$ is obtained from the n-sphere S^n by identifying diametrically opposite points $x, -x \in S^n$. Again, consider the action g of the multiplicative group $G = \mathbf{R}^* = \mathbf{R} - \{0\}$ of nonzero real numbers on the topological space $X = \mathbf{R}^{n+1} - \{0\}$ defined by

$$g : G \times X \to X, (r, x) \mapsto rx.$$

For every $x \in X$, orb x, the orbit of x represents geometrically the set of all the points except the point 0 on the straight line in $\mathbf{R}^{n+1}$ passing through the points 0 and x. This asserts that the orbit space $\mathbf{R}^{n+1}$ *mod* $\mathbf{R}^*$ is homeomorphic to the real projective space $\mathbf{R}P^n$ under the homeomorphism given by the correspondence by which the pair of antipodal points $\{x, -x\}$ is identified with the straight line passing through the points x and $-x$. The projective space is compact, since S^n is compact and $p : S^n \to \mathbf{R}P^n,\ x \mapsto [x] = \{x, -x\}$ is a continuous surjective map.

Example 2.12.21 (Complex Projective Space $\mathbf{C}P^n$) Let $S^{2n+1} = \{z = (z_0, z_1, \ldots, z_n) \in \mathbf{C}^{n+1} : \sum\limits_{i=0}^{n} |z_i|^2 = 1\}$ be the $(2n + 1)$-sphere S^{2n+1}. Consider the free action

$$\psi : S^1 \times S^{2n+1} \to S^{2n+1},\ (e^{i\theta}, z) \mapsto (e^{i\theta} z_0, e^{i\theta} z_1, \ldots, e^{i\theta} z_n), 0 \leq \theta < \pi.$$

The resulting orbit space S^{2n+1} *mod* S^1 is known as the n-dimensional complex projective space denoted by $\mathbf{C}P^n$ and is identified with the set of all complex straight lines in $\mathbf{C}^{n+1}$ passing through the origin 0.

Example 2.12.22 (Lens Space) Let $S^3 = \{z = (z_1, z_2) \in \mathbf{C}^2 : |z_1|^2 + |z_2|^2 = 1\}$ and p, q be given relatively prime integers (not necessarily prime). Define

$$f : S^3 \to S^3,\ (z_1, z_2) \to (z_1 e^{2\pi/p},\ z_2 e^{2\pi q/p}).$$

Then f generates a subgroup $G =< f >$ of the group of homeomorphisms of S^3 such that G is a cyclic group of order p, which is isomorphic to the finite cyclic group $\mathbf{Z}_p$. The orbit space $S^3\ mod\ G$ obtained by the action

$$\psi : G \times S^3,\ (f^n, z) \mapsto f^n(z)$$

is called the lens space, denoted by $L(p, q)$. It is also compact. In particular, $L(2, 1)$ is homeomorphic to $\mathbf{R}P^3$. In this sense, lens spaces are called generalization of projective spaces.

Example 2.12.23 (Klein bottle) Let T be two torus obtained as a subspace of the Euclidean space $\mathbf{R}^3$ by rotating the circle $(x - 3)^2 + z^2 = 1$ about the z-axis. Consider the homeomorphisms

$$h_1 : T \to T,\ (x, y.z) \mapsto (x, -y, -z);$$

$$h_2 : T \to T,\ (x, y.z) \mapsto (-x, -y, z);$$

and

$$h_3 : T \to T,\ (x, y.z) \mapsto (-x, -y, -z);$$

Each of these homeomorphisms is of order 2 and the group $G = \mathbf{Z}_2 = \{1_d, g\}$ act on T in three different ways and the orbit spaces corresponding to the three actions on T determined by the three homeomorphisms $h_i : i = 1, 2, 3$ are distinct.

(i) For the homeomorphism, h_1, the corresponding orbit space is homeomorphic to sphere S^2;
(ii) For the homeomorphism, h_2, the corresponding orbit space is homeomorphic to 2 torus $T^2 = T$;
(iii) For the homeomorphism, h_3, the corresponding orbit space is homeomorphic to the Klein bottle.

2.12.5 *Subgroups of the Topological Group* $(\mathbf{R}, +)$

This subsection studies subgroups of the topological group $(\mathbf{R}, +)$ and characterizes its dense topological subgroups.

The subgroups $(\mathbf{Q}, +)$(of rationals) and $(\mathbf{Z}, +)$ (of integers) of $(\mathbf{R}, +)$ have some well-known properties.

(i) The subgroup $\mathbf{Q}$ is dense in $\mathbf{R}$ but it is not cyclic;
(ii) The subgroup $\mathbf{Z}$ is cyclic but it is not dense in $\mathbf{R}$.

Remark 2.12.24 The above properties of **Q** and **Z** motivate to establish Theorems 2.12.25 and 2.12.28, which display interesting relations between algebra and topology. More precisely, Theorem 2.12.25 characterizes dense subgroups of the topological group $(\mathbf{R}, +)$ with the help of its noncyclic property. On the other hand, Theorem 2.12.28 asserts that corresponding to each cyclic subgroup of $(\mathbf{R}, +)$, there exists a dense subgroup of $(\mathbf{R}, +)$.

Theorem 2.12.25 *Let $(G, +)$ be a topological subgroup of the topological group $(\mathbf{R}, +)$. Then G is dense in* **R** *iff G is not cyclic.*

Proof Let $(G, +)$ be a dense subgroup of $(\mathbf{R}, +)$. Suppose that G is cyclic and hence $G =< x >$ for some $x \in \mathbf{R}$. Then $G = \{nx : n \in \mathbf{Z}\}$ shows that G can not be dense in $\mathbf{R}$. This contradicts our assumption that G is dense in $\mathbf{R}$. This implies that the subgroup $(G, +)$ is not cyclic.
Conversely, assume that the subgroup $(G, +)$ is not cyclic. Consider the set

$$K = \{x \in G : x > 0\}.$$

Since, since G is a group, this set K is a nonempty subset of $\mathbf{R}$, bounded below by 0. This shows that $inf\ K$ exists and is r, say. Then r is not in K, otherwise, r is the least positive element of G. Then $G = \{nr : n \in \mathbf{Z}\}$, because, $\{nr : n \in \mathbf{Z}\} \subset G$, since $r \in G$ and for any $y \in G$, if $y \notin \{nr : n \in \mathbf{Z}\}$, then there exists an $n \in \mathbf{Z}$ such that

$$nr < y < (n + 1)r,$$

which implies that

$$0 < y - nr < (n + 1) - nr = r.$$

This shows that the positive number $y - nr \in G$ is less than r. But it contradicts that r is least positive element of G. This asserts that $y \in \{nr : n \in Z\}$ and hence $G = \{nr : n \in Z\}$ implies that G is the cyclic group generated by r. But it contradicts our assumption that the group G is not cyclic. This shows that $inf\ K = r \notin K$. This implies that r is a limit of K and also a limit of G. Given an open interval (a, b), let $\epsilon = b - a > 0$. Since r is a limit point of G, the set $(r - \epsilon/2,\ r + \epsilon/2) \cap G$ is infinite. Take two distinct points $g_1, g_2 \in (r - \epsilon/2,\ r + \epsilon/2) \cap G$ such that $g_1 < g_2$ and hence

$$0 < g_2 - g_1 < \epsilon = b - a.$$

Hence there exists an integer m such that

$$m(g_2 - g_1) \in (a, b) \cap G.$$

This asserts that every open interval (a, b) in $\mathbf{R}$ intersects G and hence G is dense in $\mathbf{R}$. ❑

Corollary 2.12.26 *Let $(K, +)$ be a proper closed subgroup of the topological group $(\mathbf{R}, +)$. Then $(K, +)$ is cyclic.*

Proof By hypothesis, K is proper closed subgroup of the topological group $\mathbf{R}$. Hence, the set K can not be dense in $\mathbf{R}$. This proves the corollary by Theorem 2.12.25. ❑

An alternative proof of Corollary 2.12.26 is given in Theorem 2.12.27.

Theorem 2.12.27 *Let K be a proper closed subgroup of the topological group* $\mathbf{R}$. *Then the group K is cyclic.*

Proof By hypothesis, K is a proper closed subgroup of the topological group $\mathbf{R}$. If $K = \{0\}$, the proof is trivial. So, assume that $K \neq \{0\}$. Then for the set $\mathbf{R}^+$ of positive real numbers, $K \cap \mathbf{R}^+ \neq \emptyset$, since, all the elements of K can not be negative reals, as it is a group. Let $m = \inf\{k \in K \cap \mathbf{R}^+\}$. Then $m \neq 0$, otherwise, we have a contradiction. Because, if $m = 0$, then given an $\epsilon > 0$ and $x \in \mathbf{R}$, there exists an element $k \in (0, \epsilon) \bigcap K$ and integer $n \in \mathbf{N}$ such that for $x \geq 0$

$$nk \leq x < (n + 1)k$$

Hence, it follows that $nk \in K$ (by repeated addition) and

$$0 \leq x - nk \leq (n + 1)k - nk = k \leq \epsilon.$$

This asserts that $|x - nk| \leq \epsilon$, which implies that K is dense in $\mathbf{R}$. Proceed in a similar way for $x \leq 0$. As K is a proper closed subgroup of $\mathbf{R}$, the set K can not be dense in $\mathbf{R}$. This contradiction implies that $m > 0$. Then given $k \in K$, there exists an integer $n \in \mathbf{Z}$ such that

$$nm \leq |k| < (n + 1)m.$$

This implies that $|k| - nm \in K$ and

$$0 \leq |k| - nm \leq (n + 1)m - nm = m.$$

This implies by definition of infimum of a set that $|k| - nm = 0$, which shows that $k = +nm \ or - nm$, hence it follows that $K = < m >$.

❑

Theorem 2.12.28 determines a vast class of subgroups of $(\mathbf{R}, +)$ such that each subgroup of this class is dense in $\mathbf{R}$.

Theorem 2.12.28 *Let* $(G, +)$ *be a cyclic subgroup of* $(\mathbf{R}, +)$*, generated by some irrational number* x*. Then the subgroup* $K = \mathbf{Z} + G = \{n + g : n \in \mathbf{Z}, g \in G\}$ *is a dense subgroup of* $(\mathbf{R}, +)$*.*

Proof Clearly, $(K, +)$ is a subgroup of $(\mathbf{R}, +)$. No rational number is a generator of K, because, multiples of rational numbers are rational numbers but K contains irrational numbers. Again, no irrational number can generate K, since, a multiple of irrational number is either 0 or an irrational number, but K contains nonzero rational numbers. This asserts that the subgroup $(K, +)$ is not cyclic and hence K is dense in $\mathbf{R}$ by Theorem 2.12.25. ❑

Remark 2.12.29 Given any irrational number x, let $G(x) =< x >$ denote the cyclic subgroup of $(\mathbf{R}, +)$ generated by x. Then for two distinct irrational numbers x and y, the subgroups $K_x = \mathbf{Z} + G(x)$ and $K_y = \mathbf{Z} + G(y)$ are two different subgroups of $(\mathbf{R}, +)$. Hence, $\{K_t = \mathbf{Z} + G(t) : \forall t > 0, \textit{ and } t \in \mathbf{R} - \mathbf{Q}\}$ determines a family of dense subgroups of $(\mathbf{R}, +)$. If t is any irrational number, then the set $\mathbf{Q}(t) = \{x + yt : x, y \in \mathbf{Q}\}$ is also a subgroup of $(\mathbf{R}, +)$, which contains $\mathbf{Q}$ and hence $\mathbf{Q}(t)$ is dense in $\mathbf{R}$. But no subgroup of the family $\{K_t\}$ of subgroups contains $\mathbf{Q}$. All of them intersect $\mathbf{Q}$ at the points in the set $\mathbf{Z}$, i.e., at integral points only.

2.12.6 Generators of Connected Topological Groups

This subsection studies subgroups of connected topological groups and proves that every proper closed subgroup of the topological group $(\mathbf{R}, +)$ is cyclic.

Theorem 2.12.30 *Let* (G, τ) *be a connected topological group with identity* e*. Then every nbd of* e *forms a set of generators of* G*.*

Proof Let U be a nbd in G of the point e and $K = \langle U \rangle$ be the subgroup of G generated by the elements of the set U. To prove the theorem, it is sufficient to show that $G = K$. Now, for every $k \in K$, the nbd kU is the image of the left translation L_k on U, which is a homeomorphism. This asserts that $kU \subset K$ and K is an open set in G. Again, for any $x \in G - K$ if $xU \cap K \neq \emptyset$, then there exists an element $y \in xU \cap K$ such that $y = xu$ for some element $u \in U$. Hence it follows that $x = yu^{-1}$. But it is not possible, since both y and u^{-1} are in K but x is not in K. This implies that $xU \cap K = \emptyset$ and the nbd $L_x(U) = xU$ of x stays in $G - K$. Hence, $G - K$ is an open set in G. This shows that the given connected G is partitioned into two disjoint nonempty open sets K and $G - K$. This contradiction implies that $G = K$.

❑

Proposition 2.12.31 *Let* (G, τ) *be a topological group with its identity element* e *such that it is first countable. If* (G, τ) *is either Lindelöf or it has a dense subset, then* (G, τ) *has a countable basis.*

Proof Let $\mathcal{B} = \{B_n\}$ be a countable basis at e and the subset $X \subset G$ be countable. Then the family of sets $\mathcal{F} = \{xB_n : x \in X\}$ forms a base for the topology τ. On the other hand, if (G, τ) is Lindelöf, then select a countable set C_n for every $n \in \mathbf{N}$ such that the family $\{cB_n : n \in \mathbf{N}, c \in C\}$ forms a base for the topology τ of G. ❑

2.12.7 More Topological Applications to Matrix Algebra

This subsection gives further topological applications to the study of the set $M(n, \mathbf{R})$ of all $n \times n$ real matrices, considered as the Euclidean n^2-space $\mathbf{R}^{n^2}$, from the viewpoint of compactness and connectedness and other properties, in addition to its study in Chaps. 3 and 4. Complex matrices are also studied in an analogous way.

Proposition 2.12.32 *The orthogonal group $O(n, \mathbf{R})$ is compact but it is not connected.*

Proof $O(n, \mathbf{R})$ **is compact** : To show it consider the map

$$f : O(n, \mathbf{R}) \rightarrow O(n, \mathbf{R}), \; A \mapsto AA^t.$$

Then f is continuous and it is such that $O(n, \mathbf{R}) = f^{-1}(I_n)$. This asserts that $O(n, \mathbf{R})$ is closed in $M(n, \mathbf{R})$. For every matrix $A = (a_{i,j}) \in O(n, \mathbf{R}) \implies AA^t = I_n \implies$ each row vector of A has norm 1. This shows that every entry $a_{i,j}$ of A is such that $|a_{i,j}| \leq 1$ and, in particular,

$$a_{1i}^2 + a_{2i}^2 + \cdots + a_{ni}^2 = 1, \;\; \forall i, j = 1, 2, \ldots, n.$$

This asserts that $O(n, \mathbf{R})$ is bounded. Consequently, $O(n, \mathbf{R})$ is a closed and bounded subset of $\mathbf{R}^{n^2}$ and hence it is compact by Heine-Borel Theorem.

$O(n, \mathbf{R})$ **is no connected** : To show that $O(n, \mathbf{R})$ is not connected, consider $O(2, \mathbf{R})$. Let $A \in O(2, \mathbf{R})$. Then $det A \; is \; 1 \; or \; -1$. Consider the determinant function

$$det : O(2, \mathbf{R}) \rightarrow \{1, -1\}, \; A \mapsto det\; A.$$

For $A = I_2, \; det A = 1$ and for

$$A = \begin{pmatrix} 1 & 0 \\ 0 & -1 \end{pmatrix}$$

$det A = -1$ show that det function is a continuous surjective map. Suppose $O(2, \mathbf{R})$ is connected, then its continuous image $det\;(O(2, \mathbf{R})) = \{1, -1\}$ must be connected in $\mathbf{R}$. But is not connected in $\mathbf{R}$. This contradiction asserts that $O(2, \mathbf{R})$ is not connected. This proof in now extended in general case. For $O(n, \mathbf{R})$, consider

$$det : O(n, \mathbf{R}) \to \{1, -1\}. \ M \mapsto det \ M.$$

Since for any $M \in O(n, \mathbf{R})$, its determinant $det M = 1 \ or - 1$, the map det is well defined. Moreover, it is surjective, because, for identity matrix $I_n \in O(n, \mathbf{R}), \ det \ I_n = 1$ and for the diagonal matrix $M = (a_{i,j}) \in O(n, \mathbf{R})$, with

$$a_{1,1} = -1, \ a_{2,2} = \ a_{3,3} = \cdots = a_{n,n} = 1, \ and \ det \ M = -1.$$

Hence, the continuous image $det \ (O(n, \ \mathbf{R})) = \{1, -1\}. \implies O(n, \ \mathbf{R})$ is not connected, since $\{1, -1\}$ is not connected in $\mathbf{R}$. ❑

Corollary 2.12.33 *$O^+(n, \mathbf{R}) = \{A \in GL(n, \mathbf{R}) : det \ \ A > 0\}$ and $O^-(n, \mathbf{R}) = \{A \in GL(n, \mathbf{R}) : det \ A < 0\}$ are the connected components of $O(n, \mathbf{R})$.*

Proof It follows from the last part of Proposition 2.12.7. ❑

Corollary 2.12.34 *The special orthogonal group $SO(n.\mathbf{R})$ is a compact normal subgroup of $O(n, \ \mathbf{R})$.*

Proof Since $SO(n, \ \mathbf{R})$ is a connected component of the identity element of the topological group $O(n, \ \mathbf{R})$, it follows from Theorem 2.3.15 that $SO(n, \ \mathbf{R})$ is a closed normal subgroup of $O(n, \ \mathbf{R})$. Hence, the Corollary follows from Proposition 2.12.7 since every closed subgroup of a compact group is compact. ❑

Proposition 2.12.35 *The special linear group $SL(n, \mathbf{R})$ is not compact.*

Proof Consider the diagonal $n \times n$ matrix A_r having its diagonal entries $\{r, \frac{1}{r}, 1, \ldots, 1 : \ r > 0\}$. Then $det \ (A_r) = 1, \ \ \forall r > 0 \implies SL(n, \mathbf{R})$ is an unbounded set. This shows that $SL(n, \mathbf{R})$ is not compact. ❑

Theorem 2.12.36 *The unitary group $U(n, \mathbf{C})$ is compact.*

Proof Consider $U(n, \ \mathbf{C})$ identified with the Euclidean $\mathbf{R}^{2n^2}$-space. Proceed as in to show that $U(n, \ \mathbf{C})$ is a closed subspace of the Euclidean $\mathbf{R}^{2n^2}$-space. It is also bounded, because, for any $A = (a_{i,j}) \in U(n, \ \mathbf{C})$, the absolute value $|a_{i,j}| \leq 1$, with

$$|a_{1,i}|^2 + |a_{2,i}|^2 + \cdots + |a_{n,i}|^2 = 1.$$

Hence, it follows that $U(n, \ \mathbf{C})$ is a closed bounded set in the Euclidean $\mathbf{R}^{2n^2}$-space. This asserts that $U(n, \ \mathbf{C})$ is compact by Heine-Borel theorem. ❑

2.12.8 Special Properties Linear Functionals and Linear Transformations

This subsection studies some special properties of linear functionals and linear transformations and proves some equivalent statements on continuity appearing in analysis in the context of topological normed linear spaces.

Theorem 2.12.37 *Let X be a topological normed linear space and $\psi : X \to \mathbf{R}$ be a linear functional. Then the following statements are equivalent:*

(i) *ψ is uniformly continuous;*
(ii) *ψ is continuous;*
(iii) *for every $\epsilon > 0$, there exists an $\delta > 0$ such that*

$$|\psi(x)| < \epsilon, \forall x \in X \; with \; ||x|| < \delta.$$

(iv) *ψ is bounded.*

Proof $(i) \implies (ii)$: It follows trivially;

$(ii) \implies (iii)$. Let ψ be continuous. Since ψ is linear, $\psi(0) = \psi(x - x) = \psi(x) - \psi(x) = 0$. Then it implies (iii) by continuity of ψ at the origin.

$(iii) \implies (iv)$. Let (iii) hold but ψ is not bounded. Then for every integer $n > 0$, there is an element $x_n \in X$ such that $|\psi(x_n)| > n||x_n||$. This shows that $x_n \neq 0$. This implies that $||x_n|| > 0$. Define

$$y_n = (n||x||)^{-1}x_n \in X \; for \; every \; n \in \mathbf{N}.$$

Then it follows that

$$||y_n|| = 1/n \; and \; |\psi(y_n)| = (n||x||)^{-1}|\psi(x_n)| > 1,$$

which contradicts the hypothesis (iii) for $\epsilon = 1$.

$(iv) \implies (i)$. Let ψ be bounded. Then there exists a real number $M > 0$ such that

$$|\psi(x)| \leq M||x||, \forall x \in X.$$

Given an $\epsilon > 0$, the real number $\delta = M^{-1}\epsilon$ is such that

$$|\psi(x) - \psi(y)| = |\psi(x - y)| \leq M||x - y|| < MM^{-1}\epsilon = \epsilon, \forall x, y \in X,$$

since by hypothesis, $||x - y|| < \delta = M^{-1}\epsilon$. This asserts that ψ is uniformly continuous. ❑

Proposition 2.12.39 is proved by Hahn-Banach Theorem 2.12.38 and provides a rich supply of functionals.

Theorem 2.12.38 (Hahn-Banach Theorem) *Let X be a normed linear space and Y be a linear subspace of X. If T is a functional on Y, then it can be extended to a functional $\tilde{T_0}$ over X with the property that $||\tilde{T_0}|| = ||T||$.*

Proof Let $\mathcal{F}$ be the set of all extensions of T to functionals $\tilde{T}$ having the same norm on subspaces containing Y. Then $\mathcal{F} \neq \emptyset$. If $dom\tilde{T}$ represents the domain of $\tilde{T}$, then this domain set is partially ordered by the relation: $\tilde{T} \leq \tilde{T}'$ if $dom\tilde{T} \subset dom\tilde{T}'$ and $\tilde{T}(x) = \tilde{T}'(x)$ for all $x \in dom\tilde{T}$. Consider a chain $\{C_i : i \in \mathbf{A}\}$ in $\mathcal{F}$ and the union of any chain of members of the family $\mathcal{F}$. Then this union is an upper bound for the chain. Now use Zorn's lemma to show that there exists a maximal extension $\tilde{T}_0$. Again $dom\tilde{T} = X$, because, otherwise, $\tilde{T}_0$ can be further extended, which will contradict the maximality of $\tilde{T}_0$. ❑

Proposition 2.12.39 *Let X be a topological normed linear space over $\mathbf{R}$ and $x_0 \in X$ be a nonzero vector. If X^* denotes the set of all continuous linear functionals (real) defined on X, then there exists a functional T_0 in X^* such that $T_0(x_0) = ||x_0||$ and $||T_0|| = 1$.*

Proof Let Y be the linear subspace $Y = \{\alpha x_0 : \alpha \in \mathbf{R}\}$ of X generated by the vector x_0 and T be the functional on Y defined by $T(\alpha x_0) = \alpha||x_0||$. Then $T(x_0) = ||x_o||$ and $||T|| = 1$. Now use Hahn-Banach Theorem 2.12.38 to complete the proof. ❑

Proposition 2.12.40 also provides a rich supply of functionals.

Proposition 2.12.40 *Let X be a topological normed linear space over $\mathbf{F} = \mathbf{R}$ or $\mathbf{C}$ and Y be a closed linear subspace of X and $x_0 \in X - Y$ be a vector. Then there exists a functional f such that $f(y) = 0\ \forall\, y \in Y$ but $f(x_0) \neq 0 \in F$.*

Proof The natural projection map $p : X \to X/Y,\ x \mapsto x + Y$ is a continuous linear transformation such that $p(y) = 0,\ \forall\, y \in Y$ and $p(x_0) = x_0 + Y$. Then by Proposition 2.12.39, there exists a functional $T \in (X/Y)^*$ such that $T(x_0 + Y) \neq 0$. Define a map

$$f : X \to \mathbf{F},\ x \mapsto (T \circ p)(x)$$

Then f is a linear functional satisfying the required properties. ❑

Remark 2.12.41 Results given in Theorem 2.12.42 is analogous to Theorem 2.12.37 in the context of normed linear spaces. Theorem 2.12.37 also follows from Theorem 2.12.42.

Theorem 2.12.42 *Let X and Y be two topological normed linear spaces and $T : X \to Y$ be a* ***linear transformation****. Then the following statements are equivalent:*

(i) *T is uniformly continuous;*
(ii) *T is continuous;*
(iii) *for every $\epsilon > 0$, there exists an $\delta > 0$ such that*

$$||T(x)|| < \epsilon,\ \forall\, x \in X\ with\ ||x|| < \delta.$$

(iv) *T is bounded.*

Proof It is similar to the proof of Theorem 2.12.37. ❑

2.12.9 More Applications

This subsection communicates more applications of topological groups and topological vector spaces.

The group $(\mathbf{Z}, +)$ a topological group under the usual (Euclidean) topology σ. On the other hand, Proposition 2.12.43 shows that the group $(\mathbf{Z}, +)$ is not a topological group under the topology τ defined in this proposition.

Proposition 2.12.43 *The group $(\mathbf{Z}, +)$ is not a topological group under the topology τ obtained by declaring a subset $U \subset \mathbf{Z}$ to be open if*

$$0 \notin U \text{ or } \mathbf{Z} - U \text{ is finite.}$$

Proof Suppose $\mathbf{Z}$ is a topological group with this topology τ. Because, every left translation on a topological group is a homeomorphism by Proposition 2.2.9 and hence the left translation $L_1 : \mathbf{Z} \to \mathbf{Z}, n \mapsto 1 + n$ is a homeomorphism. Since $0 \notin \{-1\}$, the open set $\{-1\} \in \tau$ but its homeomorphic image $L_1(\{-1\}) = \{0\} \notin \tau$. This contradiction asserts that $\mathbf{Z}$ is not a topological group under the topology τ. ❑

Proposition 2.12.44 *Let G be a topological group and the subsets A, B of G be compact. Then $AB = \{ab : a \in A, b \in B\}$ is also compact.*

Proof Let G be a topological group with continuous multiplication

$$m : G \times G \to G, (x, y) \mapsto xy.$$

By hypothesis A and B are compact subsets of G. Then their product space $A \times B$ is also compact by Tychonoff theorem. Since AB is the image of $A \times B$ under the continuous multiplication m, it follows that AB is also compact. ❑

Theorem 2.12.45 *Let G be a topological group. Then*

(i) *the component C_e of the identity element e of G is a normal subgroup of G and*
(ii) *other components of G are cosets of C_e in G.*

Proof By hypothesis, G is a topological group and C_e is the component (connected) of identity element e of G.

(i) If $m : G \times G \to G$ is the continuous multiplication of G, then the continuous image $m(C_e \times C_e))$ of the connected subset $(C_e \times C_e)$ of $G \times G$ is a connected nbd of e in $G \implies m(C_e \times C_e)) \subset C_e$. Again, for $f : G \to G, x \mapsto x^{-1}$, clearly, $C_e^{-1} = f(C_e) \implies C_e^{-1}$ is also a connected nbd of e and hence $C_e^{-1} \subset C_e$. Since every translation on G onto itself is a homeomorphism, it follows that for any $g \in G$, the image gC_eg^{-1} is also a connected nbd of $e \implies gC_eg^{-1} \subset C_e \implies C_e$ is a normal subgroup of G.

(ii) For any component C_g of $x \in G$, it follows similarly that

$$gC_e \subset C_g$$

and

$$g^{-1}C_g \subset C_e.$$

It proves that $C_g = gC_e$ and hence the component C_g is a coset of C_e in G.

❑

Theorem 2.12.46 *Let X be a Hausdorff space and G a compact topological group. Suppose G acts on X. If G_x be its isotropy group at $x \in X$, and G/G_x is topolozied by quotient topology, then the map*

$$h : G/G_x \to orb\ x,\ gG_x \mapsto gx$$

is a homeomorphism.

Proof h is a bijection, since it establishes one-one and onto correspondence between the sets $\{gG_x\}$ of all cosets of G_x in G and $orb\ x$. The map h is continuous by the definition of the quotient topology on G/G_x. To show that h is a homeomorphism, use the topological result that any continuous bijective map from a compact space onto a Hausdorff space is a homeomorphism. ❑

Corollary 2.12.47 *The quotient space $O(n, \mathbf{R})/O(n-1, \mathbf{R})$ and the unit $(n-1)$ -sphere S^{n-1} are homeomorphic for $n \geq 2$.*

Proof Consider the action

$$\psi : O(n,\ \mathbf{R}) \times S^{n-1} \to S^{n-1},\ M \mapsto x^t.$$

This action is well defined, since an orthogonal matrix preserves lengths of vectors and hence $O(n,\ \mathbf{R})$ sends every unit vector $x \in \mathbf{R}^n$ to the unit vector in $x \in \mathbf{R}^n$. For example, for $n = 2$ *and* $x = (cos\ \theta, sin\ \theta) \in S^1$, the matrix

$$M = \begin{pmatrix} cos\ \theta & -sin\ \theta \\ sin\ \theta & cos\ \theta \end{pmatrix} \in O(2,\ \mathbf{R})$$

sends the vector $e_1 = (1, 0) \in S^1$ to the vector x. For any $n \geq 2$, the orthogonal group $O(n, \mathbf{R})$ is a compact subgroup of the general linear group $Gl(n, \mathbf{R})$. The sphere S^{n-1} consists of the unit vectors in $\mathbf{R}^n$. The group $G = O(n, \mathbf{R})$ acts transitively on S^{n-1} by the above group action ψ, since there is an orthogonal matrix sending any vector of unit length in any other vector in S^{n-1}. It shows that the isotropy group G_x at the point $x = (0, 0, \ldots, 1) \in S^{n-1}$ is isomorphic to the group $O(n-1, \mathbf{R})$.

Again, since this action is transitive the orbit space of this action is S^{n-1}. Hence, the quotient space $O(n, \mathbf{R})/O(n-1, \mathbf{R})$ is homeomorphic to S^{n-1} by Theorem 2.12.46. ❑

Corollary 2.12.48 *For the unitary group $U(n, \mathbf{C})$, the quotient space $U(n, \mathbf{C})/U(n-1, \mathbf{C})$ and the unit $(2n-1)$ -sphere S^{2n-1} are homeomorphic for $n \geq 2$.*

Proof Proceed as in Corollary 2.12.47 to define an action of the group $U(n, \mathbf{C})$ on S^{2n-1}. This action is transitive, since there is a unitary matrix sending any vector of unit length in any other vector in S^{2n-1}. The isotropy G_x at the point $x = (0, 0, \ldots, 1) \in S^{2n-1}$ is $U(n-1, \mathbf{C})$. Hence, the Corollary follows from Theorem 2.12.46. ❑

Corollary 2.12.49 *For the symplectic group $Sp(n, \mathbf{H})$, the quotient space $Sp(n, \mathbf{H})/Sp(n-1, \mathbf{H})$ and the unit $(4n-1)$ -sphere S^{4n-1} are homeomorphic for $n \geq 2$.*

Proof Proceed as in Corollary 2.12.47 and finally apply Theorem 2.12.46. ❑

Proposition 2.12.50 *Let (X, τ) and (Y, σ) be two topological vector spaces over the same topological field $\mathbf{F}$ and $f : X \to Y$ be a linear transformation. If Y is Hausdorff and f is continuous, then $ker f = \{x \in X : f(x) = 0_Y\} = f^{-1}(\{0_Y\})$ is closed in (X, τ).*

Proof By hypothesis, Y is Hausdorff and hence $\{O_Y\}$ is closed in (Y, σ). Again, since f is continuous, it follows that $ker f = f^{-1}(\{0_Y\})$ is closed in (X, τ). ❑

Theorem 2.12.51 characterizes continuity of linear transformation f between topological vector spaces in terms of continuity of its associated map $\tilde{f}$ defined in this theorem.

Theorem 2.12.51 *Let (X, τ) and (Y, σ) be two topological linear spaces over the same topological field $\mathbf{F}$ and $f : X \to Y$ be a linear transformation. Then f is continuous iff the map*

$$\tilde{f} : X/ker f \to Im f,\ x + ker f \mapsto f(x) \text{ is continuous.}$$

Proof Let $p : X \to ker f,\ x \mapsto x + ker f$ be the projection map. Then p is continuous. Suppose $\tilde{f}$ is continuous. Then $\tilde{f} \circ p = f$. This implies that f is also continuous, since it is the composite of two continuous maps.

Conversely, let $f : X \to Y$ be continuous and U be an open set in Imf. Then $f^{-1}(U)$ is also an open set in (X, τ). Hence, it follows that $\tilde{f}^{-1}(U) = p(f^{-1}(U))$ is also open in (X, τ), since p is continuous. This implies that $\tilde{f}$ is continuous. ❑

Proposition 2.12.52 *For every non-identity element α of a topological group X, the left translation map*

$$L_\alpha : X \to X,\ x \mapsto \alpha x$$

is a homeomorphism having no fixed point.

Proof For every non-identity element α of a topological group X, the left translation

$$L_\alpha : X \to X, \ x \mapsto \alpha x$$

is a homeomorphism and is such that

$$L_\alpha(x) = \alpha x \neq x, \ \ \forall x \in X.$$

This asserts that L_α has no fixed point. ❑

Corollary 2.12.53 *Let X be a nontrivial topological group with the identity element e. Then X has not the fixed-point property.*

Proof For every element $\alpha \neq e$ of the topological group X, the left translation

$$L_\alpha : X \to X, \ x \mapsto \alpha x$$

is a homeomorphism by Proposition 2.12.52. Hence, it follows that L_α is a continuous map such that $L_\alpha(x) = \alpha x \neq x, \ \ \forall x \in X$, which proves the Corollary. ❑

Corollary 2.12.54 *The topological group* $(\mathbf{R}, +)$ *has no fixed- point property.*

Proof It follows from Proposition 2.12.52. ❑

2.12.10 Special Properties of $O(n, \mathbf{R})$, $U(n, \mathbf{C})$ and $Sp(n, \mathbf{H})$

This subsection proves some special properties of $O(n, \mathbf{R})$, $U(n, \mathbf{C})$ and $Sp(n, \mathbf{H})$ by using Theorem 2.12.55.

Theorem 2.12.55 *Let G be a compact topological group and X be a Hausdorff space. If G acts on X, and G_x is the isotropy group at $x \in X$ with orbit space Gx, then map*

$$\psi : G/G_x \to Gx : gG_x \mapsto gx$$

is a homeomorphism.

Proof Clearly, the map ψ is continuous under the quotient topology on G/G_x. It is onto, because for any $y \in Gx$ it can be expressed as $y = g_y x$ for some $g_y \in G$ and hence $\psi(g_y G_x) = g_y x = y$. Again, for $g, h \in G, \ gx = hx$, if $g^{-1}h \in G_x$, then $gG_x = hG_x$ implies that ψ is injective. Consequently, ψ is continuous one-one and onto map from a compact space to a Hausdorff space and hence ψ is a homeomorphism. ❑

Theorem 2.12.56 *For every integer $n \geq 2$, the sphere S^{n-1} and the factor space $O(n, \mathbf{R})/O(n-1, \mathbf{R})$ are homeomorphic.*

Proof Consider the orthogonal (real) topological space $G = O(n, \mathbf{R})$ for $n \geq 2$. It is a compact topological group and $X = S^{n-1} \subset \mathbf{R}^n$ is a Hausdorff space. A matrix $M \in O(n, \mathbf{R})$ is a transformation of the Euclidean space $\mathbf{R}^n$ and it preserves lengths of the vectors and hence it is a map of S^{n-1} to itself. Again, $O(n-1, \mathbf{R})$ is regarded as the subgroup of $G = O(n, \mathbf{R})$ obtained by keeping the last coordinate fixed. Let $s_0 = (0, 0. \ldots, 1)$. This point is kept fixed by $O(n-1, \mathbf{R})$. This gives a map mapping from $O(n, \mathbf{R})$ into $X = S^{n-1}$. More precisely, define the map

$$\psi : O(n, \mathbf{R}), M \mapsto M(0, 0. \ldots, 1)^t.$$

If $A \in O(n-1, \mathbf{R})$, then $\psi(MA) = \psi(M)$ asserts that the map ψ factors via the left coset space $O(n, \mathbf{R})/O(n-1, \mathbf{R})$. The topological group $O(n, \mathbf{R})$ acts on $S^{(n-1)}$ transitively and the induced map

$$\psi^* : O(n, \mathbf{R})/O(n-1, \mathbf{R}) \to S^{n-1}$$

a continuous bijective map from a compact space to a Hausdorff space. Hence ψ^* is a homeomorphism. This proves the theorem. ❑

Theorem 2.12.57 *For every integer $n \geq 2$, the sphere S^{2n-1} and the factor space $U(n, \mathbf{C})/U(n-1, \mathbf{C})$ are homeomorphic.*

Proof Since S^{2n-1} is the set of all unit vectors in $\mathbf{C}^n$ and the unitary group $U(n, \mathbf{C})$ acts on S^{2n-1} transitively, there is a unitary matrix sending any vector of length 1 to any other vector. Now proceed as in Theorem 2.12.55 to prove the theorem. ❑

Theorem 2.12.58 *For every integer $n \geq 2$, the sphere S^{4n-1} and the factor space $Sp(n, \mathbf{H}) / Sp(n-1, \mathbf{H})$ are homeomorphic.*

Proof Since S^{4n-1} is the set of all unit vectors in $\mathbf{H}^n$ and the simpletic group $Sp(n, \mathbf{H})$ acts on S^{4n-1} transitively. because, there is a unitary matrix sending any vector of length 1 to any other vector. Now proceed as in Theorem 2.12.55 to prove the theorem. ❑

2.13 Exercises

1. Prove the following topological isomorphisms (for geometrical understanding):
   - **(i)** The quotient group $\mathbf{R}/\mathbf{Z}$ is topologically isomorphic to the circle group S^1;
   - **(ii)** the torus $T = S^1 \times S^1$ is a topological group under product topology such that the quotient topological group $(\mathbf{R} \times \mathbf{R})/(\mathbf{Z} \times \mathbf{Z})$ (called the group of Gaussian integers) is topologically isomorphic to the torus group $T = S^1 \times S^1$.

[Hint: The product of two topological groups is a topological group.]

2. Using topological tools, show that

(i) every nontrivial discrete subgroup of $(\mathbf{R}, +)$ is infinite cyclic;
(ii) every nontrivial closed subgroup of $(\mathbf{R}, +)$ with usual topology is infinite cyclic;
(iii) every nontrivial discrete subgroup of the circle group $(S^1, \cdot)$ is finite and cyclic;
(iv) every nontrivial subgroup of the circle group $(S^1, \cdot)$ with usual topology is either dense in S^1 or finite;
(v) every nontrivial closed subgroup of the circle group $(S^1, \cdot)$ with usual topology is either cyclic or finite.

3. Let X be a topological space and $\mathbf{h}omeo(X)$ be the group of homeomorphisms under usual composition of mappings. Let $\mathbf{h}omeo(X)$ be endowed with compact open topology. Show that if X is Hausdorff and locally compact, then $\mathbf{h}omeo(X)$ is a topological group.

4. Let $M, N \in O(2, \mathbf{R})$ be two matrices such that det M $= 1$ and det N $= -1$. Show that

(i) $N^2 = I_2$ (Identity matrix) and $NMN^{-1} = M^{-1}$;
(ii) every discrete subgroup of $O(2, \mathbf{R})$ is either cyclic or dihedral.

[Hint: Use (i) to prove (ii).]

5. A topological space X is said to be **regular** if for every nbd U of any point $p \in X$, there exists a nbd V of this point such that $\overline{V} \subset U$. Show that the topological space (underlying space) of every topological group is a regular space.

[Hint: Let G be a topological group with e its identity element and U be a nbd of e. As $ee^{-1} = e$, there exist a nbd V of e such that $VV^{-1} \subset U$. The closure $\overline{V} \subset U$ in G. Because, for any point $p \in \overline{V}$, its every nbd intersects V and hence pV being also a nbd of p has the property that there exists a point $y \in V$ such that $py = x \in V$. Then $p = xy^{-1} \in VV^{-1} \subset U$. This implies that $\overline{V} \subset V$.]

6. Let G be a topological group with identity element e. Given a nbd V_e of e in G and an integer $n > 0$, show that there exists a symmetric nbd U of e such that $U^n \subset V_e$.

7. Let G and H be two topological groups with e_G and e_H their respective identity elements and $\psi : G \to H$ be a group homomorphism from the topological group G to the topological group H.

Show that

(i) if ψ is continuous at the point e_G, then ψ is continuous everywhere ;
(ii) if ψ is continuous at one point $a \in G$, then ψ is continuous everywhere;
(iii) if ψ is continuous iff ψ is continuous at the point e_G.

8. Let $\{(G_a, \tau_a) : a \in \mathbf{A}\}$ be a family of topological groups. If $G = \Pi_{a \in \mathbf{A}} G_a$ is endowed with product topology, show that G is a topological group under coordinate-wise multiplication and product topology $\Pi_{:a \in \mathbf{A}} \tau_a$.

9. Let X be a left G-space with orbit space $X mod\ G$. Show that the projection map

$$\pi : X \to X mod\ G,\ x \mapsto Gx$$

is an open map.

10. Let G be a topological group. If A is a closed subspace of G, B and H are both compact subsets of G, show that

(i) AB is closed in G;
(ii) BH is compact and
(iii) moreover, if H is a normal subgroup of G, the natural projection map

$$p : G \to G/H, g \mapsto gH$$

is also closed.

11. Let G be a topological group and K be a normal subgroup of G. Show that

(i) the quotient group G/K is discrete iff K is open;
(ii) the quotient group G/K is Hausdorff iff K is closed.

12. Let G be a topological group and $g \in G$. If A, B and C be three subsets of G such that

(i) A is closed in G;
(ii) B is open in G;
(iii) C is any subset of G

show that gA, Ag and A^{-1} are closed sets in G and gB, Bg, B^{-1}, BC and CB are open sets in G.

13. Let H be a subgroup of a topological group G. Show that

(i) H is also Hausdorff;
(ii) if H is closed and G is compact, then H is also compact;
(iii) if H is closed and G is locally compact, then H is also locally compact.

14. Let G and H be two locally compact topological groups satisfying the second axiom of countability. If $f : G \to H$ is an arbitrary onto homomorphism of topological groups, show that the map f is open.

[Hint: For any open set U in the topological space G, use the local compactness and regularity properties of the topological group to show that $f(U)$ is an open set in the topological space H.]

15. Let G be a topological group and act on a topological space X. If both the spaces G and $X\ mod\ G$ are connected, show that X is also connected.

16. Let G and H be two topological groups and $f : G \to H$ be an open homomorphism of topological groups. Show that $ker\ f$ is a normal subgroup of the topological group.

17. Let G be a locally compact topological group satisfying the second axiom of countability, H be a subgroup of G and N be a normal subgroup of G. If the product space $P = HN$ is a closed subset of the topological space G and $M = H \cap N$, show that the quotient groups H/M and P/N are isomorphic as topological groups.

[Hint: Consider the map

$$f : H \to HN/N,\ h \mapsto hN.$$

Then f is an epimorphism of abstract groups. Then $ker\ f = H \cap N$. Then it follows from the theory of abstract groups that the abstract groups $H/ker\ f$ is isomorphic to the abstract group $Im(f) = HN/N$ by an isomorphism

$$\tilde{f} : H/kerf \to HN/N,\ h\,ker\ f \mapsto f(h).$$

Then $\tilde{f}$ is a homeomorphism under the given conditions.]

18. Let G be a topological group and K be a closed normal subgroup of G. Show that

(i) if G is connected, then G/K is also connected;
(ii) if both K and G/K are connected, then G is also connected.

19. **(Topological Isomorphism Theorems)** A topological isomorphism ($\cong$) between two topological groups G and H is an isomorphism (algebraically) and is also a homeomorphism between two topological groups, denoted by $G \cong H$.

(i) Let G and H be two topological groups and $f : G \to H$ be a topological group homomorphism. If f is onto, show that the quotient group $G/kerf$ and the group H are topologically isomorphic, denoted by

$$G/kerf \cong H.$$

(ii) Let G be a topological group and A, B be two closed subgroups of G such that B is normal and AB is closed. If either A or B is compact, show that

$$\frac{A}{A \cap B} \cong \frac{AB}{B}$$

[Hint: See Proposition 2.6.10 and Theorem 2.7.9]

20. Let G be a topological group satisfying the second axiom of countability. If A and B are two compact subsets of G, show that AB also compact in G.

21. Let G be a topological group and K be a normal subgroup of G. Show that if G satisfies the second axiom of countability, then the quotient group also satisfies the second axiom of countability.

22. Let G be a topological group and K be a normal subgroup of G.

(i) if G is compact, then G/K is also so;
(ii) if both K and G/K are compact, then G is also so.

23. Let G be a compact topological group act on a Hausdorff space X. Show that the orbit space $Xmod$ is a compact Hausdorff space

24. Let G be a topological group and K be a closed subgroup of G. If K and G/K are both connected spaces,

(i) show that G is also a connected space;
(ii) hence show that the special orthogonal group $SO(n, \mathbf{R}) = \{M \in O(n, \mathbf{R}) : det\ M = 1\}$ is connected;
(iii) further show that $SO(n, \mathbf{R})$ is the component of $O(n, \mathbf{R})$ containing its identity element.

25. Let G be a topological group and K be a compact normal subgroup of G. Show that the natural projection map

$$p : G \to G/K,\ g \mapsto gK$$

is closed.

26. X be a topological vector space over the field $\mathbf{C}$ and Y be a subspace of X.

If $dim_{\mathbf{C}} Y = n > 0$, show that

(i) every linear isomorphism $f : \mathbf{C}^n \to Y$ is a homeomorphism;
(ii) Y is a closed subspace of X.

27. Let X and Y be two topological vector spaces and $T : X \to Y$ be a linear transformation. If T is continuous at $0 \in X$, show that T is continuous.

[Hint: Continuity of T at 0 implies that given a nbd V of 0 in Y, there exists a nbd U of 0 in X such that $T(U) \subset V$. If $x - y \in U$, then by linearity of T it follows that $T(x) - T(y) \in V$. It shows that T maps the nbd $x + U$ of x into the nbd $T(x) + V$ of $T(x)$.]

28. Let X and Y be two topological vector spaces and $T : X \to Y$ be a linear transformation. Show that T is continuous at every point $x \in X$ iff T is continuous at some point of X.

29. Let X be a topological vector space and $\mathcal{B}$ be a local base for X. Show that every element of $\mathcal{B}$ contains the closure of some element of $\mathcal{B}$.

[Hint: Use Theorem 2.10.23 by taking $A = \{0\}$.]

30. Show that the topological product of an arbitrary family of topological vector spaces is also a topological vector space.

31. Let X be a topological vector space and the subspace $A \subset X$ be locally compact (i.e., 0 has a nbd whose closure is compact) . Show that A is closed in X.

32. Let X be a topological vector space with topology τ and a countable local base. Show that there exists a metric d on X such that

(i) $\tau = \tau_d$ (induced topology by d);

(ii) d is invariant in the sense that

$$d(x + v, y + v) = d(x, y), \ \forall\, x, y, v \in X.$$

33. Let X be a topological vector space with topology τ. If d and ρ be two invariant metrics on X such that they induce the same topology τ on X. Show that

(i) d and ρ have the same Cauchy sequences;

(ii) d is complete iff ρ is also so.

34. Let X be a topological vector space and A be a subspace (vector) having the subspace topology (inherited from X) of X. If A is an F-space, show that A is a closed subspace of X.

35. A topological space (X, τ) is said to be **completely regular** if one-pointic sets are closed in X and if for every point $p \in X$ and every closed set C not containing the point p, there exists a continuous function $f : X \to [0, 1]$ such that $f(p) = 1$ and $f(x) = 0, \ \forall\, x \in C$. Show that a Hausdorff space (X, τ) is completely regular iff the closed ideals in the topological algebra $\mathcal{C}(X)$ are in bijective correspondence with the closed subsets of X.

36. A subset S of a topological vector space X is said to be **convex** if S contains the straight line segment joining every pair of points of S in the sense that

$$rx + ty \in S, \ \forall\, x, y \in S, \ \forall\, t, s \geq 0 \ with\ t + s = 1.$$

Show that for any convex subset S of X,

(i) the Int S is also a convex subset of X;

(ii) the closure $\overline{S}$ is also a convex subset of X.

37. Let X be a locally convex topological vector space over **R** and Y be a subspace of X. Show that for a linear functional $f : Y \to \mathbf{R}$, there exists a linear functional

$$T_f \in X^* \ such\ that\ T_f = f \ on\ Y.$$

38. Let $T : X \to Y$ be a linear transformation of normed linear spaces. Show that T is continuous iff T is continuous at $0 \in X$ in the sense that $x_n \to 0$ asserts that $T(x_n) \to 0$.

39. Let G be a compact topological space and X be a topological on which G acts. If $X mod\ G$ be the corresponding orbit space, prove the following statements:

(i) If X is a normal space, then $Xmod\ G$ is also a normal space;
(ii) If X is a Hausdorff space, then $Xmod\ G$ is also a Hausdorff space;
(iii) If X is a regular space, then $Xmod\ G$ is also a regular space;
(iv) If X is a locally compact space, then $Xmod\ G$ is also a locally compact space.

40. Show that $U(1,\ \mathbf{C}) = \{z \in \mathbf{C} : |z| = 1\} = S^1 \cong \mathbf{R}/\mathbf{Z}$.
41. Show that the rotational group $SO(3, \mathbf{R})$ is a closed subgroup of $\mathbf{R}^{3\times3}$.
42. Show that

(i) the general complex linear group $GL(n, \mathbf{C})$ of all nonsingular square matrices of order n over $\mathbf{C}$ is a topological group under usual multiplication of matrices:
(ii) the topological group $GL(n, \mathbf{C})$ is not compact;
(iii) the unitary group $U(n, \mathbf{C}) \subset GL(n, \mathbf{C})$ is compact.

43. For the unitary group $U(n, \mathbf{C})$, show that topological spaces $U(n, \mathbf{C})/U(n-1, \mathbf{C})$ and S^{2n-1} are homeomorphic $\forall n \geq 2$.
44. Let X be a compact subgroup of $GL(n, \mathbf{C})$. Show that

(i) for a representation θ of the group X on a finite dimensional vector space V, there is an X-invariant positive definite hermitian form on V;
(ii) the compact subgroup X of $GL(n, \mathbf{C})$ is conjugate to a subgroup of the unitary group $U(n, \mathbf{C})$.

45. Consider the special unitary group

$$SU(n, \mathbf{C}) = \{M \in U(n, \mathbf{C}) : det\ M = 1\}.$$

Show that for all n,

(i) the group $SU(n, \mathbf{C})$ is compact;
(ii) the group $SU(n, \mathbf{C})$ is connected.

46. Let G be a connected topological group with identity element e and U be an open nbd of G containing e. Show that

$$G = \bigcup_{n=1}^{\infty} U^n, \text{ where } U^1 = U,\ U^n = U^{n-1}U.$$

[Hint: U is an open set containing e implies U^{-1} is also an open set containing $e \implies V = U \cap U^{-1}$ is also an open set containing e. Moreover, $V = V^{-1}$. Then $\bigcup_{n=1}^{\infty} V^n \subset \bigcup_{n=1}^{\infty} U^n \subset G$. Since $\bigcup_{n=1}^{\infty} V^n$ is an open set and G is connected by hypothesis, then $\bigcup_{n=1}^{\infty} V^n = G$.]
47. **(Spin group $SO(3, \mathbf{R})$)** Consider the 3-sphere S^3 identified with the set of unit quaternions

$$S^3 = \{q = t + xi + yj + zk : t, x, y, z \in \mathbf{R} \ and \ t^2 + x^2 + y^2 + z^2 = 1\}.$$

Identify $\mathbf{R}^3$ with the set of pure quaternions $\{xi + yj + zk\}$. For $q \in S^3$, the map

$$\psi_q(v) = q \cdot v \cdot q^{-1}, \ \forall \, v \in \mathbf{R}^3$$

defines an orthogonal transformation of $\mathbf{R}^3$ with determinant 1, which gives an element of $SO(3, \mathbf{R})$. Show that

(i) S^3 is a topological group;
(ii) the map

$$\psi : S^3 \to SO(3, \mathbf{R}), q \mapsto \psi_q$$

is an epimorphism of groups with kernel $\{\pm 1\}$;

$SO(3, \mathbf{R})$ is called the spin group with its generalizations $SO(n, \mathbf{R})$, $\forall \, n \geq 3$.

48. Consider the symplectic group

$$Sp(n, \mathbf{H}) = \{M \in M(n, \ \mathbf{H}) : M^*M = I\},$$

where M^* is the quaternionic conjugate transpose of M, conjugation is in the sense of reversal of all three imaginary components. Show that for all $n \geq 2$

(i) the group $Sp(n, \mathbf{H})$ is compact;
(ii) the group $Sp(n, \mathbf{H})$ is connected.

49. Let $\mathbf{R}$ be the real number space with usual topology. Show that every nontrivial closed subgroup G of $\mathbf{R}$ is an infinite cyclic group.

50. Let X be a normed linear space and $x \neq 0 \in X$. Show that there exists a continuous linear functional $f : X \to \mathbf{R}$ such that $||f|| = 1$ and $f(x) = ||x||$.

51. Let G be a topological group. Show that the inverse map

$$v : G \to G, x \to x^{-1}$$

is a homeomorphism.

[Hint: As G is a topological group, the map v is bijective and it is also continuous. Moreover, its inverse of v onto G is also continuous.]

52. Let G be a topological group with identity element e. Show that the component of e is a closed normal subgroup of G.

53. Let G be a topological group with identity element e and U be any nbd of e. Show that there exists a symmetric nbd W of e such that $W^n \subset U$ for any positive integer n.

54. Let G be a topological group with identity element e and W be any nbd of e. For any subset K of G, show that $\overline{W} \subset KW$ and also $\overline{W} \subset WK$.

55. Consider $M(n, \mathbf{R})$ identified with the Euclidean n^2-space $\mathbf{R}^{n^2}$. Show that

(i) The topological group $GL(n, \mathbf{R})$ is neither compact nor connected;
(ii) the orthogonal group $O(n, \mathbf{R})$ is compact but connected.

56. For the topological group $GL(n, \mathbf{R})$ endowed with the Euclidean topology, let $M^* = (M^t)^{-1}$. Show that each of the following maps f_1, f_2, f_3 is a homeomorphism:

(i)

$$f_1 : GL(n, \mathbf{R}) \to GL(n, \mathbf{R}), \ M \mapsto M^{-1}$$

(ii)

$$f_2 : GL(n, \mathbf{R}) \to GL(n, \mathbf{R}), \ M \mapsto M^t$$

(iii)

$$f_3 : GL(n, \mathbf{R}) \to GL(n, \mathbf{R}), \ M \mapsto M^*$$

57. Show that the symplectic group $Sp(n, \mathbf{H})$ is a compact topological group.
58. Show that

(i) the special real linear group $SL(n, \mathbf{R})$ is path connected;
(ii) the special complex linear group $SL(n, \mathbf{C})$ is path connected;
(iii) the topological orthogonal group $SO(n, \mathbf{R})$ is path connected.

59. Show that

(i) the topological group $SO(n, \mathbf{R})$ (special orthogonal group) is connected;
(ii) $S(O, 3)$ is homeomorphic to the 3-dimensional real projective space $\mathbf{R}P^3$.

60. (Pontryagin) Show that a second countable locally compact connected topological division ring is isomorphic either to the topological fields $\mathbf{R}$ or $\mathbf{C}$ or to the topological division ring $\mathbf{H}$ of real quaternions.
61. Prove that

(i) For every topological group G, its center

$$C(G) = \{x \in G : \forall y \in G, xy = yx\}$$

is a closed subset of G;
(ii) Let G be a topological group act on a topological X. Then

$$H = \{g \in G : g \cdot x = x, \ \ \forall x \in X\}$$

is a closed normal subgroup of G;
(iii) The 2- torus is obtained as an orbit space of $\mathbf{R} \times \mathbf{R}$.
(iv) The space $SO(3, \mathbf{R})$ and 3-dimensional projective space $\mathbf{R}P^3$ are homeomorphic.

62. Let $\mathcal{A}(G)$ denote the automorphism of the topological group G, i.e., every element $f \in \mathcal{A}(G)$ if is topologically a homeomorphism and algebraically an isomorphism of G onto itself. Show that

(i) the automorphism group $\mathcal{A}(\mathbf{R})$ of the additive group $(\mathbf{R}, +)$ of reals, is isomorphic to the product group $\mathbf{R} \times \mathbf{Z}_2$;
(ii) the automorphism group $\mathcal{A}(S^1)$ of circle group $(S^1, \cdot)$ is isomorphic to the group $\mathbf{Z}_2$.

[Hint: Consider the automorphism $f : \mathbf{R} \to \mathbf{R}$ such that $f(t) = tf(1),\ \forall t \in \mathbf{Q}$. Then use the result that $f(x) = xf(1),\ \forall x \in \mathbf{R}$.].

63. Let X be a subgroup of a topological group G. Show that

(i) if G is compact and X is closed, then X is compact.
(ii) if G is locally compact and X is closed, then X is also locally compact.

64. Show that every topological group G is completely regular.
65. Let G and K be two locally connected isomorphic topological groups. Show that there exists a topological group H such that

(i) G is isomorphic to the topological group G/N_1 and
(ii) K is isomorphic to the topological group G/N_2

for some discrete normal subgroups N_1 and N_2 of H.
66 (**Characterization of a connected space** in terms of its orbit space) Let a connected topological group G act on a topological space X. Show that X is connected iff the orbit space $X mod\ G$ is connected.
67. Let $X = \mathbf{R} \times \mathbf{I}$ be the product space. Determine an action

$$f : \mathbf{Z} \times X \to X$$

such that the corresponding orbit space $X mod\ \mathbf{Z}$ is homeomorphic to the Mӧbius band.
68. (**Irrational flow**) Given any irrational number β, define an action of the topological group $\mathbf{R}$ on the torus $T = S^1 \times S^1$ defined by

$$f : \mathbf{R} \times T \to T,\ (t,\ (e^{2i\pi x}, e^{2iy})) \mapsto (e^{2i\pi(x+\beta t)},\ e^{2i\pi(y+\beta t)}).$$

Show that this action is free and every orbit is a proper and dense subset in T.
69. Let G be a connected topological group. Show that every discrete normal subgroup N of G

(i) is in the center $C(G)$ of G;
(ii) is abelian.

[Hint: Given an element $g_0 \in N$, the map

$$\psi : G \to N, g \mapsto gg_0g^{-1}g_0^{-1}$$

is constant.]

70. Let R be a topological ring and M be a topological R-module. If $\Omega(M)$ denotes a family of open sets M which contain the zero element $0 \in M$ and satisfies the property that $\Omega(M)$ is a local basis at 0 in the sense that every open set in M, containing the 0 contains an element of the family $\Omega(M)$. Show that the family of sets $\mathcal{F} = \{x + U : x \in M,\ U \in \Omega(M)\}$ forms an open basis for the topological module M, called a **basis of nbds of** 0 **for** M with the properties:

(i) for any two elements (sets) $X, Y \in \Omega(R)$, their intersection $X \cap Y$ contains a third element (set) in $\Omega(R)$.
(ii) for any set $U \in \Omega(R)$, there is a set $V \in \Omega(R)$ such that

$$V - V \subset U \quad \text{and} \quad VV \subset U,$$

where $V - V$ consists of all elements $x - y$ and VV consists of all elements xy for $x, y \in V$.
(iii) for any set $U \in \Omega(R)$, any element $u \in U$ and any element $r \in R$, there is a set $V \in \Omega(R)$ such that

$$V + u \subset U \quad \text{and} \quad Vr \subset U,$$

Hence, show that for a ring R and for any family $\Omega(R)$, satisfying the above properties (i)–(iii), there exists a unique topology τ such that (R, τ) is a topological ring and the given family $\Omega(R)$ forms a basis of nbds of 0 of the topological ring (R, τ).

71. Let X be a topological vector space. Prove the following statements:

(i) X is normable iff X its origin has a convex bounded nbd;
(ii) X is of finite dimension if X is locally compact;
(iii) If X is locally bounded having the Heine- Borel property, then X is of finite dimension.

[Hint: Use Definition 2.10.20 and see Rudin (1973).]

72. Let X be a topological vector space and $A \subset X$ be a subset such that every sequence of points in A has a point of cluster in X. Show that the set A is totally bounded.

73. **(Characterization of finite dimensional topological vector spaces)** Show that a topological vector space X is of finite dimension iff there is a totally bounded nbd of its origin.

74. **(Completion theorem)** Show that every topological vector space X can be isomorphically (topological) mapped onto a dense subspace of a complete topological vector space X^*.

75. A continuous onto map $p : X \to B$ is said to have a local cross section s at a point $b \in B$, if there exist a nbd U of b in B and a continuous map $s : U \to X$ is such that $p \circ s = 1_U$ (identity map on U). Let H be a closed subgroup of a topological group G. Show that the natural projection map

$$p : G \to G/H, \; g \mapsto Hg$$

has a local cross section at every point in G/H.

2.13.1 *Multiple Choice Exercises on Topological Groups*

This section gives multiple choice exercises. Identify the correct alternative (s) (there may be more than one) from the following list of exercises:

1. Let G be a topological group with identity element e. If G_e is the connected component of e in G, then

(i) G_e is closed in G;
(ii) G_e is open in G;
(iii) G_e is a closed normal subgroup of G.

2. Let G be a topological group and X be a topological subgroup of G. Then the closure $\overline{X}$ of X is

(i) not necessarily a subgroup of G;
(ii) not necessarily a normal subgroup of G;
(iii) a normal subgroup of G if X is a normal subgroup of G.

3. Let **R** be the real number space with usual topology.

(i) The topological group $(\mathbf{R}, +)$ has no fixed- point property.
(ii) Every nontrivial closed subgroup $(G, +)$ of the topological group $(\mathbf{R}, +)$ is an infinite cyclic group.
(iii) For the cyclic subgroup $(G, +)$ of the topological group $(\mathbf{R}, +)$, generated by some irrational number, the subgroup

$$H = \{n + g : n \in \mathbf{Z}, \; g \in G\}$$

is dense in $(\mathbf{R}, +)$.

4. Let **R** be the real number space with usual topology.

(i) The additive topological group $(\mathbf{R}, +)$ with usual topology and the multiplicative topological group $(\mathbf{R}^+, \cdot)$ of positive reals with topology inherited from the usual topology on **R**, are isomorphic topological groups.

(ii) Corresponding to every cyclic subgroup of the topological group $(\mathbf{R}, +)$, there is a dense subgroup of $(\mathbf{R}, +)$.
(iii) Every dense subgroup of the topological group $(\mathbf{R}, +)$ is cyclic.

5. Let X be topological vector space over a topological field $F = \mathbf{R}$ *or* $\mathbf{C}$.

(i) $(X, +)$ is a topological group;
(ii) For an arbitrary element $x_0 \in X$, the translation

$$T_{x_0} : X \to X,\ x \mapsto x + x_0$$

is a homeomorphism;
(iii) For an nonzero element $\alpha_0 \in F$, the dilation

$$D_{x_0} : X \to X,\ x \mapsto \alpha_0 x$$

is a homeomorphism.

2.13.2 *Multiple Choice Exercises on Topological Groups of Matrices*

Consider $M(n, \mathbf{R})$ identified with the Euclidean space $\mathbf{R}^{n^2}$ with topology induced from the Euclidean space $\mathbf{R}^{n^2}$. All subsets of $M(n, \mathbf{R})$ are endowed with subspace topology induced from Euclidean topology on $M(n, \mathbf{R})$. Identify the correct alternative (s) (there may be more than one) from the following list of exercises:

1. Let $X = Gl(n, \mathbf{R})$ be the general linear topological group. Then X is

(i) open in $M(n, \mathbf{R})$;
(ii) connected in $M(n, \mathbf{R})$;
(iii) dense in $M(n, \mathbf{R})$.

2. (i) The general linear topological group $Gl(n, \mathbf{R})$ is nowhere dense;
(ii) The orthogonal group $O(n, \mathbf{R})$ endowed with subspace topology is compact;
(iii) The space orthogonal group $O(n, \mathbf{R})$ is connected.

3. (i) The general linear topological group $Gl(n, \mathbf{R})$ is compact;
(ii) The general linear topological group $Gl(n, \mathbf{R})$ is closed;
(iii) The set of all matrices $X = \{M \in M(n, \mathbf{R}) :\ \text{trace } M = 0\}$ endowed with subspace topology is compact;
(iv) The set of all matrices $X = \{M \in M(n, \mathbf{R}) :\ \text{trace } M = 0\}$ endowed with subspace topology is connected.

4. (i) The set of all matrices $X = \{M \in M(n, \mathbf{R}) :\ \det M = 1\}$ endowed with subspace topology is compact;

(ii) The set of all matrices $X = \{M \in M(n, \mathbf{R}) :$ trace $M = 1\}$ endowed with subspace topology is compact;

(iii) The set of all matrices $X=\{M \in M(n, \mathbf{R}) : M$ is symmetric and positive definite$\}$ endowed with subspace topology is connected.

5. **(i)** The set of all matrices in $M(n, \mathbf{R})$ whose eigenvalues α are such that $|\alpha| \leq 2$ is compact under the subspace topology;

(ii) The subset $\{tr A : A \in M(n, \mathbf{R})\, and\, A\, is\, orthogonal\} \subset \mathbf{R}$, endowed with Euclidean topology is compact;

(iii) The set all upper triangular matrices in $M(n, \mathbf{R})$ endowed with subspace topology is connected;

(iv) The set all nonsingular diagonal matrices in $M(n, \mathbf{R})$ endowed with subspace topology is connected.

6. **(i)** The space of all nonsingular matrices in $M(2, \mathbf{R})$ is dense;

(ii) The set of all matrices in $M(2, \mathbf{R})$ having both eigenvalues are real and endowed with subspace topology is dense;

(iii) The set of all matrices in $M(2, \mathbf{R})$ having trace 0 and endowed with subspace topology is dense.

7. **(i)** The set $X_1 = \{M \in M(n, \mathbf{R}) : M^t M = M M^t = I\}$ endowed with subspace topology is connected;

(ii) The set $X_2 = \{M \in M(n, \mathbf{R}) : tr(M) = 1\}$ endowed with subspace topology is connected;

(iii) The set $X_3 = \{M \in M(n, \mathbf{R}) : x^t M x \geq 0, \ \ \forall x \in \mathbf{R}^n\}$ endowed with subspace topology is connected.

8. **(i)** The set $X_1 = \{M \in M(2, \mathbf{R}) :$ *both eigenvalues of M are real*$\}$ and endowed with subspace topology is open;

(ii) The set $X_2 = \{M \in M(2, \mathbf{R}) :$ *neither of eigenvalues of M is real*$\}$ and endowed with subspace topology is closed;

(iii) The set $X_3 = \{M \in M(2, \mathbf{R}) :$ *both eigenvalues of M are purely imaginary*$\}$ endowed with subspace topology is closed in $M(2, \mathbf{R})$.

9. **(i)** The set $X_1 = \{M \in M(n, \mathbf{C}) : M$ *is nilpotent and* $n \geq 2\}$ endowed with subspace topology is closed in $M(n, \mathbf{C})$;

(ii) The set $X_2 = \{M \in M(n, \mathbf{C}) : n \geq 2$ *and* $M^* = M, \ M^2 = M\}$ (here M^* is the transpose conjugate of M, and M represents orthogonal projections) endowed with subspace topology is closed in $M(n, \mathbf{C})$;

(iii) The topological group $SL(n, \mathbf{C}) = \{M \in M(n, \mathbf{C}) : det\ M = 1\}$ endowed with subspace topology is compact.

10. **(i)** The special real linear topological group $SL(2, \mathbf{R})$ is path connected;

(ii) The special complex linear topological group $SL(4, \mathbf{C})$ is not path connected;

(iii) The special orthogonal topological group $SO(8, \mathbf{R})$ is path connected.

References

Adhikari, M.R.: Basic Algebraic Topology and its Applications. Springer, India (2016)

Adhikari, M.R.: Basic Topology, Volume 3: Algebraic Topology and Topology of Fiber Bundles. Springer, India (2022)

Adhikari, M.R., Adhikari, A.: Groups. Rings and Modules with Applications. Universities Press, Hyderabad (2003)

Adhikari, M.R., Adhikari, A.: Textbook of Linear Algebra: An Introduction to Modern Algebra. Allied Publishers, New Delhi (2006)

Adhikari, M.R., Adhikari, A.: Basic Modern Algebra with Applications. Springer, New Delhi, New York, Heidelberg (2014)

Adhikari, A., Adhikari, M.R.: Basic Topology, Volume 1: Metric Spaces and General Topology. Springer, India (2022)

Armstrong, M.A.: Basic Topology. Springer, New York (1993)

Bredon, G.E.: Topology and Geometry. Springer, New York (1993)

Chevalley, C.: Theory of Lie Groups. Princeton University Press. (1946)

Kelly, J.L., Namioka, I.: Linear Topological Spaces. Princeton, D Van Nostrand (1963)

Morris, P.D., Wulbert, D.E.: Functional representation of topological algebra. Pac J Math **22**(2) (1967)

Pontragin, L.: Topological Groups (Translated). Princeton University Press, Princeton, London (1946)

Rudin, W.: Functional Analysis. Mc Graw-Hill, New York (1973)

Simmons, G.: Introduction to Topology and Modern Analysis. McGraw Hill, New York (1963)

Sorani, G.: An Introduction to Real and Complex Manifolds. Gordan Breech. Science Pub, New York (1969)

Zariski, O., Samuel, P.: Commutative Algebra, vol. II. D. Van Nostrand Co., Princeton (1969)

Chapter 3
Topology and Manifolds

The main aim of this chapter is to address a systematic and comprehensive elementary approach to the topology related to manifolds with an emphasis on **differential topology** avoiding algebraic topology, except for a few isolated cases. It also studies the topology from a differential viewpoint. All manifolds studied in this chapter are by defining conditions topological manifolds in the sense that every manifold M carries a topological structure on its underlying space $|M|$. For example, differential manifolds constitute an important class of topological manifolds endowed with an additional structure which is a differential structure. Its official study starts at Sect. 3.4.

Differentiable manifolds are geometrical objects standing at the crossing of geometry, topology and analysis. On the other hand, the Lie groups (studied in Chap. 4) stand at the crossing of geometry, algebra, topology and analysis. The concept of manifolds is one of the key notions of differential topology and modern analysis and can be traced to the work of B. Riemann (1826–1866) on differential and multi-valued functions. Geometers working on differential geometry discuss curves and surfaces in ordinary space, i.e., in Euclidean plane and Euclidean space through local concepts such as curvature. But Riemann introduced a concept, known as the Riemann surface in an abstract setting in the sense that these surfaces are not defined as subsets of Euclidean space, which led to the concept of topological manifolds. This concept plays a key role in the study of modern research in topology. Manifolds as a mathematical generalizations of curves and surfaces to arbitrary dimension invite a study of their topological properties with geometric intuition needed for a deep study of manifold theory related to differential topology.

Manifolds are locally Euclidean in the sense that every point has a neighborhood, called a chart, homeomorphic to an open subset of a Euclidean space $\mathbf{R}^n$. The coordinates on a chart facilitate to study manifolds through many concepts born in $\mathbf{R}^n$ such as differentiability, tangent spaces and differential forms. Thus, the study of manifolds is facilitated by introducing a coordinate system in each of these Euclidean neighborhoods and changes of coordinates, which are continuous real-valued functions

A. Adhikari and M. R. Adhikari, *Basic Topology 2*,
https://doi.org/10.1007/978-981-16-6577-6_3

of several variables. If such functions are differentiable, then the concept of differentiable or smooth manifold (or C^∞-manifold) is introduced.

Historically, B. Riemann (1826–1866) made an extensive work generalizing the concept of a surface to higher dimensions. But before him, Carl Friedrich Gauss published a paper in 1827, where he used local coordinates on a surface carrying the concept of charts. The term "manifold" is derived from the German word "Mannigfaltigkeit," which Riemann first used. Henri Poincaré (1854–1912) used the concept of locally Euclidean spaces (while defining the concept of the topological manifold) in his monumental work when Henri Poincaré published his Analysis Situs in 1895 and its subsequent complements on homotopy and homology in the late nineteenth century. But a rapid development of manifold theory occurred in the 1930s following the definition of a manifold based on general topology.

A more deep study of manifolds was born through the search of the solution of the **Poincaré conjecture** posed in 1894. Its equivalent statement says: is a compact n-manifold homotopically equivalent to S^n homeomorphic to S^n? This conjecture is very significant in the study of the general theory of topological manifolds and is the most important topological problem of the twentieth century. For dimension $n \geq 3$, the classification problems become more complicated. Since 1894, it took more than a century till for $n = 3$, G. Perelman (1966-) proved this conjecture in 2003 by using the Ricci flow. For other values of n, it was solved by others before 1994. For more specific, for 5-dimensional or its higher dimensional manifolds, the problem was solved by Stephen Smale in 1961 and for 4-dimensional manifolds, the problem was solved by Michael Freedman in 1982. It is remarkable that William Thurston formulated in 1970 a more powerful conjecture, now known as the **Thurston geometrization conjecture** saying that every compact 3-manifold has a "geometric decomposition," in the sense that it can be cut along specific surfaces into finitely many pieces, admitting each of them one of eight highly uniform (but mostly non-Euclidean) geometric structures. Unfortunately, we fail to study them here, because of the techniques used by them need far more groundwork than we cover in this book.

A basic problem in manifold theory is the classification problem of manifolds up to topological equivalence. More attention has been paid for the classification of compact manifolds in the sense that they are considered homeomorphic to closed and bounded subsets of some Euclidean space $\mathbf{R}^n$. A complete classification of closed surfaces (2-manifold) is known. This **classification Theorem** 3.23.1 says that any compact connected surface S is either homeomorphic to a sphere or to a connected sum of tori (with $n \geq 1$ holes) or to a connected sum of $n(\geq 1)$ projective planes. No two of these three types of manifolds are homeomorphic to each other.

The basic property of a manifold which plays the key role toward various developments of manifold theory is Whitney's Embedding Theorem 3.14.20 saying that every manifold can be embedded in a Euclidean space as a closed subspace. This theorem implies that any manifold may be considered as a submanifold of a Euclidean space. It reconciles the earlier concept of manifolds with the modern abstract concept of manifolds and facilitates various development of manifold theory. The notion of regular value coupled with Sard and Brown theorem, saying that every smooth mapping has regular values, plays a key role. For simplified presentation, we take all manifolds differentiable and explicitly embedded in Euclidean space. For such a study, a small amount of general topology and of real variable theory is taken for granted. Sard's theorem is proved and it is applied to prove fundamental theorem of algebra, the Brouwer fixed-point theorem without using the tools of algebraic topology.

The original inspiration for writing this chapter was generated by the book "Topology from the Differential Viewpoint" of J. Milnor, supplemented by other books such as Adhikari (2016, 2022), Adhikari and Adhikari (2003, 2006, 2014, 2022), Apostol (1957), Bredon (1993), Brickell and Clark (1970), Donaldson (1983), Guillemin and Pollack (1974), Hirsch (1976), Lie (1880), Mandelbaum (1980), Massey (1967), Milnor (1956, 1963), Milnor and Weaver (1969), Moise (1977), Mukherjee (2015), Nakahara (2003), Singer and Thorpe (1967) and Wallace (1968) together with the paper (Whitney 1935, 1936a, b, c, 1944) and some others referred in Bibliography.

3.1 Different Types of Manifolds with Motivation

A deep insight into the structures and behavior of geometric spaces facilitates to develop the theory of manifolds intuitively by utilizing a few tools from calculus. Based on the local geometry of Euclidean spaces built by calculus, the topological spaces that look locally as some Euclidean spaces are developed. Such spaces are called manifolds. Riemann introduced a concept, known as the Riemann surface in an abstract setting in the sense that these surfaces are not defined as subsets of Euclidean space, which led to the concept of topological manifolds. Such surfaces provide workable techniques for how manifold theory can be applied to invade global problems. For example, Riemann invented the global invariant such as the **connectivity of a surface**, which is the number given by the maximum number of curves whose union does not disconnect the surface $+1$ (plus one). The connectivity of a compact space is finite. It was proved in the 1860s that compact orientable surfaces can be topologically classified by their connectivity. Henri Poincaré made the topological analysis of 3-dimensional manifolds published in "Analysis Situs" in 1895 and constructed many basic tools of algebraic topology. He used differentiable techniques at the beginning. Herman Weyl defined abstract manifolds in his book on the Riemann surfaces published in 1912. The theory of manifolds received momentum in 1936 through the work of H. Whitney. His embedding theorem is a basic result in the theory of manifolds. Thereafter, the **differential topology** was established in its own right as an important branch of mathematics because of its own problems and techniques. Its influence in other areas is also enormous. For example, manifolds appear in algebra as the Lie groups, in relativity as space-time, in mechanics as phase-spaces and in economics as indifference surfaces.

In topology, three types of manifolds are mainly studied:

(i) topological manifolds (TOP),
(ii) piecewise linear manifolds (PL) and
(iii) differentiable (or smooth) manifolds (DIFF).

All manifolds M in this chapter are assumed to be paracompact to ensure that M is a separable metric space and they have inclusions

$$DIFF \subset PL \subset TOP.$$

Remark 3.1.1 For manifolds of dimensions ≤ 3, the three concepts $DIFF$, PL and TOP of manifolds coincide [Moise, 1977]. On the other hand, $DIFF = PL$ for manifolds of dimensions ≤ 7 and there exist PL-manifolds which are not $DIFF$ [Mandelbaum, 1980].

3.1.1 Motivation of Manifolds

This subsection gives the motivation for the study of manifolds. We are familiar with the geometric objects such as curves and surfaces which are considered locally homeomorphic to $\mathbf{R}$ and $\mathbf{R}^2$, respectively. A manifold M of dimension n generalizes such geometrical objects to higher dimensional objects in the sense that M is homeomorphic to $\mathbf{R}^n$ locally. This assigns to each point $p \in M$ an ordered set of n real numbers, called the local coordinates of the point p. If M is not homeomorphic to $\mathbf{R}^n$ globally, several local coordinates are introduced. The transition from one coordinate to another one is taken to be smooth. This facilitates to apply calculus.

Like continuity in general topology, smoothness plays a key role in differential topology. An n-manifold is a Hausdorff topological space that looks locally like the Euclidean n-space $\mathbf{R}^n$, but not necessarily globally. A local Euclidean structure to the manifold by introducing the concept of a chart is utilized to use the conventional calculus of several variables. Due to the linear structure of vector spaces, for their applications in mathematics and in other areas, it needs generalization of metrizable vector spaces, maintaining only the local structure of the latter. On the other hand, every manifold can be considered as a (in general non-linear) subspace of some vector space. Both aspects are used to approach the theory of manifolds. Since the dimension of a manifold is a locally defined property, a manifold has a dimension. Our study is confined to finite-dimensional manifolds (although there are infinite-dimensional manifolds).

3.1.2 Approach to Topological Manifold

This subsection introduces the concept of a topological manifold M, which is a Hausdorff topological space that looks locally like a Euclidean space $\mathbf{R}^n$ in the sense that every point of M has a nbd homeomorphic to the Euclidean n-space $\mathbf{R}^n$; equivalently, to an open subset of $\mathbf{R}^n$, i.e., every topological manifold is locally Euclidean. In particular, the n-dimensional Euclidean space $\mathbf{R}^n$ is a topological space as well as an n-dimensional real vector space. This concept of a topological manifold was born through the geometry and theory of functions in the nineteenth century. Intuitively, it is made up of pieces of $\mathbf{R}^n$ by gluing together with the help of homeomorphisms. For a differentiable manifold, these homeomorphisms are differentiable. The sphere or torus, in which every point lies on a small curved disk which can be flattened into a disk in the Euclidean plane $\mathbf{R}^2$, is a very familiar example of manifolds; on the other hand, the cone is not a manifold, because the vertex of the cone has no nbd which looks locally like a small piece of $\mathbf{R}^2$.

The concept of a topological manifold is formulated in Definition 3.1.3. Topological manifolds form a versatile class of topological spaces having applications throughout mathematics and also beyond it.

Definition 3.1.2 An n-dimensional **topological manifold** or simply an **n-manifold** M is a Hausdorff space with a countable base such that for each point $x \in M$, there exists a homeomorphism ψ_x mapping some neighborhood of the point x onto an open subset of $\mathbf{R}^n$. It is also called a **locally Euclidean space** of dimension n. In particular, an 1-dimensional manifold is called a **curve** and a 2-dimensional manifold is called a **surface**.

Definition 3.1.3 Let M be an n-manifold and $\mathcal{C} = \{V_i : i \in \mathbf{A}\}$ be an open covering of M such that for every $i \in \mathbf{A}$, there is a homeomorphism $\psi_i : V_i \to \mathbf{R}^n$ mapping V_i onto an open subset U_i of $\mathbf{R}^n$. Each ordered pair (ψ_i, V_i) is called a **local chart** (or simply **a chart or a coordinate system**) on V_i and the family $\psi = \{(\psi_i, V_i) : i \in \mathbf{A}\}$ is called an **atlas** of the manifold M.

Example 3.1.4 There are plenty examples of topological manifolds.

(i) The Euclidean n-space $\mathbf{R}^n$ is a topological manifold of dimension n. Here, for each $x \in \mathbf{R}^n$, ψ_x is taken to be the identity map on $\mathbf{R}^n$.

(ii) The n-sphere $\mathbf{S}^n$ is a topological manifold of dimension n. Here, for the points $x, y \in \mathbf{S}^n, x \neq y$, the map ψ_x is taken to be the stereographic projection $\psi_x : S^n - \{y\} \to \mathbf{R}^n$, which is a homeomorphism.

(iii) The n-dimensional projective space $\mathbf{R}P^n$ (the space of all lines through the origin $\mathbf{0} \in \mathbf{R}^{n+1}$) is a topological manifold of dimension n (see Example 3.4.20).

(iv) Let $M(n, \mathbf{R})$ be the set of all $n \times n$ matrices over $\mathbf{R}$, identified with the Euclidean $\mathbf{R}^{n^2}$-space and $GL(n, R) = \{M \in M(n, \mathbf{R}) : det M \neq 0\}$. The set $GL(n, R)$ of nonsingular matrices is a topological manifold of dimension n^2 (see Example 3.4.20).

(v) The 2-torus $S^1 \times S^1$ with product topology is a compact and connected manifold, which is a closed topological manifold of dimension 4.

Remark 3.1.5 A topological space X is a manifold iff the connected components of X are manifolds. The only 0-dimensional connected manifold is a one-pointic space and the only 1-dimensional connected and closed manifold is the circle.

3.1.3 Approach to Differentiable Manifolds

A differentiable manifold is a topological manifold endowed with a differentiable structure, which is an additional structure other than its topological structure. A differentiable manifold defined in Sect. 3.4 is studied based on the standard differentiable structure on a Euclidean space $\mathbf{R}^n$. This approach is formally given by Hermn Weyl (1885–1955) in 1912. Historically, H. Whitney (1907–1989) during the 1930s and some others developed the theory of differentiable manifolds during the second half of the nineteenth century through differential geometry and the Lie group theory. Curves and surfaces are studied by differential geometers as subsets of Euclidean space. Their interest is on local concepts such as curvature.

Riemann made a breakthrough by his construction of an abstract manifold, now called, the Riemann surfaces, which are not subsets of Euclidean space. These surfaces are used to investigate global properties. Whitney's embedding theorem finally says that every n-dimensional manifold actually embeds in $\mathbf{R}^{2n}$ (improving his earlier own result of embedding in $\mathbf{R}^{2n+1}$) and this result unifies the original (earlier) concept of manifolds with its modern abstract concept. Thereafter, the **differential topology** was established in its own right as an important branch of mathematics because of its own problems and techniques. In a topological manifold of dimension n, every point admits a nbd homeomorphic to $\mathbf{R}^n$, and hence, in each of these Euclidean nbds, a coordinate system can be introduced. The changes of coordinates are continuous functions of several real variables. By using the standard differentiable structure on a Euclidean space, such changes of coordinates may be differentiable. This leads to the concept of differentiable manifolds (smooth or C^∞-manifold).

3.1.4 The Nature of Differential Topology

Differential topology is a subfield of topology, which studies topological problems of the theory of differentiable manifolds and differentiable functions on differentiable manifolds in general, but in particular, related to diffeomorphisms (differentiable homeomorphism), embeddings, etc., because the concepts of diffeomorphisms, embeddings and bundles play important role in differential topology. The main aim of differential topology is to study those properties of a subset of a Euclidean space which remain invariant under diffeomorphism (instead of homeomorphism). In this sense, the aim of differential topology is the study of those properties of differentiable manifolds which are invariant under diffeomorphism. Typical problems falling under this heading are the following: There is a natural question: what are the main problems of study in differential topology?

1. Under what conditions two given differentiable manifolds are said to be diffeomorphic?
2. Under what conditions a differentiable manifold is the boundary of some differentiable manifold?

3. Under what conditions a differentiable manifold is parallelizable?

The study of such problems do not fall in the premises of differential geometry, which usually studies additional structures such as a connection or a metric. The topology of the differential manifold M alone is not sufficient to invade such problems. Differential topology uses powerful tools derived from the methods of algebraic topology such as the theory of characteristic classes, where there is a shift from the manifold M to its tangent bundle $T(M)$ to obtain cohomology class in M depending on this bundle.

Differential topology saw a sea-change through the papers of Whitney. Its useful connections with algebraic topology and piecewise linear topology are found fruitful. Many problems such as embedding, immersion and classification problems by diffeomorphism are studied in differential topology. Manifolds carry some extra structure such as a Riemannian metric, a dynamical system and a binary operation. Such extra structures used as additional tools in differential topology facilitate to study global questions such as: can any manifold be embedded in another manifold? A few other problems naturally studied in differential topology are now mentioned.

1. Are homeomorphic manifolds necessarily diffeomorphic?
2. Which manifolds are the boundaries of compact manifolds?
3. Does every topological invariant of a manifold enjoy any special behavior?
4. Has every manifold a nontrivial action of some cyclic groups?

To have an answer to these questions and more others, the relevant theory is developed in differential topology. The concept of regular value coupled with the theorem of Sard and Brown saying that every smooth mapping has regular valuesplays a key role in differential topology.

3.1.5 Approach to Piecewise Linear Manifold

A topological manifold M is said to be a piecewise **linear (PL)** if M admits a piecewise linear structure defined with the help of an atlas, such that one can pass from one chart to another in M by piecewise linear transformations. An isomorphism between two piecewise linear manifolds is called a **piecewise linear homeomorphism**. This chapter does not study piecewise linear manifolds.

3.1.6 The Stiefel and Grassmann Manifolds

This subsection presents two important classes of manifolds which are the Stiefel and Grassmann manifolds. A Stiefel manifold is a natural generalization of the manifold of unit vectors tangent to spheres and a Grassmann manifold is a natural generalization of the real projective spaces.

Definition 3.1.6 (*Stiefel manifold*) Any ordered set of r $(r \leq n)$ independent vectors in the Euclidean n-space $\mathbf{R}^n$ is called an ***r*-frame**. Let $V_r(\mathbf{R}^n)$ be the set of (orthogonal) r-frames in $\mathbf{R}^n$ defined by

$$V_r(\mathbf{R}^n) = \{(v_1, v_2, \ldots, v_r) \in (S^{n-1})^r : \langle v_i, v_j \rangle = \delta_{ij}(\text{Kronecker delta})\},$$

where the **Kronecker** δ_{ij} is defined by

$$\delta_{ij} = \begin{cases} 1, & \text{if } i = j \\ 0, & \text{otherwise.} \end{cases}$$

Remark 3.1.7 Since by Definition, the Stiefel manifold $V_r(\mathbf{R}^n)$ is a closed subset of a compact space, it is also compact. Corresponding to each r-frame $(v_1, v_2, \ldots, v_r) \in (S^{n-1})^r$, there exists an associated r-dimensional subspace $\langle v_1, v_2, \ldots, v_r \rangle$ with a basis $\{v_1, v_2, \ldots, v_r\}$. Each r-dimensional subspace of $\mathbf{R}^n$ is of the form $\langle v_1, v_2, \ldots, v_r \rangle$. The manifold $V_r(\mathbf{R}^n)$ is called the Stiefel manifold of (orthogonal) r-frames in $\mathbf{R}^n$. It may be considered as the manifold of all orthogonal $(r-1)$-frames tangent to S^{n-1}. In particular, for $r = 2$, $V_2(\mathbf{R}^n)$ is the manifold of unit vectors tangent to S^{n-1} and for $r = 1$, $V_1(\mathbf{R}^n)$ is the manifold S^{n-1}.

Definition 3.1.8 (*Grassmann manifold*) The Grassmann manifold $G_r(\mathbf{R}^n)$ is a natural generalization of the real projective space $\mathbf{R}P^{n-1}$. An r-dimensional $(1 \leq r \leq n)$ linear subspace of $\mathbf{R}^n$ is called an ***r*-dimensional subspace or simply *r*-plane** of the space $\mathbf{R}^n$. Let $V_r(\mathbf{R}^n)$ be the Stiefel manifold and $G_r(\mathbf{R}^n)$ consist of all r-planes of $\mathbf{R}^n$ through the origin, with the quotient topology defined by the identification map

$$V_r(\mathbf{R}^n) \to G_r(\mathbf{R}^n), (v_1, v_2, \ldots, v_r) \mapsto \langle v_1, v_2, \ldots, v_r \rangle,$$

where the right-hand side denotes the subspace in $\mathbf{R}^n$ spanned by the vectors $\{v_i : i = 1, 2, \ldots r\}$. $G_r(\mathbf{R}^n)$ is called the Grassmann manifold.

Remark 3.1.9 In particular, for $r = 1$ in the Grassmann manifold $G_r(\mathbf{R}^n)$, it becomes $G_1(\mathbf{R}^n) = \mathbf{R}P^{n-1}$. So, the Grassmann manifold $G_r(\mathbf{R}^n)$ is called a natural generalization of the projective space $\mathbf{R}P^{n-1}$.

Clearly, $G_r(\mathbf{R}^n)$ is a compact space. The natural inclusion $G_r(\mathbf{R}^n) \subset G_r(\mathbf{R}^{n+1})$ gives rise to the topological space

$$G_r(\mathbf{R}^\infty) = \bigcup_{r \leq n} G_r(\mathbf{R}^n) \text{ with the weak topology}$$

called the **infinite-dimensional Grassmann manifold.**

3.1.7 *Surfaces*

Surfaces are important geometrical objects in topology. The early development of topology began with the study of surfaces made by L. Euler (1707–1783), E. Betti (1823–1892), B. Riemann (1826–1866), H. Hopf (1894–1971), F. Klein (1849–1925) and many others. The standard sphere S^2, the torus T^2, the projective plane $\mathbf{R}P^2$ and the Klein bottle are important examples of surfaces. In the language of manifold theory, a surface is a 2-dimensional manifold. **A complete classification of compact surfaces** initiated by A F Möbius (1790–1868) in 1861 is given in Sect. 3.23.

Example 3.1.10 The number of components of the complement of a circle in a surface is not unique. For example, the complement of the circle in the torus may have either one or two components depending on the location of the circle on the torus.

Definition 3.1.11 A **surface** is a second countable Hausdorff space S such that every point $x \in S$ has a nbd homeomorphic to $\mathbf{R}^2$.

Definition 3.1.12 A **compact surface** S is a closed and bounded subset of $\mathbf{R}^n$ such that every point in S has a nbd homeomorphic to an open ball in $\mathbf{R}^2$.

Example 3.1.13 **(i)** The Euclidean plane $\mathbf{R}^2$ is a noncompact surface.
(ii) Every open ball in $\mathbf{R}^2$ is a noncompact surface.
(iii) The torus is a connected compact surface.
(iv) The sphere S^2 is a connected compact surface.
(v) The projective plane is a connected compact surface.
(vi) The Klein bottle is a connected compact surface.

Definition 3.1.14 A surface S is called **orientable** if there exists no embedding of a Möbius strip into S. Otherwise, the surface is called **nonorientable**, i.e., if there exists an embedding of a Möbius strip into S, then S is nonorientable surface.

Example 3.1.15 The Klein bottle is a nonorientable surface.

Definition 3.1.16 A **planar model of a surface** S is a polygon in the Euclidean plane $\mathbf{R}^2$ with an identification along the edges such that the resulting identification space is S. The polygons with curved edges is permissible for the 2-sided polygon.

Remark 3.1.17 The orientation on an edge is indicated as an arrow from the first vertex to the Second, and the edges in a planar model are identified in pairs compatible with orientations. Take a vertex on the polygon and travel over the polygon representing each edge traversed as the letter a for each edge if it is traversed in the direction of the arrow and as a^{-1} if it is traversed opposite to the arrow.

Example 3.1.18 In planar model, the familiar surfaces are represented by the following standard notations:

(i) **Sphere:** aa^{-1}.
(ii) **Projective plane:** aa.
(iii) **Connected sum of** n **Projective planes:** $a_1a_1a_2a_2\cdots a_na_n$.
(iv) **Torus:** $aba^{-1}b^{-1}$.
(v) **Connected sum of** n **Tori:** $a_1b_1a_1^{-1}b_1^{-1}a_2b_2a_2^{-1}b_2^{-1}\cdots a_nb_na_n^{-1}b_n^{-1}$.
(vi) **Klein bottle:** $abab^{-1}$.

The concept of the surface is generalized to an n-dimensional topological (or n-manifold) in Sect. 3.1.2.

Example 3.1.19 **(i)** Every 0-dimensional manifold is a countable discrete space.
(ii) S^n is an n-dimensional manifold for every $n \geq 0$.
(iii) Every surface is a 2-dimensional manifold.
(iv) A connected 1-dimensional manifold is homeomorphic to either $\mathbf{R}$ or S^1 by Theorem 3.24.1.

3.2 Calculus of Several Variables Related to Differentiable Manifolds

This section communicates the fundamental concepts of calculus such as the concept of differentiable functions with some basic concepts needed for the study of differentiable manifolds, which are based on the usual calculus developed in the n-dimensional Euclidean space $\mathbf{R}^n$. Since differentiable manifolds are studied based on the standard differential structure on the Euclidean space $\mathbf{R}^n$. So, it has become necessary to convey basic relevant results on differentiable functions in $\mathbf{R}^n$, which are delighted in Sects. 3.2.3 and 3.2.4.

The space $\mathbf{R}^n$ is a topological space (endowed with Euclidean topology) as well as it is an n-dimensional real vector space under usual addition and scalar multiplication of real numbers. If $x = (x_1, x_2, \ldots, x_n) \subset \mathbf{R}^n$, then the real numbers $x_1, x_2, \ldots, x_n$ are called the coordinates of the point x. An open set in $\mathbf{R}^n$ is a set which is open in the standard metric (Euclidean) topology induced by the standard metric d on $\mathbf{R}^n$ defined by

$$d : \mathbf{R}^n \times \mathbf{R}^n \to \mathbf{R},\ (x, y) \mapsto [\Sigma_{i=1}^n (x_i - y_i)^2]^{\frac{1}{2}} : x = (x_1, x_2, \ldots, x_i, \ldots, x_n),\ y = (y_1, y_2, \ldots, y_i, \ldots, y_n) \in \mathbf{R}^n.$$

$\mathbf{R}^n$ has a natural embedding in $\mathbf{R}^{n+m}$ obtained by identifying a point $(x_1, x_2, \ldots, x_i, \ldots, x_n) \in \mathbf{R}^n$ with the point $(x_1, x_2, \ldots, x_i, \ldots x_n, 0, 0, \ldots, 0) \in \mathbf{R}^{n+m}$, which is called **natural (canonical) embedding**

$$i : \mathbf{R}^{n+m} \hookrightarrow \mathbf{R}^n,\ \ (x_1, x_2, \ldots, x_i, \ldots, x_n, 0, 0, \ldots, 0) \mapsto (x_1, x_2, \ldots, x_i, \ldots, x_n).$$

3.2.1 Derivative of a Map

This subsection introduces the concept of the derivative of a smooth map in the sense of Definition 3.2.8, which is its best linear approximation. Its more study is available in Sect. 3.3.2.

Definition 3.2.1 Let $U \subset \mathbf{R}^n$ be an open set. A real-valued function $f : U \to \mathbf{R}$ is said to be **smooth or differentiable,** if f has continuous partial derivatives of all orders with respect to the coordinates in $\mathbf{R}^n$ at every point $x = (x_1, x_2, \dots .x_n) \in U$. A differentiable function on U is also called $C^\infty(U)$-function or simply a C^∞-function, when no confusion arises.

Let f be a smooth map defined on an open set U in $\mathbf{R}^n$ into $\mathbf{R}^m$. Given a point $x \in U$ and a vector $v \in \mathbf{R}^n$, the derivative of $f : U \to \mathbf{R}^m$ at the point x in the direction v, denoted by $df_x(v)$, is defined by

$$df_x(v) = lim_{\ t\to 0}\frac{f(x+tv) - f(x)}{t}.$$

Consider df_x as a map

$$df_x : \mathbf{R}^n \to \mathbf{R}^m,\ v \mapsto df_x(v).$$

Definition 3.2.2 The vector $df_x(v) \in \mathbf{R}^m$ is called the **derivative of f at x in the direction** $v \in \mathbf{R}^n$. The map $df_x : \mathbf{R}^n \to \mathbf{R}^m$ is linear and hence it has a matrix representation with respect to the standard basis. Let $f_1, f_2, \dots, f_m$ be m differentiable real-valued functions on U and

$$f : U \to \mathbf{R}^m,\ y \mapsto (f_1(y), f_2(y), \dots, f_m(y)).$$

Jacobian matrix of f on U at x which is given by the $m \times n$ matrix

$$J(f) = \begin{pmatrix} \frac{\partial f_1}{\partial x_1} & \cdots & \frac{\partial f_1}{\partial x_n} \\ \cdots & \cdots & \cdots \\ \frac{\partial f_m}{\partial x_1} & \cdots & \frac{\partial f_m}{\partial x_n} \end{pmatrix}$$

of the first derivatives of the mapping f (as they exist by hypothesis) is called the Jacobian matrix of f, denoted by $J(f)$. If f is evaluated at a point x, then its Jacobian matrix is written as $\frac{\partial f}{\partial x}|_x$.

Definition 3.2.3 Let $U \subset \mathbf{R}^n$ be an open set. Then a map $f : U \to \mathbf{R}^m$ is said to be a C^1-map if

(i) df_x exists at every point $x \in U$;
(ii) $df : U \to L(\mathbf{R}^n, \mathbf{R}^m),\ x \mapsto df_x$ is continuous.

The map f is defined to be C^r for $1 < r < \infty$ recursively, if the map

$$df : U \to L(\mathbf{R}^n, \mathbf{R}^m) \equiv \mathbf{R}^{n+m},\ x \mapsto df_x\ is\ C^{r-1}.$$

On the other hand, f is said to be a C^∞-map, if it is a C^r-map for all $r > 1$.

Theorem 3.2.4 *Let $U \subset \mathbf{R}^n$ be an open set. Then a map*

$$f = (f_1, f_2, \dots, f_n) : U \to \mathbf{R}^n$$

is C^r iff every $f_i : U \to \mathbf{R}$ has continuous partial derivatives of all orders s such that $s \leq r$.

Proof Left as an exercise. ❑

Definition 3.2.5 Let A be a nonempty set and $f : A \to \mathbf{R}^n$ be an injective map such that $f(A)$ is an open subset of $\mathbf{R}^n$. Then f is called an n**-dimensional chart** whose domain is A. If $p_1, p_2, \dots, p_n$ denote the natural projection maps defined by

$$p_i : \mathbf{R}^n \to \mathbf{R},\ x = (x_1, x_2, \dots, x_i, \dots, x_n) \mapsto x_i,\ \ : i = 1, 2, \dots, n,$$

then for any open set $U \subset \mathbf{R}^n$, a function $f : U \to \mathbf{R}^n$ $(n \geq 1)$ is studied through the functions $f_i = p_i \circ f : i = 1, 2, \dots, n$, formulated by

$$f : U \to \mathbf{R}^n : y \mapsto (f_1(y), f_2(y), \dots, f_n(y)).$$

The set of functions $\{f_i : i = 1, 2, \dots, n\}$ is called the set of **coordinates functions in the chart** f.

Example 3.2.6 Let $X = M_{m,n}(\mathbf{R})$ be the set of all $m \times n$ real matrices. Then X can be identified with the Euclidean mn-space $\mathbf{R}^{mn}$ and hence the map

$$f : X \to \mathbf{R}^{mn},\ (x_{ij}) \mapsto (x_{11}, \dots, x_{1n}, x_{21}, \dots, x_{2n}, \dots, x_{m1}, \dots, x_{mn})$$

is injective and its image is $\mathbf{R}^{mn}$, which is an open set. This asserts that f is a chart whose domain is the entire set X.

Remark 3.2.7 Let $U \subset \mathbf{R}^n$ be an open set. Then by using the natural projection maps $p_i : \mathbf{R}^n \to \mathbf{R}$, a function $f : U \to \mathbf{R}$ can be expressed as an ordered set $\{f_1, f_2, \dots, f_n\}$ of n real-valued functions, given by

$$f : U \to \mathbf{R},\ x \mapsto p_i(f_1(x), f_2x), \dots, f_n(x)),\ \forall x \in U,$$

where

$$p_i : \mathbf{R}^n \to \mathbf{R},\ x = (x_1, x_2, \dots, x_i, \dots x_n) \mapsto x_i$$

is the ith natural projection map and $f_i = p_i \circ f$, for every $i = 1, 2, \dots., n$.

3.2.2 Smooth Maps in $\mathbf{R}^n$

This subsection introduces the concept of smooth maps (functions of multivariables) which plays a key role in the study of manifold theory.

Definition 3.2.8 Let $U \subset \mathbf{R}^n$ and $V \subset \mathbf{R}^m$ be open sets. A map

$$f : U \to V$$

is said to be **smooth or differentiable** if all of its partial derivatives

$$\partial^{nf} / \partial x_{i1} \cdots x_{in}$$

exist and are continuous. In general, let $X \subset \mathbf{R}^n$ and $Y \subset \mathbf{R}^m$ be any two subsets of Euclidean spaces (not necessarily open). A map $f : X \to Y$ is said to be smooth if f can be locally extended to a smooth map on open sets in the sense that for each point $x \in X$, there exists an open set U in $\mathbf{R}^n$ containing the point x and exists a smooth map $F : U \to \mathbf{R}^m$ such that F equals f on $U \cap X$, i.e.,

$$F(x) = f(x), \forall x \in U \cap X.$$

Remark 3.2.9 Smoothness of a map is a **local property** in the sense that a map $f : X \to \mathbf{R}^m$ is smooth if it is smooth in a nbd of every point $x \in X$. The term **global** refers to the whole space X as a unified object.

Proposition 3.2.10 *Let $U \subset \mathbf{R}^n$ be an open set. Then the set $\mathcal{D}(U)$ of all differentiable functions on U admits a ring structure under the usual compositions of pointwise addition and pointwise multiplication of functions.*

Proof For any two differentiable functions $f, g \in \mathcal{D}(U)$, define

$$f + g : U \to \mathbf{R}, x \mapsto f(x) + g(x)$$

and

$$f.g : U \to \mathbf{R}, x \mapsto f(x)g(x).$$

Then $(\mathcal{D}(U), +, \cdot)$ admits a ring structure. ❑

Definition 3.2.11 Let $U \subset \mathbf{R}^n$ be an open set and $f : U \to \mathbf{R}$ be a differentiable function. If the function f can be expanded into a convergent power series of positive radius of convergence at every point $x \in U$, then f is said to be **analytic** on U.

Example 3.2.12 Every polynomial f in the coordinates in $\mathbf{R}^n$ is differentiable on any open set $U \subset \mathbf{R}^n$, where all the derivatives of f will vanish after certain order. On the other hand, the function f defined by $f(x, y) = (1 - x^2 - y^2)^{1/2}$ fails to be differentiable on any open set U containing any point of the circle $x^2 + y^2 = 1$, but it is differentiable on every open set not containing any point of this circle.

Example 3.2.13 Let f be the real-valued function defined as

$$f : \mathbf{R} \to \mathbf{R},\ x \mapsto \begin{cases} e^{\frac{1}{x^2-1}}, & \text{if } x \in -1 < x < 1, \\ 0, & \text{if } x \leq 1,\ or\ x \geq 1. \end{cases}$$

Then f has derivatives of all orders at all points $x \in \mathbf{R}$. This shows that f is a differentiable function on the entire $\mathbf{R}$, since $lim_{\ t \to 0}\ (1/t^n)e^{-1/t} = 0$.

Remark 3.2.14 If a function is given on a set, which is not necessarily open, then we use the following definition of its differentiability.

Definition 3.2.15 Let $A \subset \mathbf{R}^n$ be any nonempty set. A function $f : A \to \mathbf{R}$ is said to be **differentiable** on A if there exists an open set U containing the set A such that f is differentiable on $U \subset \mathbf{R}^n$.

Remark 3.2.16 Differentiability of a function at a point asserts its differentiability on a nbd of the point.

Definition 3.2.17 A function $f : \mathbf{R}^n \to \mathbf{R}$ is said to of C^∞**-class at a point** $x \in \mathbf{R}^n$ if every partial derivative of f exists and is continuous on some nbd of x.

Definition 3.2.18 gives a natural extension of Definition 3.2.15 for differentiability of a function between Euclidean spaces.

Definition 3.2.18 Let $A \subset \mathbf{R}^n$ be any nonempty set and $f_1, f_2, \ldots, f_m$ be m differentiable real-valued functions on A. Then the function

$$f : A \to \mathbf{R}^m,\ x \mapsto (f_1(x), f_2(x), \ldots, f_m(x))$$

is called a differentiable function on A into $\mathbf{R}^m$.

Remark 3.2.19 For $n = 1$, the concept of differentiable function given in Definition 3.2.18 coincides with that of given in Definition 3.2.15.

Using the projection maps p_i, Definition 3.2.17 is now extended.

Definition 3.2.20 A function $f : \mathbf{R}^n \to \mathbf{R}^m$ is said to be of C^∞-class at a point $x = (x_1, x_2, \ldots, x_n) \in \mathbf{R}^n$ if every real-valued function

$$f_i = p_i \circ f : \mathbf{R}^n \to \mathbf{R}$$

is of C^∞-class at x for $i = 1, 2, \ldots m$, i.e., each partial derivative of functions f_i (with respect to its coordinates $x_1, x_2, \ldots, x_n$) exists and is continuous on some nbd of x. The function f is said to be a C^∞**-function** if it is of C^∞-class at every point $x \in \mathbf{R}^n$.

Example 3.2.21 The function f, defined by

$$f : \mathbf{R} \to \mathbf{R},\ x \mapsto \begin{cases} e^{-1/x}, & \text{if } x > 0, \\ 0, & \text{if } x \leq 0, \end{cases}$$

is a C^∞-function.

Definition 3.2.22 Let X and Y be two subsets of two Euclidean spaces and $f : X \to Y$ be a smooth map. Then f is said to be a **diffeomorphism** if it is bijective and its inverse map $f^{-1} : Y \to X$ is also smooth. If there exists a diffeomorphism $f : X \to Y$, then X and Y are said to be **diffeomorphic** and are called intrinsically equivalent.

Proposition 3.2.23 *Let $X \subset \mathbf{R}^n$ and $Y \subset X$. If $f : X \to \mathbf{R}^m$ is a smooth map on X, then its restricted map $f|Y$ is also a smooth map.*

Proof Since by hypothesis, $f : X \to \mathbf{R}^m$ is smooth, it follows that given $y \in Y \subset X$, there exist

(i) an open set U such that $y \in U \subset \mathbf{R}^n$, where $X \cap U$ is an open nbd of y in X and
(ii) a smooth map $\psi : X \to \mathbf{R}^m$ such that $\psi(x) = f(x),\ \forall\, x \in X \cap U$.

Hence, it follows that $U \cap Y$ is an open nbd of y in Y such that $\psi = f|Y$ on $Y \cap U$. ❑

Theorem 3.2.24 *Let $X \subset \mathbf{R}^n$, $Y \subset \mathbf{R}^m$ and $Z \subset \mathbf{R}^p$ be three arbitrary subsets. If the maps $f : X \to Y$ and $g : Y \to Z$ are two smooth maps, then their composite map $g \circ f : X \to Z$ is also smooth.*

Proof Given $x \in X$ if $y = f(x)$, then by the given conditions, there exist an open set V such that $y \in V \subset \mathbf{R}^m$ with $\phi : V \to \mathbf{R}^p$ is a smooth function and an open set U such that $x \in U \subset \mathbf{R}^n$ with $\psi : U \to V$ is a smooth function. Moreover,

(i) $\psi(x) = f(x),\ \forall\, x \in X \cap U$ and
(ii) $\phi(y) = g(y),\ \forall\, y \in Y \cap V$.

This shows that the composite function

$$\phi \circ \psi : U \to \mathbf{R}^p$$

is smooth and

$$(\phi \circ \psi)(x) = (g \circ f)(x),\ \forall\, x \in U \cap V,$$

because $f(X \cap Y) \subset Y \cap V$. ❑

Corollary 3.2.25 *Let $A \subset \mathbf{R}^m$ be any nonempty set and $f : A \to \mathbf{R}^n$ be a differentiable function on A. If g is a differentiable function on $f(A)$ into $\mathbf{R}^p$, then their composite function*

$$g \circ f : A \to \mathbf{R}^p$$

is also a differentiable function.

Proof It follows from Theorem 3.2.24. ❑

Corollary 3.2.26 *Let $X \subset \mathbf{R}^n$, $Y \subset \mathbf{R}^m$ and $Z \subset \mathbf{R}^p$ be three arbitrary subsets. If the maps $f : X \to Y$ and $g : Y \to Z$ are diffeomorphisms, then their composite map $g \circ f : X \to Z$ is also a diffeomorphism.*

Proof Apply Theorem 3.2.24 to the inverse map of $g \circ f : X \to Z$ to obtain a diffeomorphism. ❑

Definition 3.2.27 Let $r > 0$ be an integer and $U \subset \mathbf{R}^n$ be an open set in $\mathbf{R}^n$. Then a function $f : U \to \mathbf{R}^n$ is said to be C^r **or differentiable of class** C^r, **or a** C^r**-function** on U if it has continuous partial derivatives of all orders $\leq r$ with respect to the coordinate functions $f_1, f_2, \ldots, f_n$. On the other hand, let $U \subset \mathbf{R}^n$ be an open set. Then a map $f : U \to \mathbf{R}^m$ is differentiable if the coordinates of $f(x) = y = (y_1, y_2, \ldots, y_m) \in \mathbf{R}^m$ are differentiable functions of coordinates $x_1, x_2, \ldots, x_n$ on U.

Definition 3.2.28 Let $U \subset \mathbf{R}^n$ be an open set in $\mathbf{R}^n$. A function $f : U \to \mathbf{R}^n$ is said to be a C^0-function if it is just a continuous function on U. The function f is a C^∞**-function** if it is a C^r-function for every $r \geq 0$ on U and said to be **real analytic** on U, (i.e.,it is expandable in a power series in the coordinate functions at every point $x \in U$), then f is said to be a C^ω**- function** on U. A function of class C^r is also of class C^m for all $m < r \leq \infty$.

Definition 3.2.29 A function $f : U \to \mathbf{R}$ is said to be **locally** C^r if f is C^r on an open nbd of each point of U.

Definition 3.2.30 Let U, V be two open sets in $\mathbf{R}^n$. Then a map $f : U \to V$ is said to be C^r**-diffeomorphism** for $r \geq 1$ if

(i) f is a homeomorphism;
(ii) f is a C^r-map;
(iii) f^{-1} is also a C^r-map.

Remark 3.2.31 Let U, $V \subset \mathbf{R}^n$ be open sets. Then a map $f : U \to V$ is a diffeomorphism iff f a homeomorphism such that f is differentiable on U and f^{-1} is also differentiable on V.

Example 3.2.32 Every differentiable homeomorphism is not a diffeomorphism. For example, given an odd integer $n > 1$, the map

$$f_n : \mathbf{R} \to \mathbf{R}, \ x \mapsto x^n$$

is a differentiable homeomorphism, but its inverse is not differentiable on any nbd of the origin and hence, its inverse map is not differentiable on any nbd of the origin.

Example 3.2.33 Every smooth bijective map is not a diffeomorphism. For example, consider the map

$$f : \mathbf{R} \to \mathbf{R}, x \mapsto x^3$$

is smooth and bijective but its inverse map

$$f^{-1} : \mathbf{R} \to \mathbf{R}, x \mapsto x^{1/3}$$

is not smooth at the point $x = 0$.

Definition 3.2.34 Let $\mathbf{C}$ be the field of complex numbers and $U \subset \mathbf{C}^n$ be an open set. A complex-valued function f on U is said to be **complex analytic or holomorphic** on U if at every point $z_0 \in U$, it can be expanded into a convergent power series having positive radius of convergence.

3.2.3 *Jacobian Matrix and Jacobian Determinant*

This subsection studies the Jacobian matrix and Jacobian determinant of a differentiable function f formulated in Definition 3.2.36. They provide key tools of analysis in the study of manifolds.

Definition 3.2.35 Let $f : \mathbf{R}^n \to \mathbf{R}^n$, $x = (x_1, x_2, \ldots, x_n) \mapsto (f_1(x), f_2(x), \ldots, f_n(x))$ be a function. If

$$\frac{\partial f_j}{\partial x_i} = lim_{\ h \to 0} \frac{f_j(x_1, x_2, \ldots, x_i + h, \ldots, x_n) - f_j(x_1, x_2, \ldots, x_i, \ldots, x_n)}{h}$$

exists, $\frac{\partial f_j}{\partial x_i}$ is called the partial derivative of f_j with respect to x_i, for $i, j = 1, 2, \ldots, n$.

Definition 3.2.36 Let $A \subset \mathbf{R}^n$ be any nonempty set and $f_1, f_2, \ldots, f_m$ be m differentiable real-valued functions on A and

$$f : A \to \mathbf{R}^m, \ x \mapsto (f_1(x), f_2(x), \ldots, f_m(x))$$

be a differentiable function on A into $\mathbf{R}^m$. The **Jacobian matrix** of the differentiable function f on A is the matrix with $\frac{\partial f_i}{\partial x_j}$ in the ith row and jth column, i.e., its Jacobian matrix denoted by $J(f)$ has the form

$$J(f) = \begin{pmatrix} \frac{\partial f_1}{\partial x_1} & \cdots & \frac{\partial f_1}{\partial x_n} \\ \cdots & \cdots & \cdots \\ \frac{\partial f_m}{\partial x_1} & \cdots & \frac{\partial f_m}{\partial x_n} \end{pmatrix}$$

of the first derivatives of the mapping f (as they exist by hypothesis). If f is evaluated at a point $x_0 \in A$, then its Jacobian matrix is written as $\frac{\partial f}{\partial x}|_{x_0}$. The Jacobian matrix of the differentiable function f having $\partial f_i/\partial x_j$ at its the ij-entry determines a linear map $df : \mathbf{R}^n \to \mathbf{R}^m$ with respect to the standard bases in $\mathbf{R}^n$ and $\mathbf{R}^m$. For $n = m$, its determinant is known as the **Jacobian determinant** of f.

Proposition 3.2.37 *Let $A \subset \mathbf{R}^n$ be any nonempty set and $f_1, f_2, \ldots, f_m$ be m differentiable real-valued functions on A into $\mathbf{R}^m$*

$$f : A \to \mathbf{R}^m,\ x \mapsto (f_1(x), f_2(x), \ldots, f_m(x))$$

be a differentiable function on A into $\mathbf{R}^m$. Then the Jacobian matrix $\frac{\partial f}{\partial x}|_{x_0}$, evaluated at x_0, determines a linear transformation

$$L_f : \mathbf{R}^n \to \mathbf{R}^m,\ (x_1, x_2, \ldots, x_n) \mapsto (\alpha_1, \ldots, \alpha_m,$$

where

$$\alpha_1 = \frac{\partial f_1}{\partial x_1}|_{x_0} x_1 + \cdots + \frac{\partial f_1}{\partial x_n}|_{x_0} x_n$$

$$-----------$$

$$\alpha_m = \frac{\partial f_m}{\partial x_1}|_{x_0} x_1 + \cdots + \frac{\partial f_m}{\partial x_n}|_{x_0} x_n.$$

Proof Left as an exercise. ❑

Proposition 3.2.38 *Let $U \subset \mathbf{R}^n$ be an open set and $f : U \to \mathbf{R}^n$ be a differentiable function. If f has a differentiable inverse, then the Jacobian determinant of f is nonzero at every point of U.*

Proof Left as an exercise. ❑

Proposition 3.2.39 *Let $U \subset \mathbf{R}^n$ be an open set and $f : U \to \mathbf{R}^n$ be a differentiable function. Then there exist real numbers $\delta > 0$ and $K > 0$ such that*

$$||f(x) - f(x_0)|| \leq K||x - x_0||,\ \textit{whenever}\ ||x - x_0|| < \delta.$$

Proof Left as an exercise. ❑

Proposition 3.2.40 is proved by connectedness property of $\mathbf{R}^m$.

Proposition 3.2.40 *Let $f : \mathbf{R}^m \to \mathbf{R}^m$ be a continuously differentiable function such that $f^{-1}(C)$ is compact for every compact subset C of $\mathbf{R}^m$. If the Jacobian matrix $(\partial f_i/\partial x_j)$ has rank m everywhere, then the map f is onto.*

Proof Under the given hypothesis, $f(\mathbf{R}^m)$ is both open and closed in $\mathbf{R}^m$. As $\mathbf{R}^m$ is connected and f is continuous, then its image $f(\mathbf{R}^m)$ is also connected. Under the given conditions, since $f(\mathbf{R}^m) \neq \emptyset$, it follows that $f(\mathbf{R}^m) = \mathbf{R}^m$. ❑

3.2.4 Rank of a Differentiable Map

This subsection studies the rank of a differentiable map with the help of its Jacobian matrix.

Definition 3.2.41 Let $U \subset \mathbf{R}^n$ be an open set and $f : U \to \mathbf{R}^m$ be a map of class C^r $(r \geq 1)$. The rank of the Jacobian matrix $J(f)$ evaluated at $x_0 \in U$, abbreviated $\frac{\partial f}{\partial x}|_{x_0}$, is said to be the **rank of** f at the point x_0. This rank of f is denoted by $rank\ \frac{\partial f}{\partial x}|_{x_0}$.

Remark 3.2.42 Since the rank of a matrix $\leq$ the number of rows and columns and the rank $\frac{\partial f}{\partial x}|_{x_0}$ equals the dimension of the image of $\mathbf{R}^n$ under the linear mapping $(Df)_{x_0}$, it follows that the rank $\frac{\partial f}{\partial x}|_{x_0} \leq \ min\{n, m\}$. The points x_0 at which rank $\frac{\partial f}{\partial x}|_{x_0} = \ min\{n, m\}$ are called **regular or nonsingular points** and the points x_0 at which rank $\frac{\partial f}{\partial x}|_{x_0} < \ min\{n, m\}$ are called **singular points**.

Remark 3.2.43 Let $U \subset \mathbf{R}^n$ be an open set and $f : U \to \mathbf{R}^m$ be a map of class C^r $(r \geq 1)$. The set of regular points of f is an open set of $\mathbf{R}^n$, because of the continuity of partial derivatives of f. However, this set may be $\emptyset$.

Example 3.2.44 The differentiable map $f : \mathbf{R}^2 \to \mathbf{R}^2,\ (x, y) \mapsto (x, 0)$ has rank 1 at every point $(x, y) \in \mathbf{R}^2$.

3.2.5 The Inverse Function and Implicit Function Theorems on $\mathbf{R}^n$

Prior to the study of the topology of manifolds, this subsection recalls inverse function theorem and the implicit function theorem of real analysis which describe the local behavior of a smooth function (instead of continuous function) $f : \mathbf{R}^n \to \mathbf{R}^m$.

Definition 3.2.45 Let $U \subset \mathbf{R}^n$ be an open set and $f : U \to \mathbf{R}^n$ be a map. If $f = (f_1, f_2, \ldots . f_n)$, then f is said to be C^r map for $1 \leq r < \infty$, if each $f_i : U \to \mathbf{R}$ for $i = 1, 2, \ldots, n$ has continuous partial derivatives of all orders $\leq r$. It is said to be C^∞ if it is C^r for all r. On the other hand, the map f is called (real) analytic or C^ω, if in some nbd of every point of U, the map f is C^∞. Let U and V be two open subsets of $\mathbf{R}^n$. A C^r diffeomorphism $f : U \to V$ is a C^r homeomorphism with its inverse is also C^r.

Definition 3.2.46 Let $U \subset \mathbf{R}^n$ be an open set and $f : U \to \mathbf{R}^n$ be a smooth map. Then f is said to be a local diffeomorphism at a point $u_0 \in U$ if the inverse of f is also smooth in some neighborhood of u_0. Its generalization to smooth manifolds is given in Definition 3.9.7.

Remark 3.2.47 Let V be an open subset of $\mathbf{R}^n$ and $f : V \to \mathbf{R}^n$ be a local diffeomorphism at point $x \in V$ if there exists an open set $U \subset V$ such that

(i) $x \in U$;
(ii) $f(U) \subset \mathbf{R}^n$ is an open set and
(iii) $f|U : U \to f(U)$ is a C^r diffeomorphism.

Theorem 3.2.48 (Inverse Function Theorem) *Let $U \subset \mathbf{R}^n$ be open and $f : U \to \mathbf{R}^n$ be a smooth map. If the derivative $df_x : \mathbf{R}^n \to \mathbf{R}^n$ is nonsingular, then f maps every sufficiently small open set $U \subset \mathbf{R}^n$ about the point x diffeomorphically onto an open set $f(U) \subset \mathbf{R}^n$.*

Proof See Apostol (1957) or Dieudonné (1989). ❑

Remark 3.2.49 The map $f : U \to V$ in Theorem 3.2.48 may not be one-one if the nbd U is not sufficiently small even at the situation when every df_x is nonsingular. For example, consider the exponential map

$$exp : \mathbf{C} \to \mathbf{C}$$

of the complex plane into itself.

Remark 3.2.50 Inverse function theorem formulated in Theorem 3.2.48 has an equivalent statement in Theorem 3.2.51 in the language of local diffeomorphism. It gives a criterion for a map to be a local diffeomorphism at a point.

Theorem 3.2.51 (Another form of Inverse Function Theorem) *Let $U \subset \mathbf{R}^n$ be an open set and $f : U \to \mathbf{R}^n$ be a map. If $p \in U$, then f is a local diffeomorphism in some nbd of p iff the Jacobian determinant of f is nonzero at p.*

Proof See Apostol (1957). ❑

Theorem 3.2.52 (Implicit function theorem in surjective form) *Let $U \subset \mathbf{R}^n$ be an open set and $f : U \to \mathbf{R}^m$ be a C^r map. If $f(y) = 0$ for some $y \in U$ and*

$$df_y : \mathbf{R}^n \to \mathbf{R}^m, \ v \mapsto df_y(v)$$

is surjective, then there is a local diffeomorphism ψ of $\mathbf{R}^n$ at the origin 0 such that

(i)

$$\psi(0) = y \text{ and}$$

(ii)

$$(f \circ \psi)(x_1, x_2, \ldots, x_n) = (x_1, x_2, \ldots, x_m).$$

Proof Without loss of generality, we assume that for $f = (f_1, f_2, \ldots, f_n)$ (by a linear change of coordinates of $\mathbf{R}^m$) and $n > m$

$$\frac{\partial f_i}{\partial x_j}(y) = \delta_{ij} : i = 1, 2, \ldots, m;\ j = 1, 2, \ldots, n.$$

Define

(i)

$$g = (g_1, g_2, \ldots, g_n) : U \to \mathbf{R}^m : \ g_i = f_i \ (i = 1, 2, \ldots, m)$$

and

(ii)

$$g(x_1, x_2, \ldots, x_n) = x_{i-m}, \ \forall i = m + 1, \ldots, n.$$

This asserts that g is a C^r function and the rank of dg_y is m. Hence, it follows by Inverse Function Theorem 3.2.51 that g is a C^r local diffeomorphism at the point y. This asserts that g has a C^r inverse ψ in a nbd of 0 in $\mathbf{R}^n$. This implies that $g(\psi(x)) = x$ for x near the point 0. This proves the theorem. ❑

Theorem 3.2.53 (Implicit function theorem in injective form) *Let $U \subset \mathbf{R}^n$ be an open set and $f : U \to \mathbf{R}^m$ be a C^r map. If $y \in \mathbf{R}^n$ be such that $0 \in f^{-1}(y)$ and the derivative df_0 of f evaluated at the origin is injective. Then there is a local diffeomorphism ψ of $\mathbf{R}^m$ at the point y such that*

(i)

$$\psi(y) = 0 \textit{ and}$$

(ii)

$$(\psi \circ f)(x) = (x_1, x_2, \ldots, x_m, 0, \ldots, \ldots, 0).$$

Proof It follows from Theorem 3.2.52 by duality principle or see Apostol (1957). ❑

3.3 Manifolds in R^n

This section is a gateway to the theory of manifolds and it starts conveying the concepts of smooth manifolds in $\mathbf{R}^n$ and their smooth maps. Moreover, it studies tangent spaces of manifolds in $\mathbf{R}^n$ and derivative operations with inverse function theorem, regular and critical values for smooth maps between manifolds in Euclidean spaces. These ideas lead to facilitate the study of the general theory of manifolds, whose official study starts from Sect. 3.4.

3.3.1 *Smooth Manifolds in* $\mathbf{R}^n$

This subsection communicates the concepts of smooth manifolds based on smooth maps in $\mathbf{R}^n$ introduced in Sect. 3.2.2. Its generalization to arbitrary smooth manifolds is available in Sect. 3.5.

Proposition 3.3.1 *Let X, Y and Z be arbitrary subsets of Euclidean spaces.*

(i) *Let $f : X \to Y$ and $g : Y \to Z$ be smooth maps. Then their composite map*

$$g \circ f : X \to Z$$

is also a smooth map.

(ii) *The identity map $1_X : X \to X$ is a smooth map.*

Proof It follows from Definition 3.2.8. ❑

Remark 3.3.2 Let X and Y be arbitrary subsets of Euclidean space $\mathbf{R}^n$. A smooth map $f : X \to Y$ is a diffeomorphism if f sends X homeomorphically onto Y and f^{-1} is also a smooth map. Instead of looking at arbitrary subsets of Euclidean spaces, we feel interested to study a particular attractive useful class of subsets of $\mathbf{R}^n$, which are smooth manifolds.

Definition 3.3.3 Let M be a subset of $\mathbf{R}^m$.

(i) M is said to be a **smooth manifold** of dimension n if each point $x \in M$ has a nbd $U \cap M$ that is diffeomorphic to an open subset V of the Euclidean space $\mathbf{R}^n$.

(ii) Every diffeomorphism $f : V \to U \cap M$ is called a **parametrization** of the region $U \cap M$.

(iii) Its inverse diffeomorphism $f^{-1} : U \cap M \to V$ is called a **system of coordinates** on $U \cap M$.

Remark 3.3.4 A subset $M \subset \mathbf{R}^m$ is said to be a smooth manifold of dimension 0 if each point $x \in M$ has a nbd $U \cap M$ that precisely consists of the single point x.

Example 3.3.5 **(i)** The unit sphere $S^2 = \{(x, y, z) \in \mathbf{R}^2 : x^2 + y^2 + z^2 = 1\}$ is a smooth manifold of dimension 2.

(ii) In general, $S^n = \{x = (x_1, x_2, \ldots, x_n, x_{n+1}) \in \mathbf{R}^{n+1} : ||x|| = 1\} \subset \mathbf{R}^{n+1}$ is a smooth manifold of dimension n for every $n \geq 1$.

(iii) For $n = 0$, the manifold $S^0 \subset \mathbf{R}$ consists of only two points $\{-1, 1\}$ of $\mathbf{R}$ $(=\mathbf{R}^1)$.

Example 3.3.6 Consider the set $M \subset \mathbf{R}^2$ defined by

$$M = \{(x, y) \in \mathbf{R}^2 : x \neq 0 \text{ and } y = sin(1/x)\}.$$

Then M is a smooth manifold.

3.3.2 Tangent Spaces of Manifolds in $\mathbf{R}^n$ and Derivative Operations

This subsection continues the study of smooth maps discussed in Sect. 3.2.1 and introduces the concept of tangent spaces of smooth manifolds in Euclidean spaces and that of derivative df_x for a smooth map $f : M \to N$ at a point $x \in M$, where M and N are smooth manifolds in Euclidean spaces and studies them. Its generalization to arbitrary smooth manifolds is available in Sect. 3.5.

Definition 3.3.7 Let M and N be smooth manifolds in Euclidean spaces and $f : M \to N$ be a smooth map. The derivative df_x of f assigns to every point $x \in M \subset \mathbf{R}^n$ a linear subspace $T_x(M) \subset \mathbf{R}^n$ of dimension k, called the **tangent space** of M at the point x. Then

$$df_x : T_x(M) \to T_{f(x)}(N)$$

is a linear map. The elements $v \in T_x(M)$ are called **tangent vectors** to the manifold M at the point x.

Remark 3.3.8 The linear space $T_x(M)$ formulated in Definition 3.3.7 may be intuitively thought of as the hyperplane through the origin and parallel to the k-dimensional hyperplane in $\mathbf{R}^n$ that best approximates the manifold M near the point $x \in M$. Before studying the general case for df_x, we first consider the special case of mappings between open sets in Euclidean spaces formulated in Definition 3.3.9.

Definition 3.3.9 Let $U \subset \mathbf{R}^n$ be an arbitrary open set. The tangent space $T_x(U)$ is defined to be the whole vector space $\mathbf{R}^n$. Given any open set $V \subset \mathbf{R}^m$ and a smooth map $f : U \to V$, its derivative df_x at a point $x \in U$ is defined by

$$df_x : \mathbf{R}^n \to \mathbf{R}^m, \ h \mapsto \lim_{t \to 0} \frac{f(x+th) - f(x)}{t}, \ \forall h \in \mathbf{R}^n, x \in U.$$

$df_x(h)$ is a linear function of h, because it is the linear map that corresponds to the $m \times n$ matrix $(\partial f_i/\partial x_i)_x$ of the first partial derivatives evaluated at the point x.

3.3.3 Basic Properties of the Derivative Operation

This subsection conveys the fundamental properties of the linear function $df_x : \mathbf{R}^n \to \mathbf{R}^m$.

Theorem 3.3.10 *Let $U \subset \mathbf{R}^n$, $V \subset \mathbf{R}^k$ and $W \subset \mathbf{R}^m$ be open sets.*

(i) ***(Chain rule)*** *If $f : U \to V$ and $g : V \to W$ are smooth maps with $y = f(x)$, then*

$$d(g \circ f)_x = dg_y \circ df_x.$$

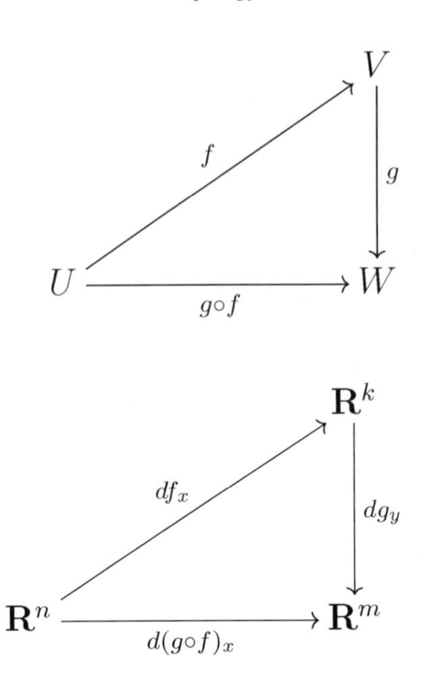

Fig. 3.1 Triangular diagram corresponding to f and g in chain rule

Fig. 3.2 Triangular diagram corresponding to df_x and dg_x in chain rule

In the mapping diagram, the chain rule can be expressed as follows: if the triangle represented in Fig. 3.1 is commutative, then the corresponding triangle in Fig. 3.2 is also commutative.

In other words, the chain rule asserts that the commutative triangular diagram of smooth maps between open sets gives rise to a commutative triangular diagram of the corresponding linear maps as shown in Fig. 3.2.

(ii) *If $I : U \to U$ is the identity map on U, then*

$$dI_x : \mathbf{R}^n \to \mathbf{R}^n$$

is the identity map on $\mathbf{R}^n$. In general, if U and U' are open sets in $\mathbf{R}^n$ and $i : U \hookrightarrow U'$ is the inclusion map, then

$$di_x : \mathbf{R}^n \to \mathbf{R}^n$$

is also the identity map on $\mathbf{R}^n$.

(iii) *If $T : \mathbf{R}^n \to \mathbf{R}^k$ is a linear map, then*

$$dT_x = T.$$

Proof It follows from Definition 3.3.9. ❑

Theorem 3.3.11 *Let $U \subset \mathbf{R}^n$ and $V \subset \mathbf{R}^k$ be open sets. If $f : U \to V$ is a diffeomorphism, then*

(i) $n = k$;
(ii) *the linear map* $df_x : \mathbf{R}^n \to \mathbf{R}^k$ *is nonsingular.*

Proof Suppose $f : U \to V$ is a diffeomorphism with $y = f(x) \in V$ for $x \in U$. Hence, it follows that the composite map

$$f^{-1} \circ f : U \to U$$

is the identity map. This implies by Theorem 3.3.10 that

$$d(f^{-1})_y \circ df_x$$

is the identity map on $\mathbf{R}^n$. Similarly,

$$df_x \circ d(f^{-1})_y$$

is the identity map on $\mathbf{R}^k$. This asserts that df_x is a both-sided inverse and hence it follows from linear algebra that $n = k$ and $df_x : \mathbf{R}^n \to \mathbf{R}^k$ is nonsingular. This proves the theorem. ❑

Remark 3.3.12 The converse of Theorem 3.3.11 is partially true in the sense formulated in Inverse Function Theorem 3.2.48. Definition 3.3.9 of tangent space is now extended in Definition 3.3.13 for an arbitrary manifold $M \subset \mathbf{R}^n$.

Definition 3.3.13 Let $M \subset \mathbf{R}^n$ be a given smooth manifold and $U \subset \mathbf{R}^m$ be open. Take a parametrization $f : U \to M \subset \mathbf{R}^n$ of a nbd $f(U)$ of a point $x \in M$, with $u = f^{-1}(x)$. Consider the map $f : U \to \mathbf{R}^n$ to define the derivative $df_u : \mathbf{R}^m \to \mathbf{R}^n$. Then the set $T_x(M) = Im\ df_u(\mathbf{R}^m)$ (image set of df_u) is an m-dimensional vector space. This construction of $T_x(M)$ is well-defined, since it does not depend on the particular choice of parametrization f as proved in Proposition 3.3.14.

Proposition 3.3.14 *The vector space* $T_x(M) = Im\ df_u(\mathbf{R}^m)$ *formulated in Definition 3.3.13 is independent of the choice of the parametrization* f.

Proof By hypothesis, $f : U \to M \subset \mathbf{R}^n$ is a parametrization of a nbd $f(U)$ of a point $x \in M$ with $u = f^{-1}(x)$. To prove the proposition, take $g : V \to M \subset \mathbf{R}^n$ as any other parametrization of a nbd $g(V)$ of x in M, with $v = g^{-1}(x)$. Then the composite map $g^{-1} \circ f$ sends some nbd U' of u diffeomorphically onto the nbd V' of v. Consider the commutative triangle of smooth maps between open sets as shown in commutative triangle in Fig. 3.3 to give rise to a commutative triangle of linear maps as shown in commutative triangle in Fig. 3.4.

This asserts that $Im\ (df_u) = Im\ (dg_u)$ and hence, the proposition is proved. ❑

Remark 3.3.15 Proposition 3.3.14 proves that $T_x(M)$ formulated in Definition 3.3.13 is well-defined and Proposition 3.3.16 proves that $T_x(M)$ is an m-dimensional vector space.

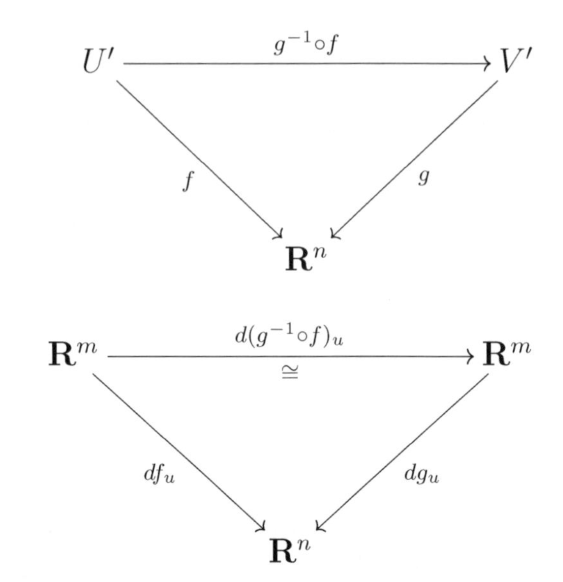

Fig. 3.3 Triangular diagram corresponding to smooth maps

Fig. 3.4 Triangular diagram corresponding to linear maps

Proposition 3.3.16 *$T_x(M)$ is an m-dimensional vector space.*

Proof Consider the tangent space $T_x(M)$ formulated in Definition 3.3.13, which is well-defined by Proposition 3.3.14. Let $f : U \to M \subset \mathbf{R}^n$ be a parametrization of a nbd $f(U)$ of a point $x \in M$, with $u = f^{-1}(x)$, where U is an open subset of $\mathbf{R}^m$. Then

$$f^{-1} : f(U) \to U$$

is a smooth map. Take an open set $W \subset \mathbf{R}^n$ containing the point x and a smooth map

$$H : W \to \mathbf{R}^m \quad \text{such that} \quad H = f^{-1}|_{W \cap f(U)}.$$

For $U_0 = f^{-1}(W \cap f(U))$, the commutativity of the triangular diagram as shown in Fig. 3.5 gives rise to the commutative triangular diagram as shown in Fig. 3.6. This implies that df_u has rank m and the vector space $T_x(M)$ is of dimension m.

This proves that $T_x(M)$ is an m-dimensional vector space over $\mathbf{R}$. ❑

Definition 3.3.17 The m-dimensional vector space $T_x(M) = Im\ df_u(\mathbf{R}^m)$ formulated in formulated in Definition 3.3.13 is called the **tangent space for the smooth manifold** $M \subset \mathbf{R}^n$ at the point $x \in M$.

Remark 3.3.18 Given two smooth manifolds M and N such that $M \subset \mathbf{R}^p$ and $M \subset \mathbf{R}^q$ and a smooth map $f : M \to N$ with $f(x) = y$, Proposition 3.3.19 considers the derivative

$$df_x : T_x(M) \to T_{y=f(x)}(N),\ v \mapsto dH_x(v),$$

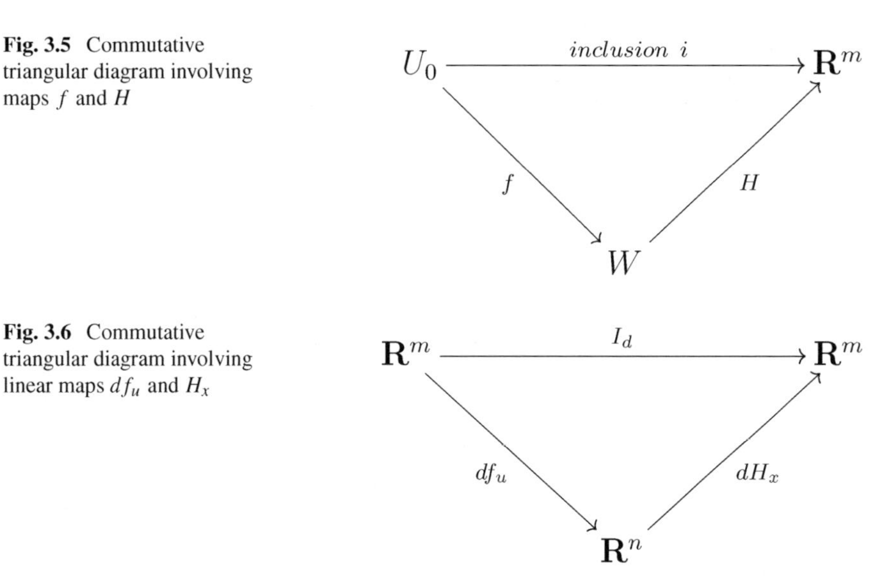

Fig. 3.5 Commutative triangular diagram involving maps f and H

Fig. 3.6 Commutative triangular diagram involving linear maps df_u and H_x

where H is defined in Proposition 3.3.16.

Proposition 3.3.19 *The derivative* $df_x : T_x(M) \to T_{y=f(x)}(N), v \mapsto dH_x(v)$ *is well-defined.*

Proof To define the derivative

$$df_x : T_x(M) \to T_{f(x)}N,$$

consider an open set W containing the point x and a smooth map

$$H : W \to \mathbf{R}^q$$

such that $f = H|_{W \cap f(M)}$ (existence of both W and H is guaranteed by the smoothness of f).

Define

$$df_x : T_x(M) \to T_{y=f(x)}(N), v \mapsto dH_x(v).$$

To show that df_x is well-defined, we have to prove that

(i) $dH_x(v) \in T_yN, \ \ \forall\, v \in T_x(M)$ and
(ii) df_x does not depend on the particular choice of H.

To prove it, take parametrizations

$$g : U \to M \subset \mathbf{R}^p \ \text{ and } \ k : V \to N \subset \mathbf{R}^q$$

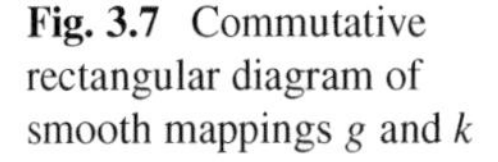

Fig. 3.7 Commutative rectangular diagram of smooth mappings g and k

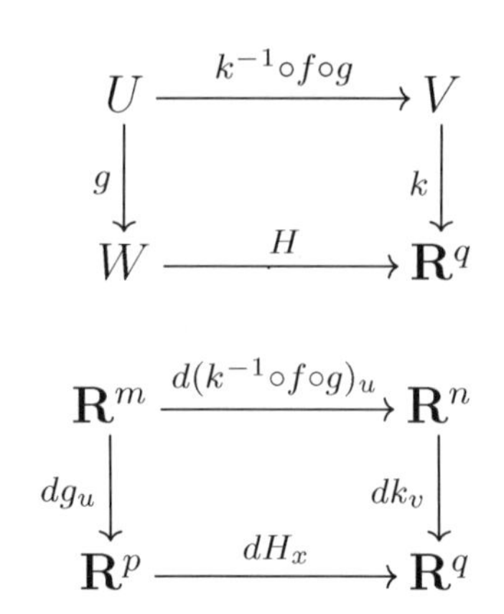

Fig. 3.8 Commutative rectangular diagram of linear mappings dg_u and dk_v

for nbds $g(U)$ of x and nbds $k(V)$ of y. Without any loss of generality, we may assume that $g(U) \subset W$ and f sends $g(U)$ into $k(V)$ (by replacing U by a smaller set if it is needed). This shows that the map

$$k^{-1} \circ f \circ g : U \to V$$

is well-defined and it is smooth. Consider the commutative rectangular diagram of smooth mappings as shown in Fig. 3.7.

It gives rise to the corresponding commutative rectangular diagram of linear maps by taking derivatives as shown in Fig. 3.8, where $u = g^{-1}(x)$ and $v = k^{-1}(y)$. Then dH_x sends $T_x(M) = Im\ dg_u$ into $T_yN = Im\ dk_v$. Moreover, the induced map df_x does not depend on the particular choice of the H, because of the commutativity of the rectangular diagram of linear mappings as given in Fig. 3.8.

This asserts that

$$df_x = dk_v \circ d(k^{-1} \circ f \circ g)_u \circ (dg_u)^{-1} : T_x(M) \to T_{f(x)}(N)$$

is a linear map.This proves that the derivative

$$df_x : T_x(M) \to T_{f(x)}N$$

is a well-defined linear map. ❑

Definition 3.3.20 Let M and N be two smooth manifolds such that $M \subset \mathbf{R}^p$ and $N \subset \mathbf{R}^q$ and $f : M \to N$ be a smooth map with $f(x) = y$. Then

$$df_x : T_x(M) \to T_{y=f(x)}(N)$$

is well-defined by Proposition 3.3.19 and it is called the **derivative** of f at the point $x \in M$.

Remark 3.3.21 Theorem 3.3.22 extends the fundamental properties of the linear function df_x formulated in Theorem 3.3.10 for manifolds in Euclidean spaces.

Theorem 3.3.22 *Let $M \subset \mathbf{R}^n$, $N \subset \mathbf{R}^k$ and $P \subset \mathbf{R}^m$ be three smooth manifolds.*

(i) ***(Chain rule)*** *If $f : M \to N$ and $g : N \to P$ are smooth maps with $y = f(x)$, then*

$$d(g \circ f)_x = dg_y \circ df_x.$$

(ii) *If $I : M \to M$ is the identity map, then*

$$dI_x : T_x(M) \to T_x(M)$$

is the identity map of $T_x(M)$.

(iii) *In general. if M is a submanifold of N with natural inclusion $i : M \hookrightarrow N$, then*

$$di_x : T_x(M) \hookrightarrow T_x(N)$$

is also natural inclusion.

Proof It follows from Definition 3.3.20. ❑

Theorem 3.3.23 *Let $M \subset \mathbf{R}^k$ be a smooth manifold of dimension n and $N \subset \mathbf{R}^p$ be a smooth manifold of dimension m. If $f : M \to N$ is a diffeomorphism, then*

$$df_x : T_x(M) \to T_x(N)$$

is an isomorphism of vector spaces. In particular, dim M = dim N.

Proof Proceed as in Theorem 3.3.11. ❑

3.3.4 Regular and Critical Values of Smooth Maps Between Manifolds of Same Dimension

This subsection introduces the concepts of regular and critical values of smooth maps between manifolds of the same dimension in Euclidean spaces. For the more general case, when their dimensions are different, see Sect. 3.17.

Definition 3.3.24 Let M and N be two manifolds of same dimension in Euclidean spaces and $f : M \to N$ be a smooth map.

(i) A point $x \in M$ is said to be a **regular point** of f, if the derivative

$$df_x : T_x(M) \to T_{f(x)}(N)$$

is nonsingular. The point $y \in N$ is called the **regular value** of f if $f^{-1}(y)$ contains only regular points (note that since df_x is nonsingular, by inverse function theorem, f sends a nbd of x in M diffeomorphically onto an open set in N).

(ii) A point $x \in M$ is said to be a **critical point** of f if the derivative

$$df_x : T_x(M) \to T_{f(x)}(N)$$

is singular. The image $f(x)$ is called a **critical value** of f.

Remark 3.3.25 Definition 3.3.24 asserts that given a smooth map $f : M \to N$, every point $y \in N$ is either a regular value or a critical value of f.

(i) $y \in N$ is a critical value of f if $f^{-1}(y)$ does contain a critical point.
(ii) $y \in N$ is a regular value of f if $f^{-1}(y)$ does not contain a critical point.
(iii) The general formulation of critical and regular values for a smooth map $f : M \to N$ between arbitrary smooth manifolds is available in Definition 3.17.1.

Proposition 3.3.26 *Let M and N be two manifolds in Euclidean spaces and $f : M \to N$ be a smooth map. If M is compact and $y \in N$ is a regular value of f, then $f^{-1}(y)$ is a finite set (may be empty).*

Proof By hypothesis, $f : M \to N$ is a smooth map, M is compact and $y \in N$ is a regular value of f. Since $f^{-1}(y)$ is a closed subset of the compact set M, it follows that the set $f^{-1}(y)$ is also compact. Again, since the map f is one-one in a nbd of every point $x \in f^{-1}(y)$, it follows that the compact set $f^{-1}(y)$ is discrete. Let card $f^{-1}(y)$ be the number points of the set $f^{-1}(y)$. Then this number card $f^{-1}(y)$ is locally constant as a function of y, where y runs over only regular values. Hence, there exists a nbd $V \subset N$ of y such that card $f^{-1}(y)$ =card $f^{-1}(y')$, $\forall\, y' \in V$. To show the existence of such a nbd V, take points $x_1, x_2, \ldots, x_n \in f^{-1}(y)$ and choose pairwise disjoint nbds $U_1, U_2, \ldots, U_n$ of the points $x_1, x_2, \ldots, x_n \in f^{-1}(y)$ which are mapped diffeomorphically onto the nbds $V_1, V_2, \ldots, V_n$ in N. The existence of our required V is proved by taking

$$V = V_1 \cap V_2 \cap \cdots \cap V_n - f(M - U_1 - U_2 - \cdots - U_n).$$

Hence, it follows that the set $f^{-1}(y)$ is finite (may be empty). ❑

3.4 Differentiable Manifolds

Every manifold is a topological manifold by definition, but there are many manifolds which are equipped with an additional structure such as differential structure, called differentiable manifolds. This additional structure makes the study of differential manifolds interesting. This section introduces the concepts of n-dimensional C^r-manifolds and differentiable (smooth) manifolds with illustrative examples, which lead to **differential topology**. Classical analysis traditionally studies real-valued functions, their continuity and differentiability in the Euclidean space $\mathbf{R}^n$. To define a continuous function between two abstract sets, it is necessary to give these abstract sets topological structures. Similarly, to define a differentiable function between two abstract sets, it is necessary to give these abstract sets differentiable structures. This generalization of a differentiable function in an abstract setting leads to the concept of **differentiable manifolds,** from which many classical results of analysis, geometry and topology are generalized.

The motivation of differentiable manifolds may be explained from the following observation: for a finite-dimensional topological manifold, a coordinate system can be introduced, and any one moving around such a manifold has to move from one set of such coordinates to the other. This produces a continuous change of coordinates described by real-valued functions of multivariables. If this change of coordinates is differentiable, then it facilitates to introduce the concept of differentiable manifolds (also called smooth or C^∞-manifold). Differentiable manifolds are studied based on the standard differential structure on the Euclidean space $\mathbf{R}^n$. So, it has become necessary to delight the basic relevant results on differentiable functions in $\mathbf{R}^n$, which are given in Sects. 3.2.3 and 3.2.4.

3.4.1 Differentiable Manifolds: Introductory Concepts

This subsection introduces the concept of abstract differentiable manifolds, formally defined by Hermn Weyl (1885–1955) in 1912 in his book on the Riemann surfaces. The theory of manifolds earned momentum through the work of H. Whitney (1907–1989) in the 1930s such as Whitney's Embedding Theorem 3.14.20 saying that every n-dimensional compact manifold can be embedded in a finite-dimensional Euclidean space. This theorem reconciles the original concept of a manifold with its modern abstract concept. Since then, differential topology witnesses a rapid development.

The concept of differentiable manifold is now considered as a basic object of study in mathematics, because of its own problems and methods. There are two quite different definitions of a differentiable manifold. The first one is traditional in the sense that it is a topological space with a structure facilitating to define differentiable functions on it. The second one says that it is a topological space constructed by gluing together open subsets of some Euclidean space in an elegant way. For example, the surface of a soccer ball can be obtained by gluing together (pasted together on overlaps) open disks (of Euclidean plane) without doing any fold or crease.

Remark 3.4.1 Convention: From now, a manifold means a smooth manifold, unless stated otherwise and the terms smooth manifold, differentiable manifold and C^∞-manifold are interchangeable.

Definition 3.4.2 Let $M \subset \mathbf{R}^m$ be a nonempty set.

(i) An ordered pair (ψ, U) consisting of a subset $U \subset M$ together with an injective map ψ from U onto an open subset in $\mathbf{R}^n$ is said to be an ***n*-coordinate pair** on M.

(ii) Two n-coordinate pairs (ψ, U) and (ϕ, V) on M are said to be C^r **related** if the maps $\psi \circ \phi^{-1}$ and $\phi \circ \psi^{-1}$ are C^r-maps having their domains of definition open sets.

(iii) A collection $\mathcal{C} = \{(\psi_i, U_i)\}$ of n-coordinate pair on M is said to be a C^r**-atlas** on M if every member of $\mathcal{C}$ is C^r related to every other member of $\mathcal{C}$, and $\bigcup U_i = M$.

(iv) A maximal collection of C^r related n-coordinate pairs is called a **maximal** C^r**-atlas** on M.

Definition 3.4.3 A topological manifold M with a given C^r-structure on M is said to be a C^r**-manifold (or a manifold of class** C^r). The dimension of the space $\mathbf{R}^n$ from which the homeomorphisms of charts act is said to be the **dimension of the** C^r**-manifold** M. In particular, if $r = 0$, then M is called a **topological *n*-manifold**.

Remark 3.4.4 Definition for C^∞ structure on M is similar. It is formulated in Definition 3.4.7.

Definition 3.4.7 of smooth maps defined on an open set is now extended over arbitrary subsets of a manifold.

Definition 3.4.5 (*Smooth map*) Let M, N be two manifolds in $\mathbf{R}^n$ and X be a subset of M. Then a map $f : X \to N$ is said to be **smooth** if f can be locally extended to a smooth map in the sense that every point $x \in X$ has an open nbd U in M and a smooth map

$$\tilde{f} : U \to N : \tilde{f}|_{X \cap U} = f.$$

Every n-coordinate pair (ψ, U) on a subset $U \subset M$ determines a set of n coordinate functions ψ_i through the function ψ such that

$$\psi_i = p_i \circ \psi : U \to \mathbf{R} \; for \; i = 1, 2, \ldots n,$$

where the projection maps p_i are defined by

$$p_i : \mathbf{R}^n \to \mathbf{R}, \; (x_1, x_2, \ldots, x_n) \mapsto x_i, \; for \; i = 1, 2, \ldots, n.$$

Clearly, each ψ_i is a C^r-map. The n-tuple of maps $(\psi_1, \psi_2, \ldots . \psi_n)$ is also called a **coordinate system** of (ψ, U).

Definition 3.4.5 is reformulated in Definition 3.4.6.

Definition 3.4.6 Let M be a manifold of dimension n and N be a manifold of dimension m. Then a continuous map

$$f : M \to N$$

is said to be C^∞ **or smooth at a point** $x \in M$ if there exist charts

(i) (ψ, U) about the point $x \in M$ and
(ii) (ϕ, V) about the point $f(x) \in N$

such that the composite map $\phi \circ f \circ \psi^{-1}$ from the open set $\psi(\phi^{-1}(V) \cap U)$ of $\mathbf{R}^m$ to $\mathbf{R}^n$ is C^∞ at the point x.

The continuous map is said to be C^∞ **or smooth** if it is C^∞ at every point $x \in M$.

Definition 3.4.7 (*Smooth manifold*) Let M be a second countable Hausdorff space. Then it is said to be an n-dimensional **smooth manifold (or differentiable manifold or C^∞-manifold)** if there exits a collection of maps $\mathcal{C} = \{\psi\}$, called **charts** such that

(i) for every chart $\psi \in \mathcal{C}$, the map $\psi : U \to V$ is a homeomorphism, where $U \subset M$ and $V \subset \mathbf{R}^n$ are both open;
(ii) every point $x \in M$ is in the domain of some chart;
(iii) given two charts $\psi_i : U_i \to V_i \subset \mathbf{R}^n$ and $\psi_j : U_j \to V_j \subset \mathbf{R}^n$ with $U_i \cap U_j \neq \emptyset$, the change of coordinates $\psi_i \circ \psi_j^{-1} : \psi_j(U_i \cap U_j) \to \psi_i(U_i \cap U_j)$ is C^∞;
(iv) the collection $\mathcal{C}$ of charts is maximal satisfying the conditions prescribed in (i), (ii) and (iii) (i.e.,in the sense that if ϕ is any homeomorphism mapping an open set in M onto an open set in $\mathbf{R}^n$ such that for every $\psi \in \mathcal{C}$, the homeomorphisms ϕ with domain U and ψ with domain V satisfy the condition: for $U \cap V \neq \emptyset$, if both the functions

$$\phi \circ \psi^{-1} : \psi(U \cap V) \to \phi(U \cap V) \text{ and } \psi \circ \phi^{-1} : \phi(U \cap V) \to \psi(U \cap V)$$

are C^∞, then $\phi \in \mathcal{C}$).
A set of charts satisfying (i), (ii) and (iii) is said to be **an atlas**.

Definition 3.4.8 Two charts $\psi_i : U_i \to V_i \subset \mathbf{R}^n$ and $\psi_j : U_j \to V_j \subset \mathbf{R}^n$ on an n-manifold are said to be C^r **compatible (or C^r overlap)** as shown in Fig. 3.9 if the coordinate changes

$$\psi_j \circ \psi_i^{-1} : \psi_i(U_i \cap U_j) \to \psi_j(U_i \cap U_j)$$

and

$$\psi_i \circ \psi_j^{-1} : \psi_j(U_i \cap U_j) \to \psi_i(U_i \cap U_j)$$

are both of C^r-class for some natural number r or they are both of C^∞-class or both of C^ω class (in the sense of real analytic). The two maps $\psi_i \circ \psi_j^{-1}$ and $\psi_j \circ \psi_i^{-1}$ are called the **transition functions** between the charts $U_i \cap U_j$. If $U_i \cap U_j = \emptyset$, then the two charts are automatically C^∞-compatible.

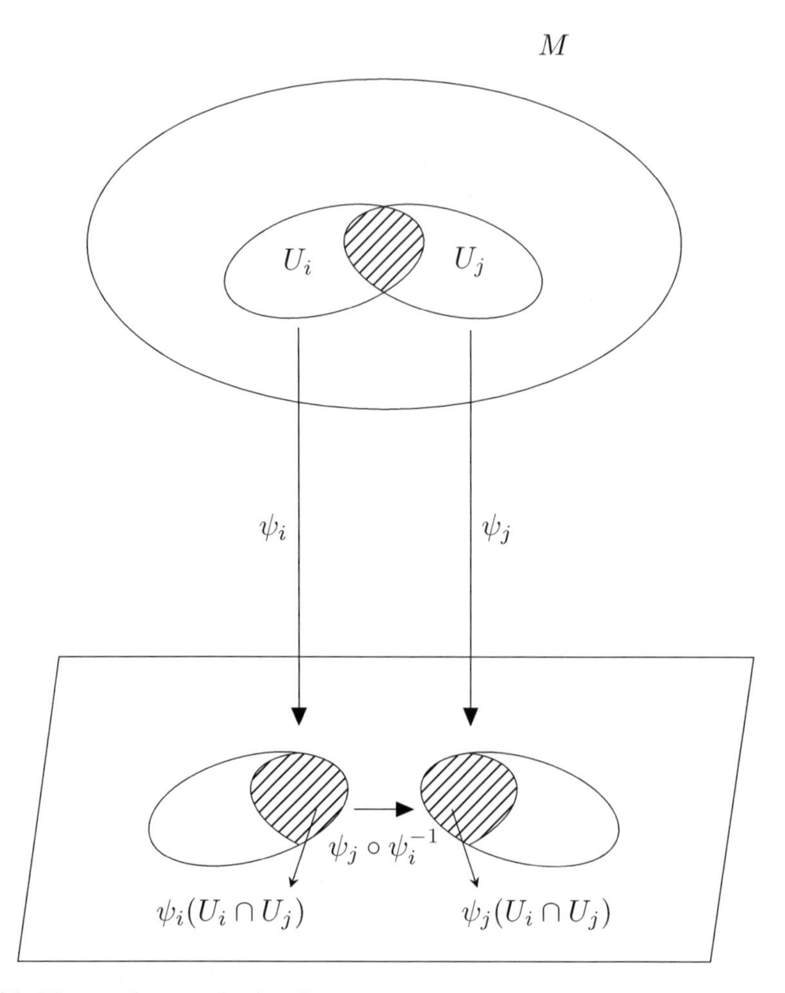

Fig. 3.9 Diagram of two overlapping charts

Remark 3.4.9 Definition 3.4.7 shows every differentiable (smooth) manifold has its own a **second countable Hausdorff topology** τ.

Definition 3.4.10 An n-dimensional differentiable manifold M is a Hausdorff topological space having a countable open covering $\{U_1, U_2, \dots\}$ such that

DM(1) for each U_i, there is homeomorphism $\psi_i : U_i \to V_i$, where V_i is an open disk in $\mathbf{R}^n$;

DM(2) if $U_i \cap U_j \neq \emptyset$, the homeomorphism

$$\psi_{ji} = \psi_j \circ \psi_i^{-1} : \psi_i(U_i \cap U_j) \to \psi_j(U_i \cap U_j)$$

is a differentiable map.

(U_i, ψ_i) is called a **local chart** of M and $\{(U_i, \psi_i)\}$ is a set of local charts of M.

Remark 3.4.11 The condition **DM(2)** asserts that ψ_{ji} is an inverse of ψ_{ij} and hence its Jacobian determinant is nonzero at every point $x \in \psi_i(U_i \cap U_j)$.

Definition 3.4.12 Two charts (ψ, U) and (ϕ, V) on a manifold M are said to be C^r**-compatible** if any one of the following conditions are satisfied:

(i) $U \cap V = \emptyset$;
(ii) if $U \cap V \neq \emptyset$, then the homeomorphism

$$\phi^{-1} \circ \psi : \psi^{-1}(U \cap V) \to \phi^{-1}(U \cap V)$$

is a C^r-diffeomorphism.

Definition 3.4.13 A set of charts $\{(\psi_i, U_i)\}$ on a manifold M is said to be a C^r**-atlas** if any two of its charts are C^r-compatible and $\bigcup_i U_i = M$.

There is a natural question: when are two atlases considered equivalent?

Definition 3.4.14 Two C^r-atlases $\{(\psi_i, U_i)\}$ and $\{(\phi_j, V_j)\}$ are said to be **equivalent** if any two charts (ψ_i, U_i) and (ϕ_j, V_j) are C^r-compatible in the sense that their union is a C^r-atlas.

Remark 3.4.15 As the relation of being equivalent of C^r-atlases on a manifold M is an equivalence relation, the set of C^r-atlases on M is partitioned into disjoint classes of equivalent atlases on M.

Definition 3.4.16 The equivalence class of C^r-atlases on a manifold M is called a C^r**-structure on** M.

Remark 3.4.17 In view of Definition 3.4.16, it is possible to restore any given C^r-structure on a manifold from any of its C^r-atlases. The union of all C^r-atlases on a manifold is again a C^r-atlas, called a **maximal** C^r**-atlas** on the manifold.

Definition 3.4.18 For $r = 1, 2, \ldots, \infty$, the C^r-structures on manifolds are called **smooth or differential structures**. A C^0-structure on manifold is known as a **topological structure** and it is unique.

Example 3.4.19 A differential manifold M may have several C^r-structures for $r \neq 0$. For example, for $M = \mathbf{R}^1$,

(i) the real line space, consider the chart (ψ, U), where $U = \mathbf{R}^1$, and $\psi = 1d : \mathbf{R}^1 \to \mathbf{R}^1$ is the identity map. It admits on $\mathbf{R}^1$ a C^∞-structure;
(ii) again for another chart $(\phi, \mathbf{R}^1)$ where $\phi : \mathbf{R}^1 \to \mathbf{R}^1, x \mapsto x^3$ also admits on $\mathbf{R}^1$ a C^∞-structure.

As these two atlases are not equivalent, it follows that $\mathbf{R}^1$ admits different C^∞-structures.

Example 3.4.20 **(i)** The Euclidean n-space $\mathbf{R}^n$ is an n-dimensional differentiable manifold. Here, the open covering $\mathcal{C}$ consists of only one open set $V = U = \mathbf{R}^n$, the whole space, and the homeomorphism $\psi : U \to V$ is taken as its identity function.

(ii) Every open set U in $\mathbf{R}^n$ has only one C^r-differential structure containing the inclusion map $i : U \hookrightarrow \mathbf{R}^n$.

(iii) The sphere $S^2 = \{(x, y, z) \in \mathbf{R}^3 : x^2 + y^2 + z^2 = 1\}$ is a 2-dimensional differentiable manifold. Here, $\mathcal{C} = \{V_1, V_2, \ldots, V_6\}$ consists of six open sets, each of them is a hemisphere. For example, here take $V_1 = \{(x, y, z) \in S^2 : z > 0\}$ to be the hemisphere $z > 0$, take the map ψ_1 sending every $(x, y, z) \in V_1$ to the point $(x, y, 0)$ and take U_1 as the unit open disk on the xy-plane with center at the origin. Then

$$\psi_1 : V_1 \to U_1,\ (x, y, z) \mapsto (x, y.0)$$

is a homeomorphism.
Similarly, take $V_2 = \{(x, y, z) \in S^2 : x > 0\}$ to be the hemisphere $x > 0$ and take the map ψ_2 sending every $(x, y, z) \in V_2$ to the point $(0, y, z)$. The collection of these image points $\{(0, y, z)\}$ forms an open set U_2 homeomorphic to the open unit disk in yz-plane and hence

$$\psi_2 : V_2 \to U_2,\ (x, y, z) \mapsto (0, y, z)$$

is a homeomorphism. Again, the map $\psi_{12} = \psi_1 \circ \psi_2^{-1}$ defined by

$$\psi_{12}(0, y, z) = ([(1 - y^2 - z^2]^{1/2}, y, 0)$$

is a differentiable. Similarly, other maps ψ_{ij} are also differentiable.

(iv) As a generalization of Example (ii), the n-sphere S^n is an n-dimensional differentiable manifold, since it is a Hausdorff topological space with a countable open covering $\mathcal{C} = \{V_1, V_2, \ldots\}$ such that

(i) for every $V_i \in \mathcal{C}$, there is a homeomorphism $\psi_i : V_i \to U_i$, where U_i is an open disk in the Euclidean space $\mathbf{R}^n$;

(ii) if $V_i \cap V_j \neq \emptyset$, the homeomorphisms ψ_i and ψ_j give rise to a homeomorphism

$$\psi_{ji} = \psi_j \circ \psi_i^{-1} : \psi_i(V_i \cap V_j) \to \psi_j(V_i \cap V_j)$$

which is a differentiable map. Hence, it follows that the n-sphere is also a differentiable manifold with dimension n.

(v) The real projective space $\mathbf{R}P^n$ is an n-dimensional differentiable manifold. The space $\mathbf{R}P^n$ is obtained as the quotient space obtained from the n-sphere S^n by identifying diametrically its opposite points. This space is also obtained as a quotient space $\mathbf{R}^{n+1} - \{0\}/\sim$, where $\sim$ is the equivalence relation defined by

$$(x_1, x_2, \ldots, x_{n+1}) \sim (y_1, y_2, \ldots, y_{n+1}) \; holds$$

iff $y_i = \alpha x_i$, for some nonzero real number α. Let (ψ_i, V_i) be the $(n+1)$ charts of $\mathbf{R}P^n$ formulated by

$$V_i = \{(x_1, x_2, \ldots, x_{n+1}) \in \mathbf{R}P^n : x_i \neq 0, \; i = 1, 2, \ldots, n+1\}$$

and

$$\psi_i : V_i \to \mathbf{R}^n, \; (x_1, x_2, \ldots, x_{n+1}) \mapsto \left(\frac{x_1}{x_i}, \frac{x_2}{x_i}, \ldots, \frac{x_{i-1}}{x_i}, \frac{x_{i+1}}{x_i}, \ldots, \frac{x_{n+1}}{x_i}\right) \in \mathbf{R}^n,$$

equivalently,

$$\psi_i : V_i \to \mathbf{R}^n, \; (x_1, x_2, \ldots, 1, \ldots x_{n+1}) \mapsto (x_1, x_2, \ldots, x_{i-1}, x_{i+1}, \ldots, x_{n+1}) \in \mathbf{R}^n,$$

where 1 occupies the ith place in the left-hand side. Then $\mathbf{R}P^n$ is an n-dimensional manifold. In particular, for $n = 2$, the real projective space $\mathbf{R}P^2$ is a surface, which is a 2-dimensional differential real manifold by Corollary 3.20.3.

(vi) Let $GL(n, \mathbf{R})$ denote the set of all nonsingular $n \times n$ matrices over $\mathbf{R}$. This set of matrices forms a locally Euclidean space of dimension n^2. Every $n \times n$ matrix over $\mathbf{R}$ can be identified with an n^2-tuple of real numbers (by stretching out its rows in a line). $GL(n, \mathbf{R})$ is a real differentiable manifold of dimension n^2. Similarly, the general complex linear group $GL(n, \mathbf{C})$ is a complex differentiable manifold of dimension n^2.

(vii) The special real linear group $SL(n, \mathbf{R}) = \{M \in GL(n, \mathbf{R}) : det M = 1\}$ is a real differential manifold of dimension $n^2 - 1$. The special complex linear group $SL(n, \mathbf{C}) = \{M \in GL(n, \mathbf{C}) : det M = 1\}$ is a complex differential manifold of dimension $n^2 - 1$.

(viii) Orthogonal group $O(n, \mathbf{R}) = \{M \in GL(n, \mathbf{R}) : MM^t = I\}$ is a real differential manifold of dimension $\frac{n^2-n}{2}$. The unitary group $U(n, \mathbf{C})$=$\{M \in GL(n, \mathbf{C}) : MM^* = I\}$, where M^* is the Hermitian conjugate, is a complex differential manifold of dimension $\frac{n^2-n}{2}$.

(xi) The special complex linear group $SL(n, \mathbf{C}) = \{M \in GL(n, \mathbf{C}) : det M = 1\}$ is a complex differential manifold of dimension $n^2 - 1$.

(x) **(Figure-eight manifold $\mathbf{F}_8$)** It consists of all points in $\mathbf{R}^2$ of the form $(sin 2t, \; sin t)$, for all $t \in \mathbf{R}$, i.e.,

$$\mathbf{F}_8 = \{(x, y) \in \mathbf{R}^2 : x = sin\, 2t, \, y = sin\, t, \; \forall t \in \mathbf{R}\}.$$

The injection

$$\psi : \mathbf{F}_8 \to \mathbf{R}, \; (sin 2t, \; sin t) \mapsto t, \; \forall t \in (0, 2\pi)$$

is onto on an open interval of $\mathbf{R}$ and hence it is a chart with domain $\mathbf{F}_8$. It admits a C^∞ atlas into $\mathbf{R}$ and it is a differentiable manifold. Consider another C^∞ atlas of $\mathbf{F}_8$ consisting of a single chart

$$\phi : \mathbf{F}_8 \to \mathbf{R},\ (sin2t,\ sint) \mapsto t,\ \forall t \in (-\pi, \pi).$$

But these two atlases ψ and ϕ are not equivalent, because the change of coordinates

$$\phi \circ \psi^{-1} : \mathbf{R} \to \mathbf{R},\ t \mapsto \begin{cases} t, \text{ for all } t \in (0, \pi) \\ t - \pi \text{ if } t = \pi \\ t - 2\pi, \text{ for all } t \in (\pi, 2\pi) \end{cases}$$

is not even continuous.

Proposition 3.4.21 *The dimension of a C^0-manifold M is an invariant in the sense that it is independent of the choice of its atlas.*

Proof Let M admit atlases $\{(U_i, \psi_i)\}$ and $\{(V_j, \phi_j)\}$, where $\psi_i : \mathbf{R}^n \to U_i$ and $\phi_j : \mathbf{R}^m \to V_j$ are homeomorphisms for $n \neq m$. Then there exist open sets U_i, V_j such that $U_i \cap V_j \neq \emptyset$ and the map

$$\phi_j^{-1} \circ \psi_i : \psi_i^{-1}(U_i \cap V_j) \to \phi_j^{-1}(U_i \cap V_j)$$

is a homeomorphism. But this contradicts the Brower dimension theorem which asserts that nonempty open sets $U \subset \mathbf{R}^n$ and $V \subset \mathbf{R}^m$ are homeomorphic only when $n = m$. ❑

Remark 3.4.22 An atlas of C^r-class on a differentiable manifold M can be conveniently defined as a collection of bijective mappings from subsets of M to open subsets of $\mathbf{R}^n$ such that all coordinate changes are C^r.

3.4.2 C^∞-Structures on Manifolds and C^∞-Diffeomorphism

This subsection conveys the concept of C^∞-structure on manifolds by using their topological structures and also the concept of C^∞-diffeomorphism, which are both basic concepts in manifold theory.

Definition 3.4.23 Let M be an n-dimensional manifold with a set of local charts $\{(\psi_i, U_i)\}$, N be an m-dimensional manifold with a set of local charts $\{(\phi_i, V_i)\}$ and $f : M \to N, x \mapsto f(x)$ be a map. Let (ψ, U) be a chart of M and (ϕ, V) be a chart of N such that $x \in U$ and $f(x) \in V$. Then f is said to be **differentiable** if the map

$$\phi \circ f \circ \psi^{-1} : \mathbf{R}^n \to \mathbf{R}^m$$

is differentiable. Since the differentiability of the map $f : M \to N, x \mapsto f(x)$ in Definition 3.4.23 does not depend on the choice of the charts of M and N, it is well-defined.

Definition 3.4.24 Let M and N be two smooth manifolds and $f : M \to N$ be a map. A chart (ψ, U) for M is said to be adapted to f by a chart (ϕ, V) for N if $f(U) \subset V$. The map

$$\phi \circ f \circ \psi^{-1} : \psi(U) \to \phi(V)$$

is well-defined and it is called the local representation of f at the point $x \in U$ in the given charts. The map f is said to differentiable (or smooth) at x if it has a local representation at x which is differentiable (or smooth). This is well-defined, because a local representation is a map between open sets in Euclidean spaces.

A basic problem in differential topology is classification problem of manifolds up to diffeomorphism. The concept of diffeomorphism is introduced in Definition 3.4.25. Two diffeomorphic manifolds are clearly homeomorphic.

Definition 3.4.25 Let M and N be two manifolds and $f : M \to N$ be a homeomorphism. If ψ and ϕ are coordinate functions such that the function

$$\phi \circ f \circ \psi^{-1} : \mathbf{R}^n \to \mathbf{R}^m$$

is invertible in the sense that there exists a differentiable function

$$\psi \circ f \circ \phi^{-1} : \mathbf{R}^m \to \mathbf{R}^n$$

such that

$$\phi \circ f \circ \psi^{-1} \textbf{ and } \psi \circ f \circ \phi^{-1}$$

are both C^∞-functions. Then f is said to be a **diffeomorphism** and the manifolds are said to be **diffeomorphic**. In particular, a homeomorphism f from an open subset of $\mathbf{R}^n$ to another open subset of $\mathbf{R}^n$ is said to be a diffeomorphism if f is a C^∞-function having C^∞ inverse.

Remark 3.4.26 Let M and N be two differentiable manifolds. A homeomorphism $f : M \to N$ between them is said to be a **diffeomorphism** if both f and its inverse f^{-1} are differentiable. The manifolds M and N are said to be **diffeomorphic** if there exists a diffeomorphism between them.

3.5 Tangent Spaces and Differentials of Smooth Maps

This section generalizes the concepts of tangent spaces of manifolds in $\mathbf{R}^n$ and derivative operations studied in Sect. 3.3.2 for arbitrary smooth manifolds. It uses the concept of derivatives to identify the linear space which gives the best approximation of a manifold M at a point $x \in U \subset M$, where U is open in M. This leads to the concept of tangent space $T_x(M)$. The concept of tangent vector formulated in Definition 3.5.7 plays a key role in the study of tangent spaces.

Definition 3.5.1 Given an n-dimensional C^r-manifold M with an atlas $\{\{\psi_i, U_i)\}$,

(i) a real-valued function

$$f : U_i \to \mathbf{R}$$

is said to be a C^k**-function** $(k \leq r)$, written as $f \in \mathcal{C}(U_i, \mathbf{R})$, if for every ψ_i, the map $f \circ \psi_i^{-1}$ is a C^k-map, mapping the image of $\psi_i \subset \mathbf{R}^n$ into $\mathbf{R}$;

(ii) on the other hand, given an open set $U \subset M$ and a point $x \in U$, a real-valued function

$$f : M \to \mathbf{R}$$

is said to be of **class** $C^k(k \leq r)$ in a nbd U of x, denoted by $f \in \mathcal{C}^k(M, x, \mathbf{R})$, if U = domain of f and $f \in \mathcal{C}^k(U, \mathbf{R})$, where U as an open set in M that admits C^r-manifold structure.

Remark 3.5.2 To define a C^k-function, we utilize the facts that

(i) a manifold M with an atlas $\{(\psi_i, U_i)\}$ looks locally like $\mathbf{R}^n$ through the coordinate systems and

(ii) $\psi_j \circ \psi_i^{-1}$ is a C^r-homeomorphism for $r \geq k$.

Definition 3.5.3 Let M be a C^r-manifold with an atlas $\{\{\psi_i, U_i)\}$, and

$$p_j : \mathbf{R}^n \to \mathbf{R},\ x = (x_1, x_2, \ldots, x_n) \mapsto x_j$$

be the jth coordinate function on $\mathbf{R}^n$. Then the jth **coordinate function** f_j on $\mathbf{R}$ for $\{\{\psi_i, U_i)\}$, defined by

$$f_j : U_i \to \mathbf{R} : f_j = p_j \circ \psi_i,$$

is a C^r-function.

Remark 3.5.4 In Definition 3.5.3, f_j is sometimes replaced by $x_j : U \to \mathbf{R}$. Then the n-tuple of functions $(x_1, x_2, \ldots, x_n)$ is also sometimes taken as a coordinate system. Let $x_1, x_2, \ldots, x_n$ represent the coordinate functions of a coordinate system ψ about a point x and $y_1, y_2, \ldots, y_n$ represent the coordinate functions of a coordinate system ϕ about a point y. Then

$$\partial/\partial x_j = \Sigma_{i=1}^n \partial(y_i)/\partial x_j \partial/\partial y_i = \Sigma_{i=1}^n \frac{\partial y_i}{\partial x_j}\frac{\partial}{\partial y_i}.$$

Definition 3.5.5 Let M be an n-dimensional C^r-manifold with a set of local charts $\{(\psi_i, U_i)\}$, N be an m-dimensional C^r-manifold with a set of local charts $\{(\phi_i, V_i)\}$

and $\psi : M \to N, x \mapsto f(x)$ be a map for some C^k function $f : N \to \mathbf{R}$. Then ψ is said to be of C^k-class ($k \leq r$), denoted by $\psi \in \mathcal{C}^k(M, N)$, if

$$f \circ \psi \in C^k(M, \mathbf{R}),\ \forall\, f \in C^k(N, \mathbf{R}).$$

Remark 3.5.6 Every vector v at a point in Euclidean space defines a map mapping each smooth function into a real number such as the directional derivative along v. Moreover, v is determined by the values of v on smooth functions. This leads to the concept of tangent vectors on manifolds.

Definition 3.5.7 Given an n-dimensional smooth manifold M with an atlas $\{U_i, \psi_i)\}$, a **tangent vector** v at a point $x \in U_i$ is a map

$$v : (U_i,\ x,\ \mathbf{R}) \to \mathbf{R},\ f \mapsto \Sigma_{i=1}^n a_i \partial/\partial x_i (f \circ \psi_i^{-1})|_{\psi(x)}.$$

Proposition 3.5.8 *Let M be a smooth manifold and v be a tangent vector at a point $x \in M$. If $T_x(M)$ denotes the set of all tangent vectors at $x \in M$, then it forms a vector space over* $\mathbf{R}$ *under the compositions defined by*

(i) $(v + w)(f) = v(f) + w(f),\ \forall\, v, w \in T_x(M)$;
(ii) $(\lambda v)(f) = \lambda(v(f)),\ \forall\, v \in TM_x,\ \forall\, \lambda \in \mathbf{R}$.

Proof Let

$$v = \Sigma_{i=1}^n a_i(\partial/\partial x_i) \text{ and } w = \Sigma_{i=1}^n b_i(\partial/\partial x_i)$$

for some n-tuples $(a_1, a_2, \ldots, a_n) \in \mathbf{R}^n$ and $(b_1, b_2, \ldots, b_n) \in \mathbf{R}^n$. This implies that

$$v + w = \Sigma_{i=1}^n (a_i + b_i)(\partial/\partial x_i) \text{ and } \lambda v = \Sigma_{i=1}^n \lambda a_i(\partial/\partial x_i).$$

Then the map

$$\psi : \mathbf{R}^n \to T_x(M),\ (a_1, a_2, \ldots, a_n) \mapsto \Sigma_{i=1}^n a_i(\partial/\partial x_i)$$

is an isomorphism of vector spaces and hence $T_x(M)$ is of dimension n with $\{\partial/\partial x_i : i = 1, 2, \ldots, n\}$ as a basis. ❑

Definition 3.5.9 Given an n-dimensional smooth manifold M, the vector space $T_x(M)$ is called the **tangent space** of M at x. In other words, the vector space $T_x(M)$ over $\mathbf{R}$ defined in Proposition 3.5.8 is called the **tangent space** to the manifold M at the point $x \in M$.

Definition 3.5.10 Given a smooth manifold M and a smooth curve $\alpha : \mathbf{R} \to M$ such that $\alpha(0) = x_0 \in M$, let U be an open nbd of x_0 in M. The directional derivative of a smooth function $f : U \to \mathbf{R}$ along the curve α at the point x_0 is defined by

$$D_\alpha(f) = \frac{d}{dt} f(\alpha(t))|_{t=0}.$$

The operator D_α is said to be the tangent vector to α at the point $x_0 \in M$. It is considered $D_\alpha = D_\beta$ if they have the same value at $x_0 \in M$ for all such functions f. The set of all tangent vectors to a smooth manifold at a point $x_0 \in M$ forms a vector space which is called the **tangent space** at x_0 to M, denoted by $T_{x_0}M$.

Definition 3.5.11 Let M be a smooth manifold and f be real-valued smooth function on M at a point $x_0 \in M$. The equivalence class of f under the equivalence relation $f \sim g$ iff $f(x) = g(x)$, $\forall\, x \in U \subset M$ for some nbd U of x_0 is called a **germ** of f.

Proposition 3.5.12 *Let M be a smooth manifold and f, g be real-valued smooth functions on M at a point $x \in M$. Using the notation D in place of D_α in Definition 3.5.10, the tangent vectors have the following properties:*

(i) $D(f+g) = D(f) + D(g)$, $\forall\, f, g \in T_xM$;
(ii) $D(rf) = rD(f)$, $\forall\, r \in \mathbf{R}. \forall\, f \in T_xM$;
(iii) $D(fg) = f(x)D(g) + D(f)g(x)$, $\forall\, f, g \in T_xM$.

Proof It follows from construction of $D = D_\alpha$ formulated in Definition 3.5.10. ❑

Definition 3.5.13 The map $D \in C^\infty(M, x, \mathbf{R})$ having the properties of Proposition 3.5.12 is called a derivation. The set of all tangent vectors at a point $x \in M$ forms an algebra, called **algebra of germs** of real-valued smooth functions at x on M.

Example 3.5.14 The tangent space $T_x(\mathbf{R}^n)$ at a point $x \in \mathbf{R}^n$ is isomorphic to the vector space $\mathbf{R}^n$ itself. Because the map

$$\psi : \mathbf{R}^n \to T_x(\mathbf{R}^n), (a_1, a_2, \ldots, a_n) \mapsto \Sigma_{i=1}^n a_i \partial/\partial y_i$$

is a vector space isomorphism.

Interpretation of Definition 3.5.13. Let M be a smooth manifold and f be real-valued smooth function on M at a point $x_0 \in M$ having $x_1, x_2, \ldots, x_n$ local coordinates at x_0. If

$$\alpha(t) = (\alpha_1(t), \alpha_2(t), \ldots, \alpha_n(t) : x_i(t) = x_i(\alpha(t)), i = 1, 2, \ldots, n,$$

then

$$D_\alpha(f) = \frac{d}{dt} f(\alpha(t))|_{t=0} = \frac{d}{dt}\Sigma_{i=1}^n f(\alpha_1(t), \alpha_2(t), \ldots, \alpha_n(t))|_{t=0} = \Sigma_{i=1}^n \frac{\partial f}{\partial x_i}\frac{d\alpha_i}{dt}|_{t=0}.$$

This asserts that

$$D_\alpha = \Sigma_{i=1}^n r_i \frac{\partial}{\partial x_i}|_{x_0} : r_i = \frac{d\alpha_i}{dt}|_{t=0},\ i = 1, 2, \ldots, n.$$

Hence, it follows that

$$\frac{\partial}{\partial x_i}|_{x_0} = D_{\beta_i} : \beta_i(t) = (0, 0, \dots, 0, t, \dots, 0) \text{ (t } \textit{stands at the ith place}).$$

This implies that the vector space $T_{x_0}(M)$ of tangent vectors of M at x_0 has $\{\frac{\partial}{\partial x_i} : i = 1, 2, \dots, n\}$ as a basis. Hence, it follows immediately that

$$(rV + sW)(f) = r(V(f)) + s(W(f), \forall V, W \in T_{x_0}(M), \ \forall s.t \in \mathbf{R}.$$

Example 3.5.15 Consider the manifold $\mathbf{R}^n$. Take a fixed point $b = (b_1, b_2, \dots, b_n) \in \mathbf{R}^n$ and a vector $v = (v_1, v_2, \dots, v_n) \in \mathbf{R}^n$. Consider the smooth curve

$$\alpha_v : \mathbf{R} \to \mathbf{R}^n, \ t \mapsto b + tv = (b_1 + tv_1, b_2 + tv_2, \dots, b_n + tv_n).$$

Let U be an open nbd of b and $f : U \to \mathbf{R}$ be a smooth function. Then D_{α_v} is the tangent vector and with respect to the basis $\{\frac{\partial}{\partial x_i}\}$, the coordinates of D_{α_v} are given by

$$D_{\alpha_v}(f) = \frac{d}{dt} f(\alpha(t))|_{t=0} = \Sigma_{i=1}^n f(\alpha_1(t), \alpha_2(t), \dots, \alpha_n(t))|_{t=0} = \Sigma_{i=1}^n \frac{\partial f}{\partial x_i}\frac{d\alpha_i}{dt}|_{t=0}.$$

This asserts that with respect to the basis $\{\frac{\partial}{\partial x_i}\}$,

$$D_{\alpha_v} = \Sigma_{i=1}^n r_i \frac{\partial}{\partial x_i} : r_i = \frac{d\alpha_i}{dt}|_{t=0} = v_i : i = 1, 2, \dots, n,$$

which implies

$$D_{\alpha_v} = \Sigma_{i=1}^n v_i \frac{\partial}{\partial x_i} =< v_1, v_2, \dots, v_n >= v.$$

This defines an isomorphism ψ of vector spaces

$$\psi : \mathbf{R}^n \to T_b(\mathbf{R}^n),$$

and hence, the dimension of $T_b(\mathbf{R}^n)$ is n.

Example 3.5.16 The vector space $T_b(\mathbf{R}^n)$ is called the **tangent space** to $\mathbf{R}^n$ at the point b and every element in $T_b(\mathbf{R}^n)$ is called a **tangent vector** to $\mathbf{R}^n$ at the point b.

Theorem 3.5.17 *Let M and N be two smooth manifolds. Suppose that M is connected and $\phi : M \to N$ is smooth. If $D\phi \equiv 0$, then ϕ is a constant map in the sense that $\phi(x) = n_0$ for some $n_0 \in N$, $\forall x \in M$.*

Proof For $n_0 \in \phi(M)$, the set $M_1 = \phi^{-1}(n_0)$ is closed in M. We claim M_1 is also open in M. Let $x_0 \in M_1$. To show it, we find an open set U in M with the following property:

$$x_0 \in U \subset M_1.$$

If V is a coordinate nbd of n_0 with coordinate functions $(y_1, y_2, \ldots, y_n)$, choose U to be a connected coordinate nbd of the point x_0 with the following property:

$$x_0 \in U \subset \phi^{-1}(V).$$

If $(x_1, x_2, \ldots, x_n)$ is the coordinate functions in U, then for every $x \in U$, the matrix corresponding to $D\phi(x)$ with respect to the basis $\{\frac{\partial}{\partial x_i} : i = 1, 2, \ldots, n\}$ for the tangent vector space $T_x(M)$ and with respect to the basis $\{\frac{\partial}{\partial y_i} : i = 1, 2, \ldots, n\}$ for the tangent vector space $T_{\phi(x)}(N)$ is given by $(\frac{\partial}{\partial x_j}(y_i \circ \phi))$. By hypothesis, $D\phi \equiv 0$. Hence, it follows that

$$\left(\frac{\partial}{\partial x_j}(y_i \circ \phi)\right) \equiv 0 \ on \ U, \forall i, j = 1, 2, \ldots, n.$$

It shows that the function $y_i \circ \phi$ is constant on U for all $i = 1, 2, \ldots, n$. Consequently,

$$\phi(x) = \phi(x_0) = n_0, \ \forall x \in U \ and \ U \subset \phi^{-1}(y_0).$$

It asserts that M_1 is an open set in M. Since by hypothesis, M is connected, it follows that $M = M_1 = \phi^{-1}(n_0)$. ❑

Definition 3.5.18 Given two smooth manifolds M and N and a smooth curve $\alpha : \mathbf{R} \to M$, the differential $d\psi$ of ψ of a smooth map $\psi : M \to N$ at a point $x \in M$ is the function defined by

$$d\psi = \psi_* : T_x(M) \to T_{\psi(x)}(N), D_\alpha \mapsto D_{\psi \circ \alpha}.$$

ψ_* is well-defined. Because, for any smooth function $k : V \to \mathbf{R}$ on a nbd V of $\psi(x)$,

$$\psi_*(D_\alpha)(k) = D_{\psi \circ \alpha}(k) = \frac{d}{dt}k(\psi(\alpha(t)))|_{t=0} = D_\alpha(k \circ \psi) \implies (\psi_*(D)(k)) = D(k \circ \psi).$$

$d\psi = \psi_*$ is called the **differential** $d\psi$ of ψ of a smooth map $\psi : M \to N$.

Proposition 3.5.19 *Given two smooth manifolds M and N and a smooth curve $\alpha : \mathbf{R} \to M$, the differential $d\psi$ of ψ of a smooth map $\psi : M \to N$ at a point $x \in M$. Then under the notations of Definition 3.5.18,*

(i) *ψ_* is a linear map;*
(ii) *for any smooth map $\phi : N \to P$,*

$$(\phi \circ \psi)_* = \phi_* \circ \psi_* : T_x(M) \to T_{(\phi \circ \psi)(x)}(P).$$

Proof **(i)** Let $X, Y \in T_x(M)$ be two tangent vectors at the point $x \in M$ and $f : U \to \mathbf{R}$ be a smooth function, where U is a nbd of x in M. Then for all $r, s \in \mathbf{R}$

$$\psi_*(rX + sY)(f) = (rX + sY)(f \circ \psi) = rX(f \circ \psi) + sY(f \circ \psi) = (r\psi_*(X) + s\psi_*(Y))(f)$$

asserts that ψ_* is a linear map.

(ii) Let $f : U \to \mathbf{R}$ be a smooth function and U be a nbd of $(\psi)(x)$ in N and $\phi : N \to P$ be an arbitrary smooth map. Since $((\phi \circ \psi)_*(D))(f) = D(f \circ \psi \circ \phi))$, it follows that

$$(\phi \circ \psi)_*(D)(f) = D(f \circ \psi \circ \phi)) = (\phi_* D)(f \circ \phi) = ((\psi_* \circ \phi_*)(D)(f), \ \forall \, D \in T_x(M).$$

This proves that

$$(\phi \circ \psi)_* = \phi_* \circ \psi_*$$

for any smooth map $\phi : N \to P$. ❑

Corollary 3.5.20 *Let M be a smooth manifold and $\alpha : \mathbf{R} \to M$ be a smooth curve. Then its differential α_* satisfies the relation*

$$\alpha_* \left(\frac{d}{dt} \right) = D_{\alpha_*}.$$

Proof Let $\alpha : \mathbf{R} \to M$ with $\alpha(0) = x_0 \in M$ be a smooth curve. Then it follows that for a smooth map $g : U \to \mathbf{R}$, where U is an open nbd of x_0 in M,

$$\alpha_* \left(\frac{d}{dt}|_{t=0} \right)(g) = \frac{d}{dt}(g \circ \alpha) = D_\alpha(g).$$

This asserts that

$$\alpha_* \left(\frac{d}{dt} \right) = D_{\alpha_*}.$$

❑

Interpretation of the differential in terms of local coordinates Let M, N be two smooth manifolds, $x_0 \in M$ be a point and $\psi : M \to N$ be a smooth map. If $x_1, x_2, \ldots, x_n$ is a local coordinate near the x_0 and $y_1, y_2, \ldots, y_n$ is a local coordinate near the $f(x_0) \in N$, then ψ can be represented as

$$\psi(x_1, x_2, \ldots, x_n) = (\psi_1(x_1, x_2, \ldots, x_n), \ldots, \psi_n(x_1, x_2, \ldots, x_n)).$$

The linear map

$$\alpha_* : T_{x_0}(M) \to T_{\psi(x_0)}(N)$$

can be expressed with respect to the bases $\{\frac{\partial}{\partial x_i}\}$ and $\{\frac{\partial}{\partial y_j}\}$ as a matrix $\{m_{i,j}\}$ such that

$$\psi_* \left(\frac{\partial}{\partial x_j}\right) = \Sigma_{i=1}^n m_{i,j} \frac{\partial}{\partial y_i}.$$

This implies that

$$\psi_* \left(\frac{\partial}{\partial x_j}\right)(y_l) = \Sigma_{i=1}^n e_{i,j} \frac{\partial}{\partial y_i}(y_l) = m_{l,j}.$$

This asserts that

$$\frac{\partial}{\partial x_j}(y_l \circ \psi) = \frac{\partial}{\partial x_j}\psi_l.$$

Consequently, the Jacobian matrix $J(\psi)$ with $(\frac{\partial \psi_l}{\partial x_j})$ at its i, jth entry represents the linear transformation ψ_*.

3.6 The Topology on a Smooth Manifold Induced by Its C^∞-Structure

This section proves that every C^∞-structure on a differentiable manifold M induces a topology τ_M on M, which is generated by its coordinate domains as basis. This topology τ_M is called an induced topology on the smooth manifold of M. If τ is own topology on M formulated in Remark 3.4.9, then these two topologies may not be the same. Example 3.6.15 shows that τ_M is not Hausdorff but τ is Hausdorff by its defining condition and hence $\tau \neq \tau_M$. Theorem 3.6.9 provides a necessary and sufficient condition under which a given topology τ on M coincides with the τ_M on M. The topology τ satisfies the first axiom of countability by Proposition 3.6.16.

3.6.1 The Topology Induced by C^∞-Structure on a Manifold

This subsection proves that every C^∞ structure on a set M induces a topology on M.

Recall that an injection $\psi : M \to \mathbf{R}^n$, whose image is an open set in $\mathbf{R}^n$, is an n-**dimensional chart**. Using the projection map $p_i : \mathbf{R}^n \to \mathbf{R}$, every such chart defines on the domain U of ψ a set of coordinate functions

$$\psi_i = p_i \circ \psi : i = 1, 2, \ldots, n$$

so that

$$\psi : U \to \mathbf{R}^n, x \mapsto (\psi_1(x), \psi_2(x), \ldots, \psi_n(x))$$

is the set of coordinates of a point $x \in U$ in the chart (U, ψ).

Definition 3.6.1 A set M endowed with a C^∞-structure of dimension n is called a **differentiable manifold of dimension** n.

Definition 3.6.2 A C^∞-atlas of a set M is called **complete** if it is not contained in any other C^∞-atlas of M. The domain of a given chart of the complete atlas of M that determines a given C^∞-structure of dimension n on the set M is called a **coordinate domain** of the manifold M.

Proposition 3.6.3 *Let M be a differentiable manifold and (ψ, U) be an n-dimensional chart. If V is a subset of U such that $\psi(V)$ is open in $\mathbf{R}^n$, then the restriction $\psi|V$ is also a chart of M.*

Proof By hypothesis, $\psi : U \to \mathbf{R}^n$ is an injection whose range is an open set in $\mathbf{R}^n$ and hence $\psi|_V$ is a chart for the set M. If (ϕ, W) is an arbitrary chart of the manifold M such that $W \cap V \neq \emptyset$, then $(\psi|_V) \circ \phi^{-1}$ is a restriction to an open set of the diffeomorphism $\psi \circ \phi^{-1}$. Hence, it follows that $\psi|_V$ is a chart of the manifold M. This defines the C^∞ structure on the set M. ❑

Proposition 3.6.4 *Let M be a manifold of dimension n. Then the family of coordinate domains of M forms a basis for a topology on the set M.*

Proof Let (ψ, U) and (ϕ, V) be two coordinate domains such that $U \cap V \neq \emptyset$. Then $\phi \circ \psi^{-1}$ is a diffeomorphism such that $\psi(U \cap V)$ is an open set in $\mathbf{R}^n$. Then its restriction $\psi|_{(U \cap V)}$ is a chart of M with domain $U \cap V$ by Proposition 3.6.3. This proves that a nonempty intersection of any two coordinate domains of the manifold M is also a coordinate domain of M. Hence, it follows that the family of coordinate domains of M forms a basis for a topology on the set M. ❑

Definition 3.6.5 (*Topology induced by C^∞ structure*) Let M be a set. The family of coordinate domains of M forms a basis for a topology τ_M on the set M. This topology τ_M is called the **topology of the manifold** M induced by the C^∞ structure of M. A subset U of M is open in this topology τ_M iff every point of U admits a coordinate nbd lying in U. In particular, this induced topology on the differentiable manifold $\mathbf{R}^n$ coincides with the standard metric topology on $\mathbf{R}^n$.

Proposition 3.6.6 *Let M be a differentiable manifold. Then using the topology τ_M, every chart of M is a homeomorphism.*

Proof Let ψ be any chart of M with domain U and range $V \subset \mathbf{R}^n$. We claim that ψ is continuous and open with respect to the topology τ_M. Take any open subset V' of $\mathbf{R}^n$. Consider its inverse image $U' = \psi^{-1}(V')$. Since $\psi(U') = V' \cap V$, it follows from Proposition 3.6.3 that $\psi|_{U'}$ is also a chart of M. This implies that U' is an open set in τ_M and hence ψ is continuous. To prove that ψ is open, take any basic open set $U' \subset U$. Consider a chart ϕ of M with domain U'. Then the change of coordinates $\phi \circ \psi^{-1}$ is a diffeomorphism in $\mathbf{R}^n$, and its domain $\psi(U \cap U') = \psi(U')$ is open in $\mathbf{R}^n$. Since any open subset of M that lies in U is a union of basic open sets, its image under ψ is open in $\mathbf{R}^n$. This asserts that ψ is open. ❑

Proposition 3.6.7 establishes a relation between a C^∞ structure on a set M and its induced topology τ_M on M.

Proposition 3.6.7 *Let $f : M \to N$ be a differentiable map at a point $x \in M$. Then under the induced topologies τ_M on M and τ_N on N, the map $f : (M, \tau_M) \to (N, \tau_N)$ is continuous at the point x.*

Proof Let ψ be chart at $x \in M$ and ϕ be chart at $f(x) \in N$. Then the representative function $g = \phi \circ \psi^{-1}$ is C^∞ and hence g is continuous at the point $\psi(x)$. On the other hand, ψ and ϕ are homeomorphisms under the induced topologies τ_M on M and τ_N on N. This proves that f is continuous at x. ❑

Remark 3.6.8 Every differentiable manifold M induces a topology τ_M on M. If τ is the given topology on M, Example 3.6.15 says that $\tau \neq \tau_M$. There is a natural problem: does there exist a situation when τ_M coincides with τ? Theorem 3.6.9 gives its positive answer. Moreover, Proposition 3.6.7 implies that for every differentiable function $f : M \to N$, the function $f : (M, \tau_M) \to (N, \tau_N)$ is continuous and hence its domains must be open subset of the manifold M.

3.6.2 A Necessary and Sufficient Condition for $\tau_M = \tau$

This subsection studies the problem raised in Remark 3.6.8: under what situation (if any) a set M endowed with some topological structure τ and the topological structure τ_M on it induced by a C^∞ structure coincide? If there exists a situation under which $\tau = \tau_M$, then Proposition 3.6.6 asserts that all the charts of the manifold M are homomorphisms under the given topology τ. This subsection solves this problem in Theorem 3.6.9 asserting that $\tau = \tau_M$ if it is true for the charts of any one atlas of M.

Theorem 3.6.9 (Manifold Structure on a Topological Space) *Let M be a set endowed with both topological structure τ and a C^∞-structure. Then the topology τ_M induced by the given C^∞-structure on M coincides with the given topology τ iff the charts of one atlas of M are homomorphisms under the given topology τ.*

Proof The necessity of the condition follows from Proposition 3.6.6. To prove the sufficiency, suppose that the charts of one atlas of M are homomorphisms under the topology τ. Consider the family $\mathcal{U}$ of open sets in the given topology τ and the family $\mathcal{V}$ of open sets in the induced topology τ_M on M. Let $\mathcal{A}$ be a given atlas of M. Take any open set $V \in \mathcal{V}$. Let ψ be a chart of the atlas $\mathcal{A}$ whose domain W is such that $W \cap V \neq \emptyset$. Then under the induced topology τ_M, the set $W \cap V$ is open, ψ is a homeomorphism and hence $\psi(W \cap V)$ is an open subset in $\mathbf{R}^n$. Since ψ is a homeomorphism under τ, the set $W \cap V \in \mathcal{U}$. Again, the set V being the union of such sets, $V \in \mathcal{U}$. This shows that every open set of $\mathcal{V}$ is also an open set of $\mathcal{U}$ and hence $\mathcal{V} \subset \mathcal{U}$. To show that $\mathcal{U} \subset \mathcal{V}$, take any open set $U \in \mathcal{U}$. Proceeding as above, it follows that $U \in \mathcal{V}$ and hence $\mathcal{U} \subset \mathcal{V}$. It proves that $\mathcal{U} = \mathcal{V}$. ❑

Corollary 3.6.10 *Let M be a compact topological space. It does not admit a C^∞-structure defined by a single chart.*

Proof Suppose M is a compact topological space such that M admits a C^∞-structure defined by a single chart.Then M is homeomorphic to an open subset V of $\mathbf{R}^n$. Since V is not compact, our supposition is not tenable. This contradiction proves the corollary. ❑

Remark 3.6.11 Let M be a set endowed with both a topological structure τ and a C^∞-structure. Does there exist any relation between the topology τ and the topology τ_M induced by the given C^∞-structure? A positive answer is available in Theorem 3.6.9 providing a necessary and sufficient condition under which the two topologies τ and τ_M on M coincide. If $\tau = \tau_M$, then Proposition 3.6.6 asserts that all the charts of the manifold M are homeomorphic in the topology τ. Theorem 3.6.9 formulates this result.

Example 3.6.12 S^n cannot admit a C^∞-structure defined by a single chart. Since S^n is a compact topological space, it follows from Corollary 3.6.10 that S^n cannot admit a C^∞-structure defined by a single chart.

3.6.3 Properties of the Induced Topology on a Differentiable Manifold

This subsection proves some special properties of the topology τ_M induced by the given C^∞-structure of the manifold M such as its T_1 property in Proposition 3.6.13, its first countability property in Proposition 3.6.16 and its locally connectedness property in Proposition 3.6.19 and showed that this topology is not necessarily Hausdorff (see Example 3.6.15). This example implies that the natural topology τ and τ_M on M are different but Theorem 3.6.9 provides a sufficient condition for coincidence: $\tau_M = \tau$.

Proposition 3.6.13 *Let M be a differentiable manifold with induced topology τ_M. Then the topology τ_M is T_1.*

Proof Let x_1 and x_2 be two distinct points of M. Consider two possibilities:

Case I: Suppose that x_1 and x_2 both lie in the domain of some chart ψ of M with range V. Then there exist two disjoint open sets V_1 and V_2 in $\mathbf{R}^n$ such that

$$\psi(x_1) \in V_1 \subset V \text{ and } \psi(x_2) \in V_2 \subset V.$$

Since ψ is continuous under the topology τ_M, it follows that $\psi^{-1}(V_1)$ and $\psi^{-1}(V_2)$ are disjoint open sets in M such that

$$x_1 \in \psi^{-1}(V_1) \in \tau_M \text{ and } x_2 \in \psi^{-1}(V_2) \in \tau_M.$$

Case II: Suppose that x_1 and x_2 do not lie in the domain of any chart ψ of M. Then there exists a chart with domain U_1 such that $x_1 \in U_1$ but $x_2 \notin U_1$ and another chart with domain U_2 such that $x_2 \in U_2$ but $x_1 \notin U_2$.

Hence, it is proved that in either case, the topology τ_M is T_1. Moreover, Case I asserts that any two points lying in the domain of one chart also satisfy the T_2 axiom. ❑

Remark 3.6.14 Proposition 3.6.13 raises the problem: is the induced topology τ_M on every differentiable manifold M Hausdorff? The answer is negative. In support, consider Example 3.6.15.

Example 3.6.15 The induced topology τ_M may not be Hausdorff. For example, consider the subset M of $\mathbf{R}^2$, which is the union of $U = \{(x, 0) \in \mathbf{R}^2\}$ (x-axis) and the one-pointic set $\{(1, 0)\}$, i.e., $M = U \cup \{(1, 0)\}$. Suppose

$$U_1 = (U - \{(0, 0)\}) \cup \{(1, 0)\}.$$

Consider two charts ψ and ψ_1 of M with domains U and U_1 into $\mathbf{R}$:

$$\psi : U \to \mathbf{R}, (x, 0) \mapsto x$$

and

$$\psi_1 : U_1 \to \mathbf{R}, (x, 0) \mapsto x \ \text{ if } x \neq 0, \ \textit{and} \ (0, 1) \mapsto 0.$$

Then the change of coordinates $\psi_1 \circ \psi^{-1}$ is the identity map on the open set $\mathbf{R} - \{0\}$. Hence, the charts ψ and ψ_1 form a C^∞-atlas and it endows the set M a differentiable manifold structure. Let τ_M be the induced topology on X. Let V be any nbd of the point $(0, 0)$ and V_1 be any nbd of the point $(0, 1)$. Then $V \cap V_1 \neq \emptyset$. Because $\psi(U \cap V)$ and $\psi_1(U_1 \cap V_1)$ are both open subsets of $\mathbf{R}$ that contain the origin 0. Hence, it follows that both of them also contain some nonzero point $p \in \mathbf{R}$. Then the point $(p, 0) \in V \cap V_1$ asserts that $V \cap V_1 \neq \emptyset$. This concludes that this induced topology τ_M is not Hausdorff.

Proposition 3.6.16 *Let M be a differentiable manifold with induced topology τ_M. Then the topology τ_M satisfies the first axiom of countability.*

Proof Let M be a differentiable manifold of dimension n and $x \in M$ be any point and ψ be a chart with domain U that contains the point x. Since the Euclidean topology of $\mathbf{R}^n$ satisfies the first axiom of countability, there exists a countable open base $\{V_i\}$ at the point $\psi(x) \in \mathbf{R}^n$. Claim that the family $\{\psi^{-1}(V_i)\}$ forms a countable open base at x. Take an arbitrary nbd W of x. Then the set $U \cap W$ open in the topology τ_M asserts the set $\psi(U \cap W)$ is a nbd of $\psi(x)$ which contains some nbd $V_k \in \{V_i\}$. This implies that $\psi^{-1}(V_i) \subset W$. ❑

Remark 3.6.17 Proposition 3.6.16 says that the topology τ_M induced by a differentiable manifold M satisfies the first axiom of countability. It is natural question: does this topology τ_M also satisfy the second axiom of countability? Its negative answer is available in Sect. 3.6.4.

Definition 3.6.18 A topological space space X is said to be **locally connected** if every point of X has a nbd base consisting of connected sets. It is said to be **path connected** if any two points $a, b \in X$ can be joined by a path in X in the sense that there exists a path $\alpha : \mathbf{I} \to X$ such that $\alpha(0) = a$ and $\alpha(1) = b$.

Proposition 3.6.19 *Let M be a differentiable manifold with induced topology τ_M. Under this topology, the space M is locally connected.*

Proof Let $x \in M$ be an arbitrary point and ψ be a chart with domain U that contains the point x. Then $\psi(U)$ is an open set in $\mathbf{R}^n$ that contains the point $\psi(x)$. Hence, there exists an open ball $B_{\psi(x)}(\epsilon)$ with center $\psi(x)$ and radius $\epsilon > 0$ such that $B_{\psi(x)}(\epsilon) \subset \psi(U)$. Since this open ball is connected, it follows that $\psi^{-1}(B_{\psi(x)})$ is a connected nbd of x. This concludes that the space (M, τ_M) is locally connected. ❑

Corollary 3.6.20 *Every connected component of a differentiable manifold M is an open set in the topology τ_M.*

Proof It follows from Proposition 3.6.19. ❑

3.6.4 Topological Restriction on a Differentiable Manifold

This subsection imposes certain topological restrictions on a manifold to solve some problems. For example, a differentiable manifold satisfying the T_2-axiom is said to be a Hausdorff manifold. This restriction is imposed, because most of the manifolds that we study are Hausdorff manifolds. So, this restriction is incorporated in defining a differentiable manifold.

Proposition 3.6.21 *Let M be an n-dimensional differentiable manifold with induced topology τ_M. Then M has a countable open base for the topology τ_M if M admits a countable atlas.*

Proof Let M admit an atlas of charts ψ_k with domains U_k for $k = 1, 2, \ldots$. Then each $\psi(U_k) = V_k$ is an open set in $\mathbf{R}^n$. Since $\mathbf{R}^n$ has the Euclidean topology with a countable basis $\mathcal{B}$, there exists a countable family of open sets $\{V_{k_j} : j = 1, 2, \ldots\}$ such that every open set in V_k is a union of open sets V_{k_j}. This implies that every open set of M lying in the open set U_k is a union of sets $\psi^{-1}(V_{k_j}) = U_{k_j}$. Hence, it follows that the countable family of sets $\{U_{k_j} : j = 1, 2, \ldots\}$ forms an open base for the topology τ_M of M, because every open set U in M is a union of open sets $U_k \cap U$ and every one of them is a union of the sets U_{k_j}. ❑

Corollary 3.6.22 *Let M be a compact n-dimensional differentiable manifold with induced topology τ_M. Then M has a countable open base for the topology τ_M.*

Proof Since any atlas of a compact manifold has a finite subatlas, the corollary follows immediately from Proposition 3.6.21. ❑

3.7 The Rank of a Smooth Map and Constant Rank Theorem

This section defines the concept of the rank of differentiable map and proves the constant rank Theorem 3.7.5 with an aim to study the local structure of a differentiable map with the help of its rank. It is used in immersion problems (see Sect. 3.9).

Rank of a Differentiable Map

This subsection defines the rank of a differentiable manifold, which plays a key role in the manifold theory.

Definition 3.7.1 Let $f : M \to N$ be a differentiable map between differentiable manifolds M and N. Given a point $x \in M$, let $\psi : U \to V$ and $\phi : W \to T$ be local coordinate systems about x and $f(x)$, respectively. Then

$$\phi \circ f \circ \psi^{-1} : V \to T$$

is differentiable map from the Euclidean open set V into the Euclidean open set T. The rank of the Jacobian matrix $\phi \circ f \circ \psi^{-1}$ at the point $f(x)$ is defined to be the rank of f at the point x. If the rank of f is the same $\forall x \in M$, then this common value is said to be the **rank of** f.

Remark 3.7.2 Since Definition 3.7.1 is independent of the choice of the local coordinates of the points x and $f(x)$, the rank of f is well-defined.

Example 3.7.3 The map $f : \mathbf{R}^2 \to \mathbf{R}^2,\ (x, y) \mapsto (x, 0)$ has rank 1 everywhere.

Remark 3.7.4 Let M be a differentiable manifold of dimension m and N be a be a differentiable manifold of dimension n. Two cases for special interest arise for a differentiable map $f : N \to M$.

(i) Case I: f has a maximal rank at a point x_0.
(ii) Case II: f has a constant rank in a nbd of a point.

For the Case I, there are three possibilities (not exclusive) such as

$$n = m, n \geq m \quad \text{or} \quad n \leq m.$$

(i) For $n = m$, it follows that f is a local diffeomorphism at the point x_0 by the inverse function theorem.

(ii) For $n \geq m$, the maximal rank of f is m and it defines a submersion at the point x_0, formulated in Definition 3.9.3, since manifolds are locally Euclidean.

(iii) For $n \leq m$, the maximal rank of f is n and it defines an immersion at the point x_0 formulated in Definition 3.9.3.

The importance of Case II is reflected through the Constant Rank Theorem 3.7.5 for manifolds.

Theorem 3.7.5 (Constant Rank Theorem) *Let M and N be two differentiable manifolds of dimensions m and n, respectively. If a differentiable map $f : N \to M$ has a constant rank r in a neighborhood of a point $x_0 \in N$, then there exist charts (ψ, U) at x_0 and (ϕ, V) in a nbd of $f(x_0) \in M$ with the property: for $(x_1, x_2, \ldots, x_n) \in \psi(U)$,*

$$(\phi \circ f \circ \psi^{-1})\,(x_1, x_2, \ldots, x_n) = (x_1, x_2, \ldots, x_r, \ldots, 0, 0),\ \forall\,(x_1, x_2, \ldots, x_n) \in \psi(U).$$

Proof By hypothesis, M and N are two differentiable manifolds of dimensions m and n, respectively. Let $f : N \to M$ be a differentiable map having a constant rank r in a neighborhood of a point $x_0 \in N$, and $(\tilde{\psi},\ \tilde{U})$ be a chart about the point $x_0 \in N$ and $(\tilde{\phi},\ \tilde{V})$ about the point $f(x_0) \in M$. Then $(\tilde{\phi} \circ f \circ \tilde{\psi}^{-1})$ is a map between two open subsets of $\mathbf{R}^n$ and $\mathbf{R}^m$. Hence, it follows by hypothesis that $(\tilde{\phi} \circ f \circ \tilde{\psi}^{-1})$ has the same constant rank r as f has in a nbd of $f(x_0)$ in $\mathbf{R}^m$. Then there is a diffeomorphism θ of a nbd of $\tilde{\psi}(x_0)$ and another diffeomorphism η of a nbd of $(\tilde{\phi} \circ f)(x_0)$ in $\mathbf{R}^m$ such that

$$(\eta \circ \tilde{\phi} \circ f \circ \tilde{\psi}^{-1} \circ \theta^{-1})((x_1, x_2, \ldots, x_n)) = (x_1, x_2, \ldots, x_r, \ldots, 0, 0).$$

Take $\psi = \theta \circ \tilde{\psi}$ and $\phi = \eta \circ \tilde{\phi}$. Then the normal form of f has no zeros if $r = m$. Hence, it follows that

$$(\phi \circ f \circ \psi^{-1})\,(x_1, x_2, \ldots, x_n) = (x_1, x_2, \ldots, x_r) \equiv (x_1, x_2, \ldots, x_r, \ldots, 0, 0).$$

❑

3.8 Manifolds With and Without Boundary

This section conveys the concept of a topological manifold with a boundary and a differentiable manifold without a boundary. The motivation of this concept comes from the observation that there are a number of geometric objects such as the n-dimensional closed disk $\mathbf{D}^n$ on which it is not possible to find a nbd of any point on its boundary $\partial\mathbf{D^n}$, which is homeomorphic to the space $\mathbf{R}^n$ or its open set. Such problems are solved by introducing the concept of manifold with boundary.

3.8.1 Topological Manifold with Boundary

This subsection introduces the concept of a topological manifold with boundary and illustrates this concept with various examples. On the other hand, Sect. 3.8.2 communicates the concept of topological manifolds without boundary.

Definition 3.8.1 An **n-dimensional topological manifold with boundary** is a second countable Hausdorff space M such that every point $x \in M$ has an open nbd homeomorphic to either $\mathbf{R}^n$ or to $\mathbf{R}^n_+ = \{x = (x_1, x_2, \ldots, x_n) : x_n \geq 0\}$.

Remark 3.8.2 Definition 3.8.1 asserts that an n-dimensional topological manifold with boundary for $n \geq 1$ is a Hausdorff space M if each point of M has a nbd homeomorphic to the open set in the subspace $x_n \geq 0$ of $\mathbf{R}^n$. The boundary of M is denoted by ∂M.

Example 3.8.3 Let $\mathbf{R}^n$ be the Euclidean n-space and $x = (x_1, x_2, \ldots, x_n) \in \mathbf{R}^n$ be an arbitrary point.

(i) The subspace $\mathbf{R}^n_+ = \{x \in \mathbf{R}^n : x_n \geq 0\} \subset \mathbf{R}^n$ is the upper-half of $\mathbf{R}^n$ and it has the boundary identified with the subspace

$$\mathbf{R}^{n-1} = \{x \in \mathbf{R}^n : x_n = 0\}.$$

This asserts that $\mathbf{R}^n_+$ is an n-dimensional manifold with $\mathbf{R}^{n-1}$ as its boundary, where the boundary $\partial\, \mathbf{R}^n_+$ is identified with hyperplane $\mathbf{R}^{n-1} \times \{\mathbf{0}\}$.
$\mathbf{R}^{n-1}$.

(ii) The subspace $\mathbf{R}^n_- = \{x \in \mathbf{R}^n : x_n \leq 0\} \subset \mathbf{R}^n$ is the lower-half of $\mathbf{R}^n$ and it has the boundary identified with the subspace

$$\mathbf{R}^{n-1} = \{x \in \mathbf{R}^n : x_n = 0\}.$$

This shows that $\mathbf{R}^n_-$ is an n-dimensional manifold with $\mathbf{R}^{n-1}$ as its boundary and hence $\mathbf{R}^{n-1}$ is the common boundary of the manifolds $\mathbf{R}^n_+$ and $\mathbf{R}^n_-$.

(iii) Let $M_1 = \{x \in \mathbf{R}^n : x_1^2 + x_2^2 + \cdots + x_n^2 \geq 1\}$ and $M_2 = \{x \in \mathbf{R}^n : x_1^2 + x_2^2 + \cdots + x_n^2 \leq 1\}$ be subspaces in $\mathbf{R}^n$. Then M_1 and M_2 are both n-dimensional manifolds with S^{n-1} as their common boundary.

3.8.2 *Topological Manifold Without Boundary*

This subsection introduces the concept of topological manifolds without boundary. On the other hand, Sect. 3.8.3 communicates the concept of differentiable manifolds with boundary.

Definition 3.8.4 Let M be an n-manifold. Then the set of points in M having a nbd homeomorphic to $\mathbf{R}^n$ is called the **interior of the manifold** M, denoted by interor M and its complement is called the boundary of the manifold M, denoted by ∂M. The manifold M is said to a **manifold without boundary** if $\partial M = \emptyset$. Moreover, if M is compact and connected, then M is said to be a closed manifold.

Example 3.8.5 **(i)** The unit disk $\mathbf{D}^{n+1}$ is an $(n+1)$-manifold with boundary unit sphere S^n.

(ii) $S^1 \times \mathbf{D}^2$ is a 3-manifold with boundary $S^1 \times S^1$.

(iii) If M is an m-manifold with boundary ∂M and N is an n-manifold with boundary ∂N, then $M \times N$ is an $(m+n)$-manifold with boundary $\partial(M \times N) = M \times \partial N \cup \partial X \times Y$.

(iv) The real projective space $\mathbf{R}P^n$ is an n-dimensional closed differentiable manifold. By Example 3.4.20, $\mathbf{R}^n$ is an n-dimensional differentiable manifold. It is a closed manifold, because it is compact and connected.

(v) The complex projective space $\mathbf{C}P^n$ is a $2n$-dimensional closed manifold. Consider the open sets

$$U_i = \{[z_1, z_2, \ldots, z_n] \in \mathbf{C}P^n : z_i \neq 0\}$$

and the homeomorphisms

$$\psi_i : U_i \to \mathbf{C}^n,\ [z_1, z_2, \ldots, z_n] \mapsto \left(\frac{z_1}{z_i}, \frac{z_2}{z_i}, \ldots, \frac{z_{i-1}}{z_i}, \frac{z_{i+1}}{z_i}, \ldots, \frac{z_n}{z_i}\right)$$

having inverses

$$\psi_i^{-1} : \mathbf{C}^n \to U_i,\ (z_1, z_2, \ldots, z_n) \mapsto [z_1, z_2, \ldots, z_{i-1}, 1, z_{i+1}, \ldots, z_n].$$

This asserts that the complex projective space $\mathbf{C}^n$ is a $2n$-dimensional manifold. It is closed, since it is compact and connected. It has no boundary.

3.8.3 *Differential Manifold with Boundary*

This subsection introduces the concept of differentiable manifolds with boundary. Recall that if $\mathcal{S} = \{(U_i, \psi_i)\}$ is a set of local charts of a differentiable manifold M, then $\mathcal{S}$ is said to be a differentiable structure on M. Every subset of $\mathcal{S}$ which satisfies the condition $M = \bigcup \{U_i\}$, and also the axioms **DM(1)** and **DM(2)**, is called a basis for the differential structure $\mathcal{S}$ of the manifold M.

Definition 3.8.6 An n-dimensional differentiable manifold with boundary is a topological space X with a subspace Y and a countable open covering $\mathcal{C} = \{U_1, U_2, \ldots\}$ together with homeomorphisms $\psi_1, \psi_2, \ldots$ such that

(i) each $U_n \in \mathcal{C}$ is either contained in $X - Y$ with the property that there is a homeomorphism $\psi_i : U_i \to V_i$, where V_i is an n-disk in $\mathbf{R}^n$, or there is a homeomorphism $\psi_i : U_i \to V_i$, where V_i is a hemisphere given by $\{x = (x_1, x_2, \ldots, x_n) \in \mathbf{R}^n : \Sigma_{i=1}^n x_i^2 < 1,\ x_i \geq 0\}$, with the property that $\psi_i(U_i \cap Y) = V_i$ for which $x_n = 0$;

(ii) given $U_i,\ U_j \in \mathcal{C}$ and homeomorphisms $\phi_i : U_i \to V_i, \phi_j : U_j \to V_j$ described in (i), if $U_i \cap U_j \neq \emptyset$, then the map

$$\psi_{ij} = \psi_i \circ \psi_j^{-1} : \psi_j(U_i \cap U_j) \to \psi_i(U_i \cap U_j)$$

is onto and differentiable.

Example 3.8.7 Let $\mathbf{R}_+^m$ be the upper closed half-space

$$\mathbf{R}_+^m = \{(x_1, x_2, \ldots, x_m) \in \mathbf{R}^m : x_m \geq 0\}.$$

Its boundary $\partial\, \mathbf{R}_+^m$ is defined to be the hyperplane $\mathbf{R}^{m-1} \times \{\mathbf{0}\} \subset \mathbf{R}^m$.

Remark 3.8.8 Definition 3.8.6 asserts that a subset $X \subset \mathbf{R}^n$ is a smooth m-dimensional manifold with boundary if every point $x \in X$ has a nbd $X \cap U$ diffeomorphic to an open subset $\mathbf{R}_+^n \cap V$ of $\mathbf{R}_+^n$. The boundary ∂X of X consists of precisely all points in X which correspond to points of $\mathbf{R}_+^n$ under such a diffeomorphism. The boundary ∂X is well-defined and it is a smooth manifold of dimension $m - 1$ with the interior $X - \partial X$, which is a smooth manifold of dimension m. The tangent space $T_x(X)$ is an m-dimensional vector space for all points $x \in X$ (even the point x stays on its boundary ∂X).

Example 3.8.9 Let $\mathbf{R}^n$ be the Euclidean n-space and $x = (x_1, x_2, \ldots, x_n) \in \mathbf{R}^n$ be an arbitrary point.

(i) The set

$$S_1 = \{x \in \mathbf{R}^n : \Sigma_{i=1}^n x_i^2 \leq 1\}$$

is an n-dimensional C^∞-manifold with boundary S^{n-1}.

(ii) The set

$$S_2 = \{x \in \mathbf{R}^n : \Sigma_{i=1}^n x_i^2 \geq 1\}$$

is also an n-dimensional C^∞-manifold with boundary S^{n-1}.

Proposition 3.8.10 *Let M be an m-manifold with boundary ∂M and N be an n-manifold with $m > n$. If $f : M \to N$ is a smooth map and $y \in N$ is a regular value of both f and $f|_{\partial M}$, then*

(i) *$f^{-1}(y) \subset M$ is a smooth manifold of dimension $m - n$ with boundary and*
(ii) *its boundary $\partial(f^{-1}(y))$ is same as $f^{-1}(y) \cap \partial M$.*

Proof **(i)** It is sufficient to consider the special case of a map $f : \mathbf{R}^m_+ \to \mathbf{R}^n$ with regular value $y \in \mathbf{R}^n$, since it is required to prove a local property. Suppose $x' \in f^{-1}(y)$. If x' is an interior point, then $f^{-1}(y)$ is a smooth manifold in the nbd of the point x'. If x' is a boundary point, consider a smooth map $g : U \to \mathbf{R}^n$ defined on a nbd of x' in $\mathbf{R}^m$ and coinciding with the restriction $f|_{U \cap \mathbf{R}^m_+}$. Without any loss of generality, we may assume that h has no critical points, because U may be replaced by a smaller nbd (if it is necessary). This asserts that $f^{-1}(y)$ is a smooth manifold of dimension $m - n$. This proves the first part.

(ii) To prove the second part, let

$$p : g^{-1}(y) \to \mathbf{R},\ (x_1, x_2, \ldots, x_m) \mapsto x_m$$

be the coordinate projection. Then 0 is a regular value of p. Because the tangent space of $g^{-1}(y)$ at a point $x \in p^{-1}(0)$ is the same as the null space of the linear map

$$dg_x = df_x : \mathbf{R}^m \to \mathbf{R}^n.$$

By hypothesis, the restriction of f over the boundary $\partial\mathbf{R}^m_+$ of $\mathbf{R}^m_+$ is regular at the point x. This ensures that the null space of dg_x cannot completely reside in $\mathbf{R}^{m-1} \times \{0\}$. Hence, the set

$$g^{-1}(y) \cap \mathbf{R}^m_+ = f^{-1}(y) \cap U = \{x \in g^{-1}(y) : p(x) \geq 0\}$$

is a smooth manifold by Exercise 63 of Sect. 3.27, with boundary same as $p^{-1}(0)$ by Exercise 63. This proves the second part. ❑

3.8.4 Surfaces with Boundaries

This subsection continues the study of Sect. 3.8 by discussing in particular, surfaces which are manifolds of dimension two. The spheres, ellipsoid, paraboloids, circular cylinders and the hyperboloid of one or two sheets are familiar examples of surfaces in $\mathbf{R}^3$.

Definition 3.8.11 **A surface** S is a Hausdorff space such that each point of S has a nbd homeomorphic to either $\mathbf{R}^2$ or the closed upper-half-space $\mathbf{R}^2_+$. The interior of S is precisely the set of points having a nbd homeomorphic to $\mathbf{R}^2$.

Definition 3.8.12 The boundary of a surface S consists of those points $x \in S$ for which there is a nbd V, and a homeomorphism $f : \mathbf{R}^2_+ \to V$ such that $f(0) = x$, where $\mathbf{R}^2_+$ is the closed upper-half-space of $\mathbf{R}^2$.

Remark 3.8.13 The intuitive concepts of interior and boundary of a surface agree with their formal concepts. A point cannot stay both in interior and on the boundary of a surface.

Example 3.8.14 (*The Möbius strip or band*) It is a surface with only one side and only one boundary. The Mö*bius* strip has the mathematical property of being nonorientable.

3.9 Immersion, Submersion and Transversality Theorem

This section generalizes the inverse function theorem related to Euclidean spaces to the corresponding theorem for smooth manifolds. It conveys the concept of immersion of a manifold with illustrative examples. The concepts of immersions and submanifolds are closely related which are used to obtain the solution of some problems.

Definition 3.9.1 Let M and N be two differentiable manifolds and $f : N \to M$ be a differentiable map.

(i) f is said to be an immersion if, at each point of its domain, its rank coincides with the dimension of the manifold N. If in particular, if the domain of f is the whole space of N, then f is said to be an **immersion** of N into M. In other words, if M is of dimension m and N is of dimension n, then a smooth map $f : N \to M$ is said to be an immersion at a point $p \in N$ if $m \leq n$ and rank $f = m$ at p.
(ii) f is said to be a **submersion** at p if $m \geq n$ and rank f = n at p.
(iii) The map f is said to be called an immersion, or a submersion, if it is so at every point $p \in N$.

Example 3.9.2 The natural inclusion map

$$i : S^2 \hookrightarrow \mathbf{R}^3$$

is an immersion, since at each point of S^2, the map i has rank 2 and the dimension of S^2.

Definition 3.9.1 is reformulated in Definition 3.9.3 in terms of derivative map df_x of $f : M \to N$.

Definition 3.9.3 Let M and N be smooth manifolds and $f : M \to N$ be a smooth map. If

$$df_x : T_x(M) \to T_{f(x)}(N)$$

is its derivative map at $x \in M$, then

(i) f is said to be an **immersion** at the point $x \in M$ if its derivative map df_x is a monomorphism of tangent spaces. If f is an immersion at every point, it is simply called an **immersion**.
(ii) f is said to be a **submersion** at the point $x \in M$ if its derivative map df_x is an epimorphism of tangent spaces. If f is a submersion at every point, it is simply called a **submersion**.
(iii) (M, f) is said to be a **submanifold** if f is an immersion and one-one.
(iv) f is said to be an **embedding** if (M, f) is a submanifold and $f : M \to f(M)$ is a homeomorphism under relative topology on $f(M)$. If

$$f : M \to N$$

is an embedding, then $f(M)$ is said to be an **embedded submanifold** of N.

In terms of dimensions of manifolds, the concepts of immersion and submersion given in Definition 3.9.3 are reformulated in Definition 3.9.4

Definition 3.9.4 Let M be a smooth manifold of dimension n and N be a smooth manifold of dimension m. Then a smooth $f : M \to N$

(i) is said to be an **immersion** at a point $x \in M$ if $n \leq m$ and rank of f at x is n. It is said to be an immersion if it is an immersion at every point $x \in M$;
(ii) is said to be a **submmersion** at a point $x \in M$ if $n \geq m$ and rank of f at x is m. It is said to be a submersion if it is a submersion at every point $x \in M$;
(iii) is an **embedding** if it is an immersion and a homeomorphism onto its image $f(M)$;
(iv) is a **diffeomorphism** if $n = m$ and f is a surjective embedding.

Example 3.9.5 Let $\mathbf{Sy(n, R)}$ be the set of all symmetric matrices of order n over $\mathbf{R}$ and $M(n, \mathbf{R})$ be the set of all square matrices of order n over $\mathbf{R}$, which is identified with the n^2-Euclidean $\mathbf{R}^{n^2}$ (see Chap. 2). Then $\mathbf{Sy(n, R)}$ has a standard C^∞ structure described by a global chart

$$\psi : Sy(n, \mathbf{R}) \to M(n, \mathbf{R}), (m_{i,j}) \to (m_{11}, \ldots, m_{1n}, m_{22}, \ldots, m_{2n}, m_{33}, \ldots, m_{nn}).$$

Its range is the Euclidean $n(n+1)/2$-space $\mathbf{R}^{n(n+1)/2}$. Hence, it follows that ψ is an immersion with this structure.

Remark 3.9.6 Let M, N be two smooth manifolds and $\psi : M \to N$ be a smooth map with its induced linear homomorphism $\psi_* : T_x(M) \to T_{\psi(x)}(N)$.

(i) If ψ is an embedding or an immersion, then a tangent vector $v \in T_x(M)$ is identified with the vector $\psi_*(v) \in T_{\psi(x)}N$.
(ii) The product manifold $M \times N$ is also a smooth manifold with charts as the products of charts for M and N. Moreover,

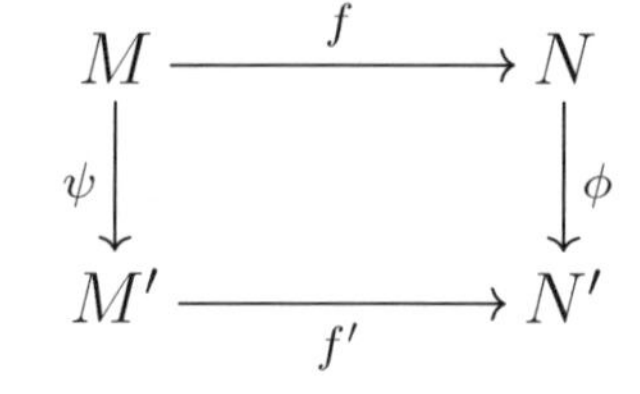

Fig. 3.10 Diagram for equivalent maps f and f'

$$T_{(x,y)}(M \times N) \approx T_x(M) \times \times T_{\psi(x)}(N).$$

If a smooth map $f : M \to N$ is an embedding (or an immersion), a tangent vector $v \in T_x(M)$ is identified with the vector $df_x(v) \in T_{f(x)}(N)$.

Definition 3.9.7 Let M and N be two smooth manifolds of the same dimension n. A smooth map $f : M \to N$ is said to be a **local diffeomorphism** at a point $x \in M$, if its derivative map

$$df_x : T_x(M) \to T_{f(x)}(N)$$

is an isomorphism of tangent spaces. A local diffeomorphism at a point $x \in M$ carries a nbd of x diffeomorphically onto a nbd of $f(x) \in N$.

Remark 3.9.8 Definition 3.9.7 embodied in Inverse Function Theorem is really remarkable and valuable. Because df_x is a single linear map, which can be represented by a matrix M_{df_x}, and it is nonsingular if the determinant det M_{df_x} of the matrix M_{df_x} is nonzero. This says that the problem: whether a map $f : M \to N$ sends a nbd of x in M demographically onto a nbd of $f(x)$ in N is reduced to examining whether the determinant det M_{df_x} is nonzero.

Definition 3.9.9 Let M and N be two smooth manifolds. A smooth map $f : M \to N$ is said to be proper if the inverse image of every compact set in N is also compact in X. An immersion which is both injective and proper is said to be an **embedding**.

Definition 3.9.10 Two smooth maps $f : M \to N$ and $f' : M' \to N'$ are said to be **equivalent** up to diffeomorphism if there exist a diffeomorphism $\psi : M \to M'$ and a diffeomorphism $\phi : N \to N'$ such that $\phi \circ f = f' \circ \psi$, i.e., make the rectangle in the diagram in Fig. 3.10.

Remark 3.9.11 **(i)** Since a diffeomorphism $f : M \to N$ is a bijection and both f and f^{-1} are smooth maps, it is an equivalence relation on smooth manifolds and hence it classifies smooth manifolds into disjoint classes such that any two smooth manifolds are in the same class if one can be deformed into the other smoothly. In this sense, two diffeomorphic manifolds are considered to be the same smooth manifold. Clearly, every diffeomorphism is a homeomorphism. There is a natural question: is every homeomorphism a diffeomorphism? Its answer is negative, in view of the supporting examples for smooth manifolds of dimensions > 3.

(ii) The differential structure on manifolds may not be unique. There exist manifolds having multiple differentiable structures. It was believed before 1956 that a topological space may admit only one differentiable structure. Examples show that the differentiable structure of a topological space may not be unique. For example, John Willard Milnor (1931-) proved in 1956 that **the 7-sphere S^7 admits 28 different differentiable structures** Milnor (1956). Milnor was awarded the Fields Medal in 1962 for his work in differential topology. He was also awarded the Abel Prize in 2011. For another example, Sir Simon Kirwan Donaldson (1957-) proved in 1983 that **$\mathbf{R}^4$ admits an infinite number of different differentiable structures** Donaldson (1983). He was awarded the Fields Medal in 1986.

3.10 The Inverse Function, Local Immersion and Local Submersion Theorems on Smooth Manifolds

This section proves three key theorems: The inverse function Theorem 3.10.1, local immersion Theorem 3.10.5 and local submersion Theorem 3.10.6 on smooth manifolds.

Theorem 3.10.1 (Inverse function theorem on smooth manifold) *Let M and N be two smooth manifolds of the same dimension n and $U \subset M$ and $V \subset N$ be two open sets. If $f : U \to V$ is a smooth map such that rank $f = n$, then for every point $u \in U$, there exists an open nbd W of u such that $f|W$ is a diffeomorphism onto an open nbd $f(u)$ in V.*

Proof The theorem is true for the particular case, when $M = N = \mathbf{R}^n$ by Theorem 3.2.51. Again, by the given conditions, any local representation of f

$$h = \phi \circ f \circ \psi^{-1}$$

(using notation of Definition 3.7.1) has rank n at the point $\psi(u)$. Hence, there exists an open set W' of $\psi(u)$ on which the map h is a diffeomorphism. Thus, it asserts that the restricted map $f|W = \psi^{-1}(W')$ is also a diffeomorphism, which proves the general case. ❑

3.10.1 Canonical Embedding and Canonical Submersion

This subsection presents examples of canonical embedding and canonical submersion and shows that an injective immersion is not necessarily an embedding.

Example 3.10.2 Given integers $m \geq n$, the standard inclusion map

$$i : \mathbf{R}^n \hookrightarrow \mathbf{R}^m, \ x = (x_1, x_2, \ldots, x_n) \mapsto (x_1, x_2, \ldots, x_n, 0, \ldots, 0) = (x, 0)$$

is an embedding, called the **canonical embedding**.

Example 3.10.3 Given integers $m \leq n$, the standard projection map

$$s : \mathbf{R}^n \to \mathbf{R}^m, \ x = (x_1, x_2, \ldots, x_m, \ldots, x_n) \mapsto (x_1, x_2, \ldots, x_m)$$

is a submersion, called the **canonical submersion**.

Example 3.10.4 Every injective immersion is not necessarily an embedding. For example, consider the map

$$f : \mathbf{R} \to S^1 \times S^1, \ t \mapsto (e^{2\pi a t}, \ e^{2\pi b t}),$$

where $\frac{a}{b}$ is irrational, is injective, but it is not an embedding. f is injective, because if $f(t) = f(t')$, then both $a(t - t')$ and $b(t - t')$ are integers. This is possible only when $t = t'$. Clearly, f is an immersion, because $\frac{df}{dt} \neq 0$. Again, f is not an embedding, because the image $f(\mathbf{R})$ is an everywhere dense curve winding around the 2-torus $S^1 \times S^1$ but it contradicts Proposition 3.11.10 saying that if M is an n-submanifold of an m-manifold for $n < m$, then M is not a dense subset of N. The above discussion asserts that f is an injective immersion but it is not an embedding of manifolds.

3.10.2 Local Immersion and Local Submersion Theorems

This subsection proves local immersion and local submersion theorems for smooth manifolds. Theorem 3.10.5 proves that every immersion is locally equivalent to the canonical embedding i formulated in Example 3.10.2. On the other hand, Theorem 3.10.6 proves that every immersion is locally equivalent to the canonical submersion s formulated in Example 3.10.3.

Theorem 3.10.5 (Local Immersion Theorem) *If M is a smooth manifold of dimension m and N be a smooth manifold of dimension n and if $f : M \to N$ is an immersion at a point $a \in M$, then there is a local representation of f at the point a such that it is the canonical embedding*

$$i : \mathbf{R}^m \hookrightarrow \mathbf{R}^n, \ x = (x_1, x_2, \ldots, x_m) \mapsto (x_1, x_2, \ldots, x_m, 0, \ldots, 0) = (x, 0).$$

Proof Suppose M is an m-dimensional smooth manifold with a set of local charts $\{(\psi_i, U_i)\}$ and N is an n-dimensional smooth manifold with a set of local charts $\{(\phi_i, V_i)\}$. Let (ψ, U) be a chart of M and (ϕ, V), be a chart of N. Consider a local representation $g = \phi \circ f \circ \psi^{-1}$ of f corresponding to a pair (ψ, ϕ) of coordinate systems. We assume without loss of generality that $\psi(a) = 0$ and $\phi(f(a)) = 0$ and the matrix representation of g at 0 takes the form (if necessary permuting the coordinates in ϕ)

$$J(g)_0 = \begin{pmatrix} A \\ B \end{pmatrix},$$

where A is an $m \times m$ nonsingular matrix . By changing the coordinates in $\mathbf{R}^n$ by a linear transformation

$$T : \mathbf{R}^n \to \mathbf{R}^n,$$

the matrix representation $M(T)$ of T takes the form

$$M(T) = \begin{pmatrix} A^{-1} & 0 \\ -BA^{-1} & 1_{n-m} \end{pmatrix},$$

where 1_{n-m} is the identity matrix and 0 is the null matrix. Then the matrix $J(g)_0$ assumes the following form:

$$\begin{pmatrix} A^{-1} & 0 \\ -BA^{-1} & 1_{n-m} \end{pmatrix} \begin{pmatrix} A \\ B \end{pmatrix} = \begin{pmatrix} 1_m \\ 0 \end{pmatrix}.$$

Let U be the domain of g. Define a map

$$k : U \times \mathbf{R}^{n-m} \to \mathbf{R}^n, \ (x, y) \mapsto g(x) + (0, y).$$

Then $k \circ i = g$ and the Jacobian matrix $J(h)_0$ is the identity matrix 1_n. Hence, it follows by inverse function Theorem 3.10.1 that k is a local diffeomorphism at $0 \in \mathbf{R}^n$ and

$$\phi \circ f \circ \psi^{-1} = g = k \circ i.$$

It implies that

$$(k^{-1} \circ \phi) \circ (f \circ \psi^{-1}) = i.$$

Hence, it follows that the local representation of f at the point $a \in M$ corresponding to the pair of coordinate systems $(\psi, \ k^{-1} \circ \phi)$ is i, which is the local embedding

$$i : \mathbf{R}^m \hookrightarrow \mathbf{R}^n.$$

❑

Theorem 3.10.6 (Local submersion theorem) *Let M be a smooth manifold of dimension m and N be a smooth manifold of dimension n. If $f : M \to N$ is a submersion at a point $a \in M$, then there is a local representation of f at the point a such that it is the canonical submersion s.*

Proof Suppose M is an m-dimensional smooth manifold with a set of local charts $\{(\psi_i, U_i)\}$ and N is an n-dimensional smooth manifold with a set of local charts $\{(\phi_i, V_i)\}$. Let (ψ, U) be a chart of M and (ϕ, V) be a chart of N. Consider a local representation $g = \phi \circ f \circ \psi^{-1}$ of f corresponding to a pair (ψ, ϕ) of coordinate systems as in proof of Theorem 3.10.5 such that $\psi(a) = 0$ and $\phi(f(a)) = 0$ and the Jacobian matrix $J(g)$ of g at 0 (on a linear change of coordinates in $\mathbf{R}^m$) is

$$J(g)(0) = (1_n \quad 0).$$

Then the map

$$k : U \to \mathbf{R}^m, x \mapsto (g(x), x_{n+1}, \dots, x_m)$$

has the Jacobian matrix 1_m at $x = 0$ and hence $g = s \circ k$ implies that

$$(\phi \circ f) \circ (k \circ \psi)^{-1}.$$

z is the canonical submersion s (as formulated in Example 3.10.3). ❑

3.11 Submanifolds of Smooth Manifolds

This section studies the concept of submanifolds which is itself a manifold lying in a manifold in such a way that the local coordinate systems of these two manifolds are consistent. Its precise definition is now given in Definition 3.11.1. The concepts of immersions and submanifolds are closely linked and they are specially used for the solution of some problems.

Definition 3.11.1 Let M be an n-dimensional differentiable manifold and $N \subset M$ be nonempty. Then N is said to be a **submanifold** of M of dimension $m \leq n$ if it satisfies the following conditions:

(i) N is an m-dimensional differentiable manifold;
(ii) if for every point $y \in N$, there is a local coordinate nbd U in N such that the local coordinate system $\psi : U \to V$, where V is an open cell in $\mathbf{R}^n$ with the property that $\psi(U \cap N)$ is a subset of V and $x_{m+1} = x_{m+2} = \cdots = x_n = 0$;
(iii) $\psi|_{(U \cap N)}$ is a local coordinate system for N around the point $y \in N$.

Remark 3.11.2 Definition 3.11.1 of a submanifold N of a smooth manifold M asserts that

(i) the local coordinate $x_1, x_2, \dots, x_m$ on N around the point $y \in N$ is set up on M such that N satisfies the equations (local)

$$x_{m+1} = x_{m+2} = \cdots = x_n = 0;$$

(ii) the local coordinate systems on N are inherited by the local coordinate systems on M;
(iii) N lies in a tubular type nbd in M in the sense of Definition 3.22.4.

Definition 3.11.1 is reformulated in Definition 3.11.3.

Definition 3.11.3 Let M be an n-dimensional differentiable manifold and $N \subset M$ be nonempty. Then N is a **submanifold** of M of dimension m if it satisfies the following conditions:
for every point $y \in N$, there is a local coordinate nbd U in M such that the local coordinate system $\psi : U \to V$, where V is an open cell in $\mathbf{R}^n$ with the property that ψ maps $U \cap N$ onto an open subset of $\mathbf{R}^m \subset \mathbf{R}^n$, where $\mathbf{R}^m$ is considered as the subspace of the first m coordinates in $\mathbf{R}^n$,

$$\mathbf{R}^m = \{(x_1, x_2, \ldots, x_m, x_{m+1}, \ldots, x_n) \in R^n : x_{m+1} = x_{m+2} = \cdots = x_n = 0\}.$$

Then the collection

$$\{(N \cap U, \psi|_{U \cap N}) : (U, \psi) \text{ is a chart in } M \quad \text{and } N \cap U \neq \emptyset\}$$

forms a smooth atlas of N.

Example 3.11.4 S^2 is a submanifold of the manifold $\mathbf{R}^3$.

Example 3.11.5 Example 3.9.5 implies that $Sy(n, \mathbf{R})$ is a submanifold of $M(n, \mathbf{R})$.

Example 3.11.6 Let M and N be two differentiable manifolds. If N is a submanifold of M, then the natural inclusion map $i : N \hookrightarrow M$ is an immersion.

Example 3.11.7 **(i)** Let $M = \mathbf{R}^n$ and $x = (x_1, x_2, \ldots, x_m, x_{m+1}, x_{m+2}, \ldots, x_n) \in \mathbf{R}^n$ be an arbitrary point. Then the Euclidean subspace N defined by

$$N = \{x \in \mathbf{R}^n : x_{m+1} = x_{m+2} = \cdots = x_n = 0, m < n\}$$

is a submanifold of M.
(ii) Let $M = S^2$ be the 2-sphere in $\mathbf{R}^3$ and N be the circle at which the sphere S^2 meets the plane $z = 0$. Then z can be taken as one of the local coordinates on S^2 at any point $y \in N$, and the other coordinates may be taken either x or y. This implies that the circle N is a submanifold of $M = S^2$.
(iii) Let $M = \mathbf{R}^3$ be the Euclidean 3-space and $N = S^2$ be the 2-sphere with the equation $x^2 + y^2 + z^2 = 1$ in $\mathbf{R}^3$. Take the coordinates x, y, z as the local coordinates around any point $p \in \mathbf{R}^3$ and two of the coordinates x, y, z can be used as local coordinates of S^2 around any point on S^2. Then $N = S^2$ is a differentiable manifold and it is a submanifold of $M = \mathbf{R}^3$.

Proposition 3.11.8 *Let M be a smooth manifold of dimension n and N be a submanifold of M of dimension m. Then for every point $y \in N$, there exists an open nbd U of y in N and a submersion*

$$f : U \to \mathbf{R}^{n-m} : f^{-1}(0) = N \cap U.$$

Proof To prove the proposition, we use that $\mathbf{R}^n = \mathbf{R}^m \times \mathbf{R}^{n-m}$. It follows from Definition 3.11.3 that there is a coordinate chart $\psi : U \to \mathbf{R}^n$ about the point y in N such that

$$\psi^{-1}(\mathbf{R}^m \times \{0\}) = N \cap U.$$

Let $p : \mathbf{R}^n \to \mathbf{R}^{n-m}$ be the standard projection onto the second factor. Then

$$f = p \circ \psi : U \to \mathbf{R}^{n-m}$$

is a submersion with $f^{-1}(0) = N \cap U$. ❑

Proposition 3.11.9 *Let M be a manifold of dimension n and N be a manifold of dimension m $(m \leq n)$. Then a subset X of N is a submanifold of N iff X is the image of a smooth embedding $f : M \to N$.*

Proof First suppose that $f : M \to N$ is a smooth embedding such that $X = f(M)$. Using local immersion Theorem 3.10.5, it follows that for every point $a \in M$, there is a coordinate system $y_1, y_2, \ldots, y_n$ in an open nbd V of $f(a)$ such that

$$X \cap V = \{p \in V : y_{m+1}(p) = y_{m+}(p) = \cdots = y_n(p) = 0\}$$

and the restrictions of the remaining coordinate functions

$$\{y_i|_{X \cap Y} : i = 1, 2, \ldots, m\}$$

constitute a local chart on X at the point $f(a)$. This asserts that X is a submanifold of N.

Conversely, suppose that X is a submanifold of N. Then the inclusion map

$$i : X \hookrightarrow N$$

is a smooth embedding, which follows from the natural smooth structure on X derived from the smooth structure of N. ❑

Proposition 3.11.10 *Let M be an n-submanifold of an m-manifold N $(n \leq m)$. Then M is not a dense subset of N.*

Proof Let (U, ψ) be a coordinate chart of N such that $V = M \cap U \neq \emptyset$ and $\psi(V) \subset \mathbf{R}^n \times \{0\}$. Then

$$W = \psi^{-1}(\mathbf{R}^n \times (\mathbf{R}^{m-n} - \{0\})) \neq \emptyset$$

is an open set of N such that $W \subset U$ and $W \cap V \neq \emptyset$. This proves that M is not dense in N. ❑

3.12 Germs of Differentiable Functions

This section introduces the concept of germs at a point x of a differentiable manifold M and proves the following properties of germs of differentiable functions.

(i) Proposition 3.12.2 proves that corresponding to every germ at $x \in M$, there exists a commutative algebra $\mathcal{A}_x$ (see Proposition 3.12.2).
(ii) Theorem 3.12.3 proves that corresponding to every pair of points $x, y \in M$, there exists an algebra isomorphism

$$h : \mathcal{A}_x \to \mathcal{A}_y.$$

Recall that given a point at x in a differentiable manifold M, a function f is said to be differentiable near the point x if there is an open nbd U of x such that f is differentiable on U. Let $\mathcal{F}(x)$ be the set of all differentiable functions near $x \in M$.

Definition 3.12.1 Two functions $f, g \in \mathcal{F}(x)$ are said to be equivalent denoted by $f \sim g$ if they assume the same value on a nbd of x. The equivalence class $[f]$ of f is said to be the **germ of** f at x. The quotient set, denoted by $\mathcal{F}(x)/\sim$, forms a commutative algebra.

Proposition 3.12.2 *For every differentiable manifold M, the quotient set* $\mathcal{F}(x)/\sim$ *forms a commutative algebra denoted by* $\mathcal{A}_x$ *for every* $x \in M$ *under the compositions*

(i)

$$[f] + [g] = [f + g];$$

(ii)

$$[f] \cdot [g] = [f \cdot g];$$

(iii)

$$c[f] = [cf], \ \forall [f], [g] \in \mathcal{A}_x,$$

where c is a real or complex constant.

Proof It follows from Definition 3.12.1 of the germ $[f] \in \mathcal{A}_x$. ❑

Theorem 3.12.3 *Given a differentiable manifold M and a pair of points* $x, y \in M$, *there exists an algebra isomorphism*

$$h : \mathcal{A}_x \to \mathcal{A}_y,$$

where $\mathcal{A}_x$ *is the commutative algebra of germs at* x.

Proof Let (ψ, U) and (ϕ, V) be two local coordinate charts of the differentiable manifold M containing the point x and the point y, respectively. Suppose that $\psi(x) = \phi(y)$. Let $\alpha \in \mathcal{A}_x$ be represented by f. Then $\alpha = [f]$ and the function f is differentiable on some open set in U. Hence, the function $g = f \circ \psi^{-1} \circ \phi$ defined on a nbd of y is also differentiable near the point y, because $g \circ \phi^{-1} = f \circ \psi^{-1}$ is a differentiable function on a nbd $\phi(y)$. Consider the map

$$h : \mathcal{A}_x \to \mathcal{A}_y, \ \alpha \mapsto \beta = [g].$$

The map h is well-defined, because it is independent of the choice of the representative of the class. To show it, let $f' \in \alpha$. Then $f - f' = 0$ on some nbd of x. If $g' = f' \circ \psi^{-1} \circ \phi$, then

$$g - g' = (f - f') \circ \psi^{-1} \circ \phi = 0$$

on a nbd of y shows that $[g] = \beta = [g']$. Clearly, h is an isomorphism of commutative algebras. ❑

3.13 Product and Quotient Manifolds

This section studies the product and quotient manifolds and their corresponding topologies induced by their differential structure.

3.13.1 Product Manifold and Its Induced Topology

This subsection studies finite product of C^r-manifolds having finite dimension and the induced topology of the product manifold. Proposition 3.13.5 relates the induced topology of the product manifold to the product of the induced topologies of its factor manifolds.

Proposition 3.13.1 facilitates to formulate the concept of product manifold in Definition 3.13.2.

Proposition 3.13.1 *Let M be an n-dimensional C^r-manifold and N be an m-dimensional C^r-manifold. Then the cartesian product space of $M \times N$ is an $(n+m)$-dimensional C^r-manifold.*

Proof Let M be an n-dimensional C^r-manifold with an atlas $\{(\psi_i, U_i)\}$ and N be an m-dimensional C^r-manifold with an atlas $\{(\phi_j, V_j)\}$. Consider the function

$$\psi_i \times \phi_j : M \times N \to \mathbf{R}^n \times \mathbf{R}^m \approx \mathbf{R}^{n+m}.$$

It is an injective map such that its range $\psi_i(U_i) \times \phi_i(V_i)$ is an open set in $\mathbf{R}^{n+m}$. This shows that it is a chart for the product space $M \times N$ with domain $U_i \times U_j$. Again, if $\psi_i' \times \phi_j'$ is another chart having domain $U_i' \times V_j'$ that interests $U_i \times V_j$, then the change of coordinates

$$(\psi_i' \times \phi_j') \circ (\psi_i \times \phi_j)^{-1} = (\psi_i' \times \phi_j') \circ (\psi_i^{-1} \times \phi_j^{-1}) = (\psi_i' \circ \psi_i^{-1}) \times (\phi_j' \circ \phi_j^{-1})$$

is a diffeomorphism. This asserts that a set of all such charts which is a C^r-atlas defines a C^r-structure on $M \times N$ with dimension $n + m$. ❑

Definition 3.13.2 Given an m-dimensional C^r-manifold M with an atlas $\{(\psi_i, U_i)\}$ and an n-dimensional C^r-manifold N with an atlas $\{(\phi_j, V_j)\}$, their **product C^r-manifold** $M \times N$ is an $(n + m)$-manifold with atlas $\{((\psi_i \times \phi_j), (U_i \times V_j)\}$, where $M \times N$ is the cartesian product space of M and N.

Remark 3.13.3 Definition 3.13.2 can be extended to any finite number of C^r-manifolds. The coordinate function (ψ_i, ϕ_j) acts on a point $(x, y) \in M \times N$ obtaining the point $(\psi_i(x), \phi_j(y)) \in \mathbf{R}^{m+n}$, and the product $M \times N$ with $\{((\psi_i \times \phi_j), (U_i \times V_j)\}$ satisfies all the conditions of a manifold.

Example 3.13.4 **(i)** Let $f : \mathbf{R}^n \to \mathbf{R}^m$ be a C^r-map and its graph

$$G_f = \{(x, f(x))\} \subset \mathbf{R}^n \times \mathbf{R}^m$$

is an n-dimensional submanifold in $\mathbf{R}^{n+m}$ of C^r-class. In particular, for

$$f : \mathbf{R} \to \mathbf{R},\ x \mapsto sin(1/x) : x \neq 0,$$

its graph G_f is a 1-dimensional manifold in $\mathbf{R}^2$ with C^r-class.

(ii) The cylinder $S^1 \times \mathbf{R}$ is a product of manifolds.

(iii) The torus $T^2 = S^1 \times S^1$ is a C^r-manifold of dimension 2.

(iv) The n-torus $T^n = S^1 \times S^1 \times \cdots \times S^n$ (the product of n-circles) is a C^r-manifold of dimension n.

Proposition 3.13.5 studies topologies induced by differential structures given in Definition 3.6.5 and relates the induced topology of the product manifold to the product of the induced topologies of its factor manifolds.

Proposition 3.13.5 *The induced topology of the product manifold is the product of the induced topologies of its factor manifolds.*

Proof Let M and N be two C^r-manifolds. To prove the proposition by using Theorem 3.6.9, it is sufficient to show that the charts ψ_i and ϕ_j (formulated in Proposition 3.13.1) are also homeomorphisms in the product topology of $M \times N$. Since ψ_i and ϕ_j are both homeomorphisms in the corresponding topologies of M and N, then their products $\psi_i \times \phi_j$ are also homeomorphisms. ❑

3.13.2 Quotient Manifold and Its Induced Topology

The aim of this subsection is to define the quotient manifold as a submersion and also to study its associated topologies. Proposition 3.13.11 gives a positive answer of the problem: whether the induced topology of a quotient manifold is the quotient topology?

Definition 3.13.6 Let M be a manifold and $\sim$ be an equivalence relation on M. Then $M/\sim$ is the quotient set with natural projection

$$p : M \to M/\sim, x \mapsto [x],$$

which sends every point $x \in M$, to its equivalence class $[x]$. If the set $M/\sim$ is endowed with the structure of differentiable manifold such that

$$p : M \to M/\sim, x \mapsto [x]$$

is a submersion, then the differential manifold $M/\sim$, denoted by $Q(M)$, is called a **quotient manifold** of M. In other words, the quotient space $M/\sim$ is a quotient manifold of M if $M/\sim$ admits the structure of a smooth manifold such that

$$p : M \to M/\sim, x \mapsto [x]$$

is a submersion.

Example 3.13.7 Let $\mathbf{R}_*^{(n+1)} = \mathbf{R}^{(n+1)} - \{0\}$ and $\sim$ be an equivalence relation on $\mathbf{R}_*^{(n+1)}$ be defined by

$$z \sim w \Leftrightarrow z = tw \quad \text{for some } t \in \mathbf{R}_* = \mathbf{R} - \{0\}.$$

Then there exists a bijection

$$f : \mathbf{R}_*^{(n+1)}/\sim \ \to \mathbf{R}P^n, \ z \mapsto p_z^1$$

between the quotient set $\mathbf{R}_*^{(n+1)}/\sim$ and the real projective space $\mathbf{R}P^n$ obtained by sending z to the 1-dimensional plane p_z^1 of $\mathbf{R}^{n+1}$ containing z. Hence, f defines a C^∞-structure on the quotient set $\mathbf{R}_*^{(n+1)}/\sim$ and it admits a quotient manifold structure of dimension n.

Example 3.13.8 Let M and N be two smooth manifolds and $f : M \to N$ be a submersion. Define an equivalence relation $\sim$ on M by the rule

$$x_1 \sim x_2 \Leftrightarrow f(x) = f(y).$$

Define a bijection

$$g : M/\sim \ \to N, x \mapsto f(x).$$

Then g defines a C^∞-structure on the quotient set $M/\sim$. Hence, it admits a quotient manifold structure such that it is diffeomorphic to the manifold N.

Example 3.13.9 The topology induced on a submanifold of a differentiable manifold by its C^∞-structure may not be its subspace topology. For example, consider the Figure-eight $\mathbf{F}_8$. It admits a C^∞-structure (see Example 3.4.20) making it a submanifold of $\mathbf{R}^2$. It is bounded and hence it is compact in the subspace topology induced from natural topology on $\mathbf{R}^2$. On the other hand, there is another C^∞-atlas on $\mathbf{F}_8$ determined by a single chart which implies that $\mathbf{F}_8$ is not compact.

Remark 3.13.10 Example 3.13.9 raises the problem: whether the induced topology of a quotient manifold is the quotient topology? Proposition 3.13.11 solves this problem.

Proposition 3.13.11 *Let M be a smooth manifold. Then the induced topology of a quotient manifold $Q(M)$ of M is the same as the quotient set topology.*

Proof Consider the natural projection

$$p : M \to Q(M) = M/\sim,\ x \mapsto [x].$$

Then p is continuous, open, surjective and is also differentiable. If a quotient set has a topology τ in which p is both continuous and open, then the topology τ is the same as the quotient set topology. This implies that the induced topology on $Q(M)$ is the quotient set topology. ❑

There is a natural question: if a quotient space $Q(M) = M/\sim$ of a smooth manifold M admits a quotient manifold structure, is this structure unique? Proposition 3.13.12 gives its affirmative answer.

Proposition 3.13.12 *If a quotient space $Q(M) = M/\sim$ of a smooth manifold M admits a quotient manifold structure, then this structure is unique.*

Proof To prove the proposition, take two quotient manifold structures $\mathcal{M}_1$ and $\mathcal{M}_2$ on $M/\sim$. Now, consider their associated canonical projections

$$p : M \to (M/\sim,\ \mathcal{M}_1)$$

and

$$p : M \to (M/\sim,\ \mathcal{M}_2).$$

Then they coincide and are related by $p = 1_{(M/\sim)} \circ p$ and hence $1_{(M/\sim)}$ is a smooth map, which is a diffeomorphism. This implies that the quotient manifold structures $\mathcal{M}_1$ and $\mathcal{M}_2$ on $M/\sim$ are identical. This proves the uniqueness of this structure. ❑

Example 3.13.13 Let M and N be two smooth manifolds. Then every surjective immersion $\psi : M \to N$ determines an equivalence relation on M defined by

$$x \sim y \Leftrightarrow \psi(x) = \psi(y).$$

This determines a bijective map

$$f : M/\sim \ \to N, [x] = p(x) \to \psi(x),$$

where

$$p : M \to M/\sim, \ x \to [x]$$

is the canonical projection. Carrying the given differential structure of N onto $M/\sim$, it follows that $M/\sim$ admits a quotient manifold structure diffeomorphic to N.

3.14 Embedding of Compact Manifold in Euclidean Spaces and Whitney's Theorem

This section proves a basic theorem on embedding of a finite-dimensional compact manifold as a closed subspace in a finite-dimensional Euclidean space in Theorem 3.14.20, which is known as **Whitney's embedding theorem**. This theorem implies that any manifold may be considered as a submanifold of a Euclidean space and it conveys one of the important properties of a manifold, which facilitates various development of manifold theory. Historically, the concept of manifolds was introduced in this way. But Whitney's Theorem 3.14.20 reconciles this earlier concept of manifolds with the modern abstract concept of manifolds. Here, a manifold means a smooth manifold (it may have a boundary).

3.14.1 Smooth Partition of Unity

This subsection uses the concept of smooth partition of unity to solve the problem: under what conditions a manifold can be embedded in some finite-dimensional Euclidean space? For its answer, see Sect. 3.14.2. It is one of the main tools of manifold theory in transferring its local structure to global structure. The main reason for restricting the topology of a smooth manifold M is to guarantee the existence of a specialized family of real-valued functions on M, known as the partition of unity. The concept of partition of unity subordinate to an open covering of an n-dimensional C^r-manifold is analogous to that of a topological space defined in Chap. 5 of **Basic Topology, Volume 1** of the present series of books.

Definition 3.14.1 Let M be an n-dimensional C^r-manifold (or smooth manifold). For the function $f : M \to \mathbf{R}$, the closure $\overline{S_f}$ of the set

$$S_f = \{x \in M : f(x) \neq 0\}$$

is called the **support** of f, denoted by $supp\ f$.

Definition 3.14.2 Let M be an n-dimensional smooth manifold and $\mathcal{C} = \{U_i\}_{i \in \mathbf{A}}$ be an open covering of M. Then a family of smooth functions

$$\mathcal{F} = \{\psi_i : M \to \mathbf{R}\}_{i \in \mathbf{A}}$$

is said to be a **smooth partition of unity** subordinate to the covering $\mathcal{C}$ if

(i) each support set $supp\ S_{\psi_i} = \overline{\psi_i} \subset U_i,\ \forall\, i \in \mathbf{A}$;
(ii) $0 \leq \psi_i(x) \leq 1,\ \ \forall\, x \in A$;
(iii) the family $\mathcal{F}$ forms a local covering of M, i.e., every point $x \in M$ has a nbd on which all functions ψ_i excepting finitely many are identically zero and
(iv) $\Sigma_j \psi_j(x) = 1$ for every $x \in M$, which is well-defined by condition (iii).

Remark 3.14.3 Every smooth manifold with an open covering $\mathcal{U} = \{U_i : i \in \mathbf{A}\}$ admits a smooth partition of unity subordinate to $\mathcal{U}$ (see Theorem 3.14.15).

Remark 3.14.4 Bump function formulated in Definition 3.14.5 is an important smooth function in differential topology. For example, it is used to study smooth homotopy and normalized homotopy (see Definition 3.19.4).

Definition 3.14.5 A smooth function $\mathcal{B} : \mathbf{R} \to \mathbf{R}$ is said to be **bump function** if it satisfies the conditions

$$\mathcal{B}(x) = \begin{cases} 0, & \text{for all } x \leq 0 \\ 1, & \text{for all } x \geq 1 \end{cases}$$

and

$$0 < \mathcal{B}(x) < 1 \text{ for all } x \ \text{ such that } \ 0 < x < 1.$$

Example 3.14.6 To have a bump function, consider the smooth function $f : \mathbf{R} \to \mathbf{R}$ defined by

$$f(x) = \begin{cases} e^{\frac{1}{x(x-1)}}, & \text{if } \ 0 < x < 1 \\ 0, & \text{otherwise}. \end{cases}$$

Clearly, f is a non-negative smooth function. It is non-vanishing for $0 < x < 1$. Then the function

$$\mathcal{B} : \mathbf{R} \to \mathbf{R},\ x \mapsto \frac{\int_0^x f(t)dt}{\int_0^1 f(t)dt}$$

is a bump function.

Proposition 3.14.7 *Let M be a smooth manifold of dimension n. If K is compact and U is open such that $K \subset U \subset M$, then there is a smooth function $f : M \to [0, \infty)$ such that*

(i) $f(x) > 0, \ \forall x \in K$ *and*
(ii) *the support* f *is contained in* U, *i.e.,* $\overline{S_f} \subset U$.

Proof Define a smooth function

$$f : \mathbf{R} \to \mathbf{R}, \ x \mapsto \psi(1 - |x|),$$

where $\psi : \mathbf{R} \to \mathbf{R}$ is a bump function given in Example 3.14.6. Then

$$f(x) \begin{cases} > 0, & \text{if } |x| < 1 \\ = 0, & \text{if } |x| \geq 1. \end{cases}$$

For each $k \in K$, we define a smooth function $F_k : M \to [0, \infty)$ such that

(i) $F_k(k) > 0$ and
(ii) *supp* $F_k \subset U$

as follows: take local coordinates $(x_1, x_2, \ldots, x_n)$ about the point k such that

$$B_r = \{(x_1, x_2, \ldots, x_n) : |x| < r\} \subset U$$

for some $r > 0$. Define

$$F_k : M \to [0, \infty) \text{ by } x \mapsto \begin{cases} f\left(\frac{x_1}{r}\right) \cdots f\left(\frac{x_n}{r}\right), & \text{if } x \in B_r \\ 0, \text{ otherwise.} \end{cases}$$

Then the family $\mathcal{U} = \{x \in M : F_k(x) > 0\}_{k \in K}$ forms an open covering of K. As by hypothesis, K is compact, and $\mathcal{U}$ has a finite subcovering. Then the required function $F : M \to [\,0, \infty)$ is defined by taking the sum of the corresponding finite number of functions F_k. ❑

Proposition 3.14.8 *Let* $\mathcal{U} = \{U_a\}_{a \in \mathbf{A}}$ *and* $\mathcal{V} = \{U_b\}_{b \in \mathbf{B}}$ *be two open coverings of a smooth manifold* M. *If* $\mathcal{U}$ *refines* $\mathcal{V}$ *and* $\mathcal{U}$ *has a smooth partition of unity subordinate to the covering* $\mathcal{U}$, *then* $\mathcal{V}$ *has also so.*

Proof Suppose that $\{\psi_a\}_{a \in \mathbf{A}}$ is a smooth partition of unity subordinate to the covering $\mathcal{U}$ and $f : \mathbf{A} \to \mathbf{B}$ be a map such that

$$U_a \subset V_{f(a)}, \ \forall a \in \mathbf{A}.$$

Define

$$F_b : M \to \mathbf{R}, \ x \mapsto \Sigma_{a \in f^{-1}(b)} \psi_a(x).$$

Then the family $\{F_b\}_{b \in \mathbf{B}}$ forms a smooth partition of unity. Hence, the proposition follows. ❑

Definition 3.14.9 Let $\mathcal{C} = \{U_i\}$ be an open covering of a topological space X. Then this covering is said to be **locally finite** if every point of X has an open nbd which meets only finitely many members of $\mathcal{C}$. If $\mathcal{C}' = \{U_i'\}$ is another open covering of the topological space X, then $\mathcal{C}'$ is said to be a refinement of $\mathcal{C}$ if every member of $\mathcal{C}'$ is in some member of $\mathcal{C}$. A Hausdorff space X is **paracompact** if its every open covering has an open locally finite refinement.

Theorem 3.14.10 proves the paracompactness property of a smooth manifold by showing that every smooth manifold admits a locally finite refinement which is also countable.

Theorem 3.14.10 *Every smooth manifold is paracompact.*

Proof Let M be a smooth manifold of dimension n. Then it is locally homeomorphic to either $\mathbf{R}^n$ or $\mathbf{R}^n_+$, which are locally compact, because every point of either of them has a compact nbd. Hence, every point $x \in M$ has an open nbd U such that $\overline{U}$ is compact, because if V is an arbitrary nbd of x and W is a compact nbd of x, then

$$U = V \cap Int\ W \subset W.$$

It implies that $\overline{U}$ is compact, because it is a closed subset of the compact set W. Again, since by hypothesis, M is second countable, it has a countable basis $\{U_j\}$ with the property that each $\overline{U}_j$ is compact. Consider the ascending sequence of compact subsets of M

$$W_1 \subset W_2 \subset \cdots \subset W_j \subset \cdots$$

such that

(i) $M = \bigcup_j W_j$ and
(ii) $W_j \subset Int\ W_{j+1},\ \forall\, j.$

Taking $W_1 = \overline{U}_1$ and defining W_j inductively, such that if $m > j$ is the smallest integer with the property

$$W_j \subset U_1 \cup U_2 \cup \cdots \cup U_m.$$

Take

$$W_{j+1} = \overline{U}_1 \cup \cdots \cup \overline{U}_m = \overline{U_1 \cup \cdots \cup U_m}.$$

Suppose $\mathcal{U} = \{U_i\}$ is an open covering of M. Take a locally finite refinement $\mathcal{V}$ defined as follows: Take $W_{-1} = W_0 = \emptyset$ and for every integer $j > 1$, consider the open sets

$$(Int\ W_{j+2} - W_{j-1}) \cap U_i : U_i \in \mathcal{U}.)$$

Since these open sets form a covering of the compact set $W_{j+1} - Int\ W_j$, there is a finite subcovering $\mathcal{V}_j = \{V_1^j, \ldots, V_k(j)^j\}$ for some integer $k(j)$. Since the collection

$\{W_{j+1} - Int\ W_j\}$ of open sets forms a covering of M, it follows that $\mathcal{V} = \mathcal{V}_1 \cup \mathcal{V}_2 \cup \cdots \cup \mathcal{V}_j \cup \cdots$ has a covering of M, which is also a refinement of the covering $\mathcal{U}$, since every $V_i^j \subset U_i$ for some U_i. It is also locally finite, since if $x \in W_j$ for some j, then $Int\ W_{j+1}$ is a nbd of x which meets no member of $\mathcal{V}_r$ for $r > j+1$. This proves that M is paracompact. ❑

Theorem 3.14.11 proves the metrizability property of a smooth manifold. The role of this metric in the study of locally Euclidean structure on the manifold is not well-known.

Theorem 3.14.11 *Every smooth manifold is metrizable.*

Proof It follows from Urysohn's metrization theorem which asserts that every second countable normal space is topologically embeddable in the infinite-dimensional Hilbert coordinate space, because every paracompact space is normal and every smooth manifold is second countable. ❑

Proposition 3.14.12 *Let M be an n-dimensional smooth manifold and $\mathcal{U} = \{U_i\}$ be a covering of M. Then $\mathcal{U}$ has a locally finite countable refinement by every smooth coordinate nbds such that every one of them has a compact closure.*

Proof Take U_i, V_i and W_i as in Theorem 3.14.10 and choose the covering of each compact set $W_{j+1} - Int\ W_j$ properly. Then every point x of the open set

$$(Int\ W_{j+2} - W_{j-1}) \cap V_i$$

has a coordinate nbd U_x and a homeomorphism ψ_x from U_x into $\mathbf{R}^n$ or into $\mathbf{R}^n_+$ with the properties

(i) $U_x \subset (Int\ W_{j+2} - W_{j-1}) \cap V_i$ and
(ii) $\psi_x(U_x)$contains a closed nbd $B_{\psi_x(x)}(n)$ with the choice

$$K_x = \psi_x^{-1}(Int\ B_{\psi_x(x)}(n)),$$

where $B_{\psi_x(x)}(n))$ denotes the closed ball with center $\psi_x(x)$ and radius n in $\mathbf{R}^n$ (or in $\mathbf{R}^n_+$). Then $\overline{K_x}$ is compact and hence there is a finite number of K_x which cover $W_{j+1} - Int\ W_j$. To complete the proof, proceed as in Theorem 3.14.10. ❑

Definition 3.14.13 Let M be a smooth manifold and $\mathcal{U} = \{U_i\}$ be an open covering of M. Then a family $\mathcal{V} = \{V_i\}$ of open sets is said to be a **shrinking** of the covering of $\mathcal{U}$ if $\overline{V}_i \subset U_i$, for all i.

Lemma 3.14.14 (Shrinking lemma) *Let M be a smooth manifold and $\mathcal{U} = \{U_i : i = 1, 2, \ldots\}$ be a countable family of locally finite open covering of M. Then corresponding to $\mathcal{U}$, there is another open covering $\mathcal{V} = \{V_i : i = 1, 2, \ldots\}$ such that*

$$\overline{V}_i \subset U_i,\ \forall i \geq 1.$$

Proof Suppose that M is connected. For each positive integer m, define the open set $\mathcal{U}_m$ by the rule

$$\mathcal{U}_m = \bigcup_{i \geq m} U_i.$$

Construct inductively the family of open sets $\{V_i\}$ as follows:

Since the closed set $C_1 = U_1 - \mathcal{U}_2$ is contained in the open set U_1, it follows that $M = C_1 \cup \mathcal{U}_2$. Again, since M is paracompact by Theorem 3.14.10, it is also normal. Now choose an open set V_1 with the property

$$C_1 \subset V_1 \subset \overline{V}_1 \subset U_1,$$

which implies that

$$M = V_1 \cup \mathcal{U}_2.$$

Choose the open sets $V_1, V_2, \ldots, V_{m-1}$ such that

$$\overline{V}_i \subset U_i, \; \forall i = 1, 2, \ldots, m-1 \text{ and } M = V_1 \cup V_2 \cdots \cup V_{m-1} \cup \mathcal{U}_m.$$

Then the closed sets C_m have the following property:

$$C_m = U_m - (V_1 \cup V_2 \cdots \cup V_{m-1} \cup \mathcal{U}_{m+1}) \subset U_m.$$

It asserts that

$$M = V_1 \cup V_2 \cup \cdots \cup V_{m-1} \cup C_m \cup \mathcal{U}_{m+1}.$$

Again, choose the open set V_m such that

$$C_m \subset V_m \subset \overline{V}_m \subset U_m.$$

Hence, it follows that

$$M = V_1 \cup V_2 \cup \cdots \cup V_{m-1} \cup V_m \cup \mathcal{U}_{m+1}.$$

The family $\{V_i\}$ of open sets forms an open covering of M. Because covering $\mathcal{U}$ is locally finite, for any point $x \in M$, there exists the greatest integer k such that $x \notin U_m, \; \forall m \geq k$. Then $x \notin U_k$. Finally, it follows that

$$x \in V_1 \cup V_2 \cup \cdots \cup U_{k-1},$$

since

$$M = V_1 \cup V_2 \cup \cdots \cup V_{k-1} \cup \mathcal{U}_k.$$

This proves that $\mathcal{V} = \{V_i\}$ forms an open covering of M. ❑

Theorem 3.14.15 provides a basic result in differential topology, and it is used to covert many global problems into local forms.

Theorem 3.14.15 *Every smooth manifold M with an open covering $\mathcal{U} = \{U_a\}_{a\in\mathbf{A}}$ has a smooth partition of unity subordinate to $\mathcal{U}$.*

Proof Suppose that the given covering $\mathcal{U} = \{U_a\}_{a\in\mathbf{A}}$ is countable and locally finite such that every member U_a in $\mathcal{U}$ is coordinate nbd with compact closure by Proposition 3.14.12. Then there is another covering $\mathcal{V} = \{V_a\}$ of B such that $\overline{V}_a \subset U_a$ by Shrinking Lemma 3.14.14. Following the method described in Proposition 3.14.7, construct the smooth function $f_a : M \to \mathbf{R}$ such that

(i) $f_a(x) > 0$ on $\overline{V}_a$ and
(ii) *support* $f_a \subset U_a$.

Then the function $\Sigma_{a\in\mathbf{A}} f_a$ is well-defined and it is a positive smooth function. Hence, the family of functions

$$\{\psi_a\} = \frac{f_a}{\Sigma_{a\in\mathbf{A}} f_a}$$

gives the required smooth partition of unity subordinate to $\mathcal{U}$. ❑

Proposition 3.14.16 *Let M be a smooth manifold and $\{\psi : U_i \to \mathbf{R}\}$ be a smooth partition of unity on M. If $\{f_i : U_i \to \mathbf{R}\}$ is a family of smooth functions, then the function*

$$f : M \to \mathbf{R},\ x \mapsto \Sigma_i(\psi_i f_i)(x) = \Sigma_i \psi_i(x) f_i(x)$$

is smooth.

Proof Consider the smooth function $\psi_i f_i : U_i \to \mathbf{R}, x \mapsto \psi_i(x) f_i(x)$. Since it vanishes on a nbd of $M - U_i$, it can be extended over M by using the zero function on $M - U_i$. This asserts that the sum function

$$f = \psi_i f_i : M \to \mathbf{R},\ x \mapsto \Sigma_i(\psi_i f_i)(x) = \Sigma_i \psi_i(x) f_i(x),$$

is smooth. ❑

Theorem 3.14.17 characterizes a smooth map with the help of partition of unity.

Theorem 3.14.17 *Let U be an open set in $\mathbf{R}^n_+$ and $f : U \to \mathbf{R}^m$ be a given map. Then f is smooth iff there is an open set V in $\mathbf{R}^n$ and a smooth map $H : V \to \mathbf{R}^n$ such that*

(i) $V \cap \mathbf{R}^n_+ = U$ *and*
(ii) $H|_U = f$.

Proof First suppose that $f : U \to \mathbf{R}^m$ is a smooth map. Then for each $x \in U$, there is an open nbd V_x of x in $\mathbf{R}^n$ and a smooth map $g_x : V_x \to \mathbf{R}^m$ such that $f = g_x$ on $U \cap V_x$. If $K = \Sigma_{x \in U} V_x$, then K is an open set in $\mathbf{R}^n$ such that $U \subset K$. The manifold K has a partition of unity $\{\psi_x\}$ subordinate to the covering $\{V_x\}$. Consider the map

$$F = \Sigma_{x \in U} \psi_x g_x : K \to \mathbf{R}^m.$$

Then it follows by using Proposition 3.14.16 that H is a smooth map. Again, there is another open set V_1 of $\mathbf{R}^n$ such that $V_1 \cap \mathbf{R}^n_+ = U$. Then take $V = K \cap V_1$ and $H = F|_V$ to prove the first part of the theorem. The converse part is trivial. ❑

Proposition 3.14.18 *Let M be a smooth manifold and $f : M \to \mathbf{R}$ be a continuous function such that $f(x) > 0, \ \forall x \in M$. Then there is a smooth function $h : M \to \mathbf{R}$ such that*

$$0 < h(x) < f(x), \ \forall x \in M.$$

In particular, if M is compact, then h taken as a constant function serves the purpose.

Proof Consider a locally finite covering $\mathcal{U} = \{U_i\}$ of M and another open covering $\mathcal{V} = \{V_i\}$ of M such that $\overline{V_i}$ is compact and $\overline{V_i} \subset U_i$, as in the proof of Theorem 3.14.15. Let $\{\psi_i\}$ be a smooth partition of unity such that

(i) $\psi > 0$ on $\overline{V}_i$ and
(ii) support of ψ_i is a subset of U_i.

Let δ_i be a positive real number such that it is smaller than $inf \ f$ on the compact set $\overline{V}_i$. Define

$$f : M \to \mathbf{R}, \ x \mapsto \Sigma_i \delta_i \psi_i(x).$$

This asserts that f is smooth. Moreover, $h(x) < f(x)$, because $\Sigma_i \psi_i(x) = 1$ and the maximum of the corresponding $\delta_i < f(x)$. Since all $\delta_i > 0$, it follows that $h(x) > 0$. ❑

Remark 3.14.19 For more result on paracompactness of a manifold, see Exercise 20 of Sect. 3.27.

3.14.2 Embedding of Compact Manifold in Euclidean Spaces

This subsection proves Whitney's Embedding Theorem 3.14.20 (Whitney 1936a, c, b), which gives a basic property of a manifold. This theorem plays the key role toward various developments of manifold theory and unifies the original (earlier) concept of manifolds with the modern abstract concept making a revolution in the theory of manifolds. This theorem is the most powerful result of the manifold theory that opens up various beautiful developments of this theory.

Theorem 3.14.20 (Whitney's Embedding Theorem) *Every n-dimensional compact smooth manifold can be embedded in a finite-dimensional Euclidean space.*

Proof Let M be an n-dimensional compact smooth manifold. Since M is compact, there exists a finite open covering $\mathcal{V} = \{V_1, V_2, \ldots, V_n\}$ of M such that each open set V_i can be embedded in $\mathbf{R}^k$ for some positive integer k by embedding

$$f_i : V_i \rightarrow \mathbf{R}^k : \; for\ i = 1, 2, \ldots, n.$$

Since by hypothesis, M is Hausdorff and compact, it is also normal. Then there is a partition of unity $\mathcal{P} = \{g_1, g_2, \ldots, g_n\}$ subordinate to this covering $\mathcal{V}$. Let $M_i = \ support\ of\ g_i,\ for\ 1, 2, \ldots, n$, which is a closed set in M. Consider the map

$$\psi_i : M \rightarrow \mathbf{R}^k : x \mapsto \begin{cases} g_i(x).f_i(x), & \text{if } x \in V_i \\ 0, & \text{if } x \in M - V_i, \end{cases}$$

for $i = 1, 2, \ldots, n$, where the right-hand product is product of the vector $f_i(x)$ with the scalar which is the real number $g_i(x)$. Each map ψ_i is well-defined by its defining conditions. Clearly, each ψ_i is continuous. Construct the continuous map

$$H : M \rightarrow \mathbf{R} \times \mathbf{R} \times \cdots \times \mathbf{R}\ (n\ times) \times \mathbf{R}^k \times \mathbf{R}^k \times \cdots \times \mathbf{R}^k\ (n\ times),$$

which maps

$$x\ to\ (g_1(x), g_2(x), \ldots, g_n(x), \psi_1(x), \psi_2(x), \ldots, \psi_n(x)),$$

i.e.,

$$H(x) = (g_1(x), g_2(x), \ldots, g_n(x), \psi_1(x), \psi_2(x), \ldots, \psi_n(x)), \ \forall\, x \in M.$$

Injectivity of H follows by the compactness property of M. To prove it, let $H(x) = H(y)$. Then by construction of H, it follows that $g_i(x) = g_i(y)$ and $\psi_i(x) = \psi_i(y)$ for each $i = i = 1, 2, \ldots, n$. Again, $\Sigma g_i(x) = 1$ asserts that $g_i(x) > 0$ for some index m say. This implies that $g_i(y) > 0$ for the same index m and $x, y \in V_m$. Now, $g_m(x) \cdot f_m(x) = \psi_m(x) = \psi_m(y) = g_m(y) \cdot f_m(y)$. Since $g_m(x) = g_m(y)$ is a positive real number, it follows that $f_m(x) = f_m(y)$. Finally, since the map $f_m : V_i \rightarrow \mathbf{R}^k$ is injective, it follows that $x = y$. This proves that

$$H : M \rightarrow \mathbf{R}^k$$

is an embedding for some positive integer k. ❑

Proposition 3.14.21 *Every compact Hausdorff locally Euclidean space M is homeomorphic to a subspace of some Euclidean space.*

Proof Proceed as in the proof of Theorem 3.14.20. ❑

Remark 3.14.22 Theorem 3.14.20 is a weak form of the theorem that every compact n-manifold can be embedded in $\mathbf{R}^{2n+1}$. It appears from the proof of Theorem 3.14.20 that $k > 2n + 1$.

Proposition 3.14.23 (Chain rule) *Let M, N and P be three smooth manifolds. If $f : M \to N$ and $h : N \to P$ be two smooth maps and $U \subset M$ is open, then*

$$d(h \circ f)_x = dh_{f(x)} \circ df_x, \; \forall\, x \in U.$$

Proof It follows from Definition $df_x : T_x(M) \to T_x(N)$. ❑

An alternative form of the smooth Urysohn Lemma 3.14.24 is given in Exercise 28 of Sect. 3.27.

Lemma 3.14.24 (Smooth Urysohn Lemma) *Let M be a smooth manifold, U be an open set in M and A be a closed set such that $A \subset U \subset M$. Then there exists a smooth function $f : M \to \mathbf{R}$ such that*

(i) $f(x) = 1, \; \forall\, x \in A$;
(ii) $0 \leq f(x) \leq 1 \; \forall\, x \in M$ *and*
(iii) *support of $f \subset U$.*

Proof Suppose that $U_1 = U$ and $U_2 = M - A$. Then $\{U_1, U_2\}$ forms an open covering of M. If $\psi_1 : U_1 \to \mathbf{R}$ and $\psi_2 : U_2 \to \mathbf{R}$ form a smooth partition of unity subordinate to the open covering $\{U_1, U_2\}$, then ψ_1 can be extended over M by the zero function outside U_1. It determines a smooth function f satisfying the required properties of f. ❑

Theorem 3.14.25 (Whitney's weak embedding theorem) *Let M be a compact smooth n-dimensional manifold. Then there is an embedding*

$$f : M \to \mathbf{R}^k, \; \textit{where } k = m(n+1) \textit{ for some integer } m > 0.$$

Proof By hypothesis, M is a compact smooth manifold. Then there exists an open finite covering of M having coordinate charts $\{(\psi_i, U_i) : i = 1, 2, \ldots, m\}$ and $\{(\psi_i, V_i) : i = 1, 2, \ldots, m\}$ such that $\overline{V}_i \subset U_i, \; \forall\, i$. Moreover, by the smooth Urysohn Lemma 3.14.24, there exists a smooth function

$$\lambda_i : M \to \mathbf{R}$$

such that

(i) $\lambda_i | \overline{V_i} = 1$ in the sense that $\lambda_i(x) = 1, \ \forall x \in \overline{V_i}$ and
(ii) support $\lambda_i \subset U_i : i = 1, 2, \ldots, m$.

Define a map

$$\psi_i : M \to \mathbf{R}^k : x \mapsto \begin{cases} \lambda_i(x).\psi_i(x), & \text{if } x \in U_i \\ 0, & \text{if } x \in M - U_i. \end{cases}$$

Define another map

$$f : M \to \mathbf{R}^{m(n+1)}, \ x \mapsto (\psi_1(x), \psi_2(x), \ldots, \psi_m(x), \lambda_1(x), \lambda_2(x), \ldots, \lambda_m(x)).$$

Then the Jacobian matrix $J(f)$ has rank n at every point $x \in M$. Moreover, f is injective. This shows that f is an injective immersion of the compact space M to a Hausdorff space $f(M)$ and hence it follows that f is a homeomorphism from M onto its image $f(M)$. It proves that f is an embedding. ❑

3.15 Vector Bundles Over Smooth Manifolds and Their Homotopy Properties

This section is devoted to the study of vector bundles over smooth manifolds and their homotopy properties. This theory provides a convenient language to describe many ideas in manifolds. A vector bundle is a bundle with an additional vector space structure on each of its fibers. H. Whitney gave the first general definition of fiber of a bundle over a base space. The concept of a bundle is very important in topology. It is a union of fibers parameterized by its base space and glued together by the topology of the total space. A fiber bundle is a topological space which likewise looks locally as a direct product of two topological spaces.

3.15.1 Vector Bundles on Smooth Manifolds

The work of H. Whitney, H. Hopf and E. Stienfel established the importance of fiber bundles for applications of topology to geometry around 1940. Since then, this topic has created general interest for its finest application to other fields such as general relativity and gauge theories and has promised many more. It also makes a return of algebraic topology to its origin and revitalized this topic from its origin in the study of classical manifolds (see **Basic Topology, Volume 3** of the present series of books).

This subsection introduces the concept of vector bundles over smooth manifolds and conveys the similarity between the concepts of vector bundles and smooth manifolds.

Definition 3.15.1 formulates the concept of a vector bundle over **R**. All the definitions of vector bundles on smooth manifolds over fields $\mathbf{R} and \mathbf{C}$ and the division ring **H** are similar. Let **F** represent either **R**, **C** or **H**.

Definition 3.15.1 (*Vector bundle on a smooth manifold*) A (real) vector bundle $\xi = (X, p, B)$ of dimension n is a triple consisting of a pair of smooth manifolds X and B with a smooth surjective map

$$p : X \to B$$

satisfying the following conditions:

1. **($\mathbf{VB_1}$)** For every $b \in B$, the inverse image $X_b = p^{-1}(b)$ is an n-dimensional vector space over **R**;
2. **($\mathbf{VB_2}$)** For every $b \in B$, there exists an open nbd U of b and a diffeomorphism

$$\psi : U \times \mathbf{R}^n \to p^{-1}(U)$$

such that

(i) the triangle in Fig. 3.11 is commutative, where

$$p_1 : U \times \mathbf{R}^n \to U \ (y, v) \mapsto y$$

is the projection map onto the first factor;

(ii) for every $y \in U$, there exists a linear isomorphism

$$\psi_y : \mathbf{R}^n \to p^{-1}(y), \ v \mapsto \psi(y, v).$$

There is a natural question: does there exist any similarity between the concepts of a vector bundle and a manifold. Remark 3.15.2 gives its answer.

Remark 3.15.2 Similarity between a vector bundle and a manifold.

(i) A vector bundle is analogous to a manifold in the sense that both of them are built up from elementary objects by gluing together by specified maps. For

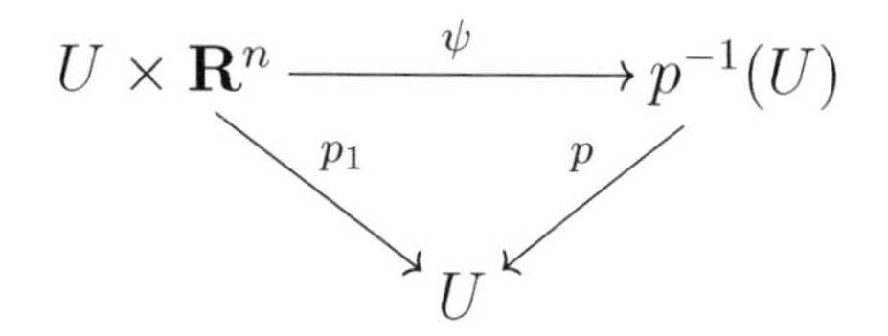

Fig. 3.11 Commutative diagram of vector bundles over smooth manifolds

example, the elementary objects for vector bundle are trivial bundles $U \times \mathbf{R}^n$ and the gluing maps are morphisms of the form

$$\psi : U \times \mathbf{R}^n \to U \times \mathbf{R}^n, (x, y) \mapsto (x, g(x)) \text{ for some } g : U \to GL(n, \mathbf{R}).$$

(ii) On the other hand, the elementary objects for manifolds are open subsets of $\mathbf{R}^n$ and the gluing maps are homeomorphisms.

(iii) In more general, a vector bundle over a topological space B consists of a family $\{X_b\}_{b\in B}$ of disjoint vector spaces parametrized by the space B. The union $X = \bigcup_{b\in B} X_b$ of these vector spaces is a space X and the map

$$p : X \to B, X_b \mapsto b$$

is continuous and it is locally trivial in the sense that X looks locally like the product $U \times \mathbf{R}^n$.

(iv) On the other hand, given a manifold M of dimension n and an open covering $\{U_i : i \in \mathbf{A}\}$, a vector bundle of over M is constructed by gluing together a family of product bundles $\{U_i \times \mathbf{R}^n\}_{i\in\mathbf{A}}$ by an action of the group of $GL(n, \mathbf{R})$ on $\mathbf{R}^n$.

Definition 3.15.3 Let $\xi = (X, p, M)$ be a vector bundle over a smooth manifold M.

(i) The smooth manifold X is called the **total space**,

(ii) M is called the **base space** of ξ and $p : X \to M$ is called its **projection**,

(iii) For any $b \in M$, the inverse image $X_b = p^{-1}(b)$ is called the **fiber over** b,

(iv) On the other hand, $\mathbf{VB_2}$ in Definition 3.15.1 is called the **local triviality** condition and the pair (ψ, U) is called a **vector bundle chart** and

(v) U is called a trivializing open set. The family $\{\psi_i, U_i)\}$ of vector bundle charts is called a **vector bundle atlas**.

Remark 3.15.4 Definition 3.15.1 asserts that a vector bundle of dimension n over a smooth manifold B is a product bundle $B \times \mathbf{R}^n$ given by gluing together a family of product bundles $\{U_i \times \mathbf{R}^n\}$, where $\{U_i\}$ is an open covering of B and is obtained by an action of the linear group $GL(n, \mathbf{R})$ on $\mathbf{R}^n$. The dimension of a vector bundle $\xi = (X, p, B, \mathbf{F}^n)$ is actually the dimension of its fibers (instead of dimension of the manifolds X or B). It is well-defined, because, for any fiber F_x over $x \in B$, the map

$$\psi : B \to \mathbf{R}, x \mapsto dim\ \mathbf{F}_x$$

is locally constant and it is constant on each component of B. If ψ is constant on the entire B, then $dim\ \mathbf{F}_x$ is constant for all $x \in B$ and this value is the dimension of ξ.

Example 3.15.5 Given any topological space B, the trivial bundle $(B \times \mathbf{F}^n, p, B, \mathbf{F}^n)$ is an n-dimensional $\mathbf{F}$-vector bundle.

Example 3.15.6 $G_r(\mathbf{F}^n)$ be the Grassmann manifold of r-dimensional subspaces of $\mathbf{F}^n$. Define

$$X = \{(V, y) \in G_r(F^n) \times F^n : p(V, y) \text{ is the orthogonal projection of } y \text{ into } V\}.$$

Then the $\gamma_r^n = (X, p, G_r(\mathbf{F}^n), \mathbf{F}^n)$ is an n-dimensional F-vector bundle.

3.15.2 The Whitney Sum of Vector Bundles Over a Smooth Manifold

Definition 3.15.7 Let $\xi_1 = (X_1, p_1, M)$ and $\xi_2 = (X_2, p_2, M)$ be two vector bundles over the same manifold M. Define the total space $X_1 \oplus X_2$ and the map $p : X_1 \oplus X_2 \to M$ as follows:

(i) $X_1 \oplus X_2 = \{(x_1, x_2) \in X_1 \times X_2 : p_1(x_1) = p_2(x_2)\}$ and
(ii) $p : X_1 \oplus X_2 \to M, (x_1, x_2) \mapsto p_1(x_1) = p_2(x_2)$.

Then $(X_1 \oplus X_2, p, B)$ is vector bundle over the manifold M with fiber over $b \in M$, the direct sum $p_1^{-1}(b) \oplus p_2^{-1}(b)$. The vector bundle $(X_1 \oplus X_2, p, B)$ is called the **Whitney sum** of the vector bundles ξ_1 and ξ_2 and is denoted by $\xi_1 \oplus \xi_2$.

3.15.3 Equivalence of Vector Bundles

This subsection studies equivalence of vector bundles and proves in Theorem 3.15.10 that a homomorphism $f : \xi \to \xi_1$ between two vector bundles is a bundle equivalence iff f is an isomorphism in the sense of Definition 3.15.9.

Definition 3.15.8 Let $\xi = (X, p, M)$ and $\xi_1 = (X_1, p_1, M_1)$ be two vector bundles over smooth manifolds. Then a **vector bundle homomorphism** or a **morphism** or a **bundle map** $(f, g) : \xi \to \xi_1$ consists of a pair of smooth maps

$$f : X \to X_1 \text{ and } g : M \to M_1$$

such that

(i) the rectangular diagram in Fig. 3.12 is commutative in the sense that $p_1 \circ f = g \circ p$;
(ii) the restriction f_b of f to each fiber $p^{-1}(b)$

$$f_b = f|p^{-1}(b) : X_b \to p_1^{-1}(g(b)) = (X_1)_{g(b)}$$

is a linear map for each $b \in M$, where X_b denotes the fiber $p^{-1}(b)$ over $b \in B$ and $(X_1)_{g(b)}$ denotes the fiber $p_1^{-1}(g(b))$ over $g(b)$ in ξ_1;

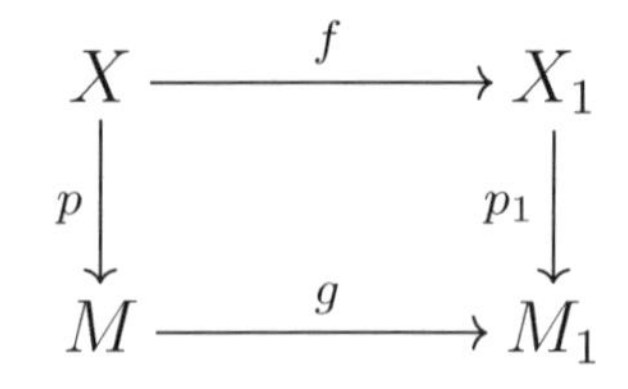

Fig. 3.12 Morphism of vector bundles over manifolds M_1 and M_2

(iii) in particular, for $M_1 = M$, a vector bundle homomorphism over the same base space M is a pair of maps $(f, 1_M) : (X, p, M) \to (X_1, p_1, M)$ satisfying the above condition, sometimes denoted by $f : \xi \to \xi_1$.

Definition 3.15.9 Let $\xi = (X, p, M)$ and $\xi_1 = (X_1, p_1, M)$ be two vector bundles over the same smooth manifold M.

(i) A vector bundle homomorphism $f : \xi \to \xi_1$ over the same base space M is said to be a **monomorphism** or an **epimorphism** according as each $f_b = f|p^{-1}(b)$ is a monomorphism or an epimorphism of linear map for each $b \in M$.
(ii) In particular, a vector bundle homomorphism $f : \xi \to \xi_1$ is said to be a vector **bundle equivalence** if $f_b = f|p^{-1}(b)$ is an isomorphism and it is also a diffeomorphism. In this case, $f_b = f|_{X_b}$ is a linear isomorphism and its inverse is the restriction to $f^{-1}|_{(X_1)_b}$. The equivalence is denoted by $\xi \approx \xi_1$.

Theorem 3.15.10 characterizes bundle equivalence over a smooth manifold in terms of isomorphism of bundles.

Theorem 3.15.10 *Let $\xi = (X, p, M)$ and $\xi_1 = (X_1, p_1, M)$ be two vector bundles over the same smooth manifold M. Then a homomorphism $f : \xi \to \xi_1$ is a bundle equivalence iff f is an isomorphism.*

Proof To prove the theorem, it is sufficient to show that a smooth map $f : X \to X_1$, which maps every fiber isomorphically onto a fiber, is a diffeomorphism. Consider a map

$$g : X_1 \to X,\ x \mapsto f_b^{-1}(x) : p_1(x) = b.$$

We claim that g is a smooth map. Consider a vector bundle chart for X

$$\psi : U \times \mathbf{R}^n \to p^{-1}(U)$$

and a vector bundle chart for X_1

$$\psi_1 : U \times \mathbf{R}^n \to p_1^{-1}(U)$$

corresponding to a common trivializing open set $U \subset M$. Then the map $\psi_1^{-1} \circ f \circ \psi$ takes the form

$$\psi_1^{-1} \circ f \circ \psi : U \times \mathbf{R}^n \to U \times \mathbf{R}^n, (b, v) \mapsto (b, k(b)v)$$

for some $k : U \to \mathcal{T}(\mathbf{R}^n, \mathbf{R}^n)$ (set of all linear transformation $T : \mathbf{R}^n \to \mathbf{R}^n$). This implies that the map

$$f : p^{-1}(U) \to p_1^{-1}(U) \;\text{ is smooth over }\; U \Leftrightarrow k \;\text{ is smooth.}$$

Again, f is an isomorphism on each fiber iff $Im\ K \subset GL(n, \mathbf{R})$. Moreover, if k is smooth, then

$$k^{-1} : U \mapsto GL(n, \mathbf{R}), b \mapsto k(b)^{-1}$$

is also smooth, because the map

$$h : GL(n, \mathbf{R}) \to GL(n, \mathbf{R}), A \mapsto A^{-1}$$

is smooth. From the above discussion, it follows that $f|_{p^{-1}}(U)$ is smooth $\Longrightarrow$ $\psi_1^{-1} \circ f \circ \psi$ is smooth $\Longrightarrow k$ is smooth $\Longrightarrow k^{-1}$ is smooth $\Longrightarrow \psi^{-1}$ $\circ\, g \circ \psi_1$ is smooth $\Longrightarrow g|_{p_1^{-1}}(U)$ is smooth. This asserts that g is smooth, because it is true for every common trivializing open set U. ❑

Definition 3.15.11 A vector bundle ξ is said to be **trivial** if it is equivalent to a product bundle.

Example 3.15.12 Let $\xi_k = (S^n \times \mathbf{R}^k, p, S^n)$ be the product bundle. Then there is a bundle equivalence

$$\mathcal{T}(S^n) \oplus \xi_1 \approx \xi_{n+1},$$

because the map

$$f : \mathcal{T}(S^n) \oplus \xi_1 \to S^n \times \mathbf{R}^{n+1}, (x, (v, \alpha)) \mapsto (x, \alpha x + v), v \in \mathcal{T}(S^n)_x,\ \alpha \in \mathbf{R}$$

is a bundle isomorphism with its inverse

$$g : S^n \times \mathbf{R}^{n+1} \to \mathcal{T}(S^n) \oplus \xi_1,\ (x, v) \mapsto (x, (v - \langle x, v\rangle \cdot x, \langle x, v\rangle)), v \in \mathbf{R}^{n+1}.$$

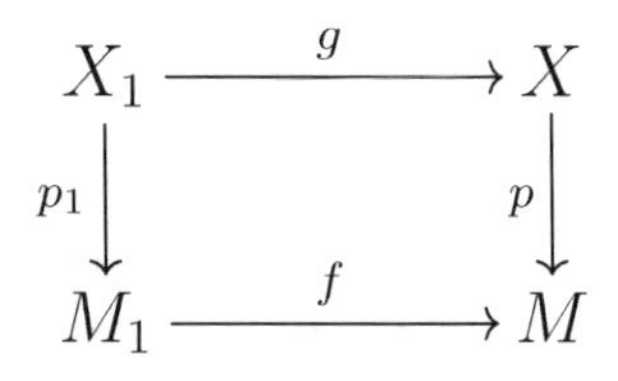

Fig. 3.13 Morphism of vector bundles over smooth manifolds

Proposition 3.15.13 *The projection map*

$$p : p^{-1}(U) \to U$$

given in Definition 3.15.1 is a submersion.

Proof It follows from commutativity of triangle in Fig. 3.11 that $p = p_1 \circ \psi^{-1}$ holds locally. Since ψ^{-1} is a diffeomorphism and p_1 is a submersion and hence p is a submersion. ❑

Proposition 3.15.14 *Let $\xi = (X, p, M)$ be a vector bundle over a smooth manifold M and M_1 be a smooth manifold. If $f : M_1 \to M$ is a smooth map, then there is a vector bundle $\xi_1 = (X_1, p_1, M_1)$, uniquely determined up to equivalence, and a bundle morphism*

$$(g, f) : \xi_1 \to \xi$$

such that for every $b \in M_1$, the map

$$g_b : p^{-1}(b) \to p^{-1}(f(b))$$

is a linear isomorphism.

Proof Under the given condition, define

$$X_1 = \{(b, x) \in M_1 \times X : f(b) = p(x)$$

and

$$p_1 : X_1 \to B_1, (b, x) \mapsto b \text{ and } g : X_1 \to X, \ (b, x) \mapsto x.$$

Consider the commutative diagram in Fig. 3.13.

For local triviality of ξ_1, take a vector bundle chart for ξ

$$\psi : U \times \mathbf{R}^n \to p^{-1}(U).$$

Define

$$\psi_1 : f^{-1}(U) \times \mathbf{R}^n \to p_1^{-1}(f(U)), (b, v) \mapsto (b, \psi(f(b), v).$$

Then ψ_1 is a vector bundle chart for ξ_1 with its inverse formulated by

$$\psi_1^{-1}(b, x) = (b, p_2 \circ \psi^{-1}(x)),$$

where $b \in f^{-1}(U) and\ x \in p^{-1}(U)$ so that $f(x) = p(x)$ and p_2 is the projection map onto the second factor. This implies that g is a diffeomorphism and also an isomorphism on every fiber. ❑

Definition 3.15.15 The bundle ξ_1 formulated in Proposition 3.15.14 is called the **induced bundle**, induced from the bundle ξ by f and is denoted by $f^*\xi$ or by $f^*(\xi)$. The induced bundle $f^*(\xi)$ is also called the **pull-back** of ξ by f. The pair of maps (g, f) formulated in the same definition is called the **canonical bundle map** of the induced map $f^*(\xi)$.

Definition 3.15.16 Let $\xi = (X, p, M)$ be a vector bundle over a smooth manifold M. Then $\xi_1 = (X_1, p_1, M)$ is said to be a **subbundle** of ξ if X_1 is a submanifold of X and $p_1 = p|_{X_1}$. On the other hand, if M_1 be a submanifold of M, the **restricted bundle** of ξ to M_1, denoted by $\xi|_{M_1}$, is bundle (X_1, p_1, M_1), where $X_1 = p^{-1}(M_1)$ and $p_1 = p|_{X_1}$.

Proposition 3.15.17 *Let $\xi = (X, p, M)$ be a vector bundle over a smooth manifold M.*

- **(i)** *If M_1 be a submanifold of M and $f : M_1 \hookrightarrow M$ is the inclusion map, then the restricted vector bundle $\xi|_{M_1}$ and the induced vector bundle $f^*\xi$ are isomorphic.*
- **(ii)** *If $f_1 : M_1 \to M$ and $f_2 : M_2 \to M_1$ are smooth maps, then the induced bundles $(f_1 \circ f_2)^*(\xi)$ and $f_2^*(f_1^*(\xi))$ are isomorphic.*
- **(iii)** *If $1_d : M \to M$ is the identity map on M, then its induced bundle $(1_M)^*(\xi)$ is isomorphic to ξ.*

Proof It follows from the definition of induced vector bundle. ❑

Proposition 3.15.18 *Let $\xi_1 = (X_1, p_1, M_1)$ and $\xi_2 = (X_2, p_2, M_2)$ be two vector bundles over smooth manifolds. Then every vector bundle morphism*

$$(g, f) : \xi_1 \to \xi_2$$

can be expressed as a product

$$(g, f) = (h, f) \circ (k, 1_{M_1}) : g = h \circ k,$$

where $(k, 1_{M_1})$ is a homomorphism and (h, f) is the canonical bundle map of the induced bundle $f^*(\xi_2)$.

Proof Let $f^*(\xi_2) = (X, p, M_1)$. Consider the map

$$h : X \to X_2, (b, x) \mapsto x \quad \text{and} \quad k : X_1 \to X : x \mapsto (p_1(x), g(x)).$$

Then $Im\ K \subset X$, because $f \circ p_1 = p_2 \circ g$. This implies that k is linear on every fiber and $g = h \circ k$. ❑

Proposition 3.15.19 *Given any smooth manifold M, let N be submanifold of M such that N is a closed subset of M. If $\xi = (X, p, M$ is a vector bundle over M, then every smooth section s of the restricted bundle $\xi|N$ can be extended to a smooth section of ξ.*

Proposition 3.15.20 *Let $\xi = (X, p, M)$ and $\xi' = (X', p', M)$ be two vector bundles of same dimension over a smooth manifold M and N be a closed submanifold of M. Then every isomorphism*

$$\psi : \xi|N \to \xi'|N$$

has an extension to an isomorphism

$$\tilde{\psi} : \xi|U \to \xi'|U$$

over an open nbd U of N.

3.15.4 Homotopy of Smooth Maps and Homotopy Property of Vector Bundles

This subsection communicates the concept of smooth homotopy between smooth maps. Its more properties are studied in Sect. 3.19.1. It also studies homotopy properties of vector bundles and proves that smooth homotopic maps induce the same vector bundle up to equivalence in Theorem 3.15.23. More results on smooth homotopy are given in Exercises 74 and 75 of Sect. 3.27.

We now use Definition 3.15.8 of a bundle map for smooth manifolds.

Definition 3.15.21 Let $f, g : M \to N$ be two smooth maps. They are said to be **smoothly homotopic** if there exists a smooth map

$$F : M \times \mathbf{R} \to N : F(x, 0) = f(x) \text{ and } F(x, 1) = g(x), \forall x \in M.$$

This gives rise to a family of smooth maps

$$F_t : M \to N, x \mapsto F(x, t), \forall t \in \mathbf{R}.$$

The smooth map F is called a **smooth homotopy** between smooth maps f and g. In particular, if the map F is just continuous, the maps f and g are said to be continuously homotopic or simply homotopic.

Remark 3.15.22 Geometrically, two smooth maps are smoothly homotopic if one can be deformed to the other through smooth maps. For a continuously homotopy F between two continuous maps $f, g : M \to N$, the map F is taken as a continuous map from $M \times \mathbf{I} \to N$; on the other hand, for a smooth homotopy F between smooth maps $f, g : M \to N$, the map F is taken as a smooth map from $M \times \mathbf{R} \to N$. The technical reason is that if the manifold M has a boundary, then the product space $M \times \mathbf{I}$ is not a smooth manifold. For a detailed study of homotopic maps between topological spaces, **Basic Topology, Volume 3** of the present book series is referred.

Theorem 3.15.23 proves that smooth homotopic maps induce the same vector bundle up to equivalence.

Theorem 3.15.23 *Let $f, g : M \to N$ be two smooth homotopic maps and ξ be a vector bundle over M. Then their induced vector bundles $f^*(\xi) \cong g^*(\xi)$.*

Proof By hypothesis, $f, g : M \to N$ be two smooth homotopic maps. Let $F : M \times \mathbf{R} \to N$ be a smooth homotopy between f and g and $p : M \times \mathbf{R} \to M$ be the projection map. Then there is an isomorphism between vector bundles $F^*(\xi)$ and $p^* F_t^*(\xi)$ over the closed subinterval $M \times \{t\}$, because $F = F_t \circ p$ on $M \times \{t\}$, which is closed in the product topology. Hence, by Exercise 75 of Sect. 3.27, there exists an isomorphism

$$\psi : F^*(\xi) \to p^* F_t^*(\xi)$$

over some vertical strip $M \times (t - \delta, t + \delta)$. This implies that the isomorphism class of $F_t^*(\xi)$ is a locally constant function of t and it is constant, because the real line space $\mathbf{R}$ is connected. This proves that $f^*(\xi) \cong g^*(\xi)$. ❑

Definition 3.15.24 A smooth manifold M is said to be **contractible** if the identity map $1_M : M \to M$ is homotopic to some constant map $c : M \to b_0 \in M$.

Corollary 3.15.25 *Every vector bundle over a contractible manifold is trivial.*

Proof Let $\xi = (X, p, M)$ be a vector bundle over a contractible manifold M and $b_0 \in M$. Then the identity map $1_M : M \to M$ is homotopic to the constant map $c : M \to b_0$. Hence, it follows by Theorem 3.15.23 that $f^*(\xi) \cong g^*(\xi)$. But the vector bundle $g^*(\xi)$ is trivial, because it is a vector bundle over a point. This proves that the vector bundle $\xi = (X, p, M)$ over a contractible manifold M is trivial. ❑

3.16 Tangent, Normal and Cotangent Bundle of a Smooth Manifold

This section continues the study of bundles by introducing the concepts of tangent, normal and cotangent bundle of smooth manifolds.

3.16.1 *Tangent Bundle of a Smooth Manifold*

This subsection introduces the concept of tangent bundle over a smooth manifold and studies it. If M is an n-dimensional smooth manifold and $T_x(M)$ be tangent space at $x \in M$, then

$$\mathcal{T}(M) = \bigcup_{x \in M} \{T_x(M)\}$$

(disjoint union) together with a projection $p : \mathcal{T}(M) \to M$ forms a tangent bundle of the manifold M. It is formalized in Definition 3.16.1.

Definition 3.16.1 (*The tangent bundle of a manifold*) The tangent bundle of a smooth manifold M is the bundle $(\mathcal{T}(M), p, M)$, where

(i) the total space $\mathcal{T}(M)$ is the disjoint union of all tangent spaces $T_x(M)$ as x runs over M. This is the set of all ordered pairs (x, v) such that $x \in M$ and $v \in T_x(M)$;

(ii) the map

$$p : \mathcal{T}(M) \to M, \ (x, v) \to x$$

is called the **projection map** of the tangent bundle.
$T_x(M)$ is called the **tangent plane** at x and $v \in T_x(M)$ is called the **tangent vector** with initial point x.

More precisely, let M be an n-dimensional smooth manifold, $T_x(M)$ be the vector space of all tangent vectors to M at any point $x \in M$ and $\mathcal{T}(M) = \bigcup_{x \in M} \{T_x(M)\}$ (disjoint union). Then the set $\mathcal{T}(M)$ is the disjoint union set of all tangent spaces $T_x(M)$ and there is projection map

$$p : \mathcal{T}(M) \to M, \ (x, v) \mapsto x.$$

The subspace $F_x = p^{-1}(x) = T_x(M)$ with topology inherited from M is called the **fiber** over x. If M is a smooth manifold of dimension n, then $\mathcal{T}(M)$ is a $2n$-dimensional manifold (see Theorem 3.16.3). A **cross section of the tagent bundle** $(\mathcal{T}(M), p, M)$ is a smooth map $s : M \to \mathcal{T}(M)$ such that $p \circ s = 1_M$ and hence $(p \circ s)(x) = x, \ \forall x \in M$.

Remark 3.16.2 If M is a smooth manifold of dimension n, then $\mathcal{T}(M)$ is the set of all tangent vectors (x, v) at all points $x \in M$. If p assigns to every vector (x, v) its initial point x, then $T_x(M) = p^{-1}(x) = F_x$ is the tangent plane at the point x, which is a linear space. A cross section of the tangent bundle $\mathcal{T}(M)$ of M is a **vector field** over M.

Theorem 3.16.3 gives a unique differentiable structure on the manifold $\mathcal{T}(M)$.

Theorem 3.16.3 *Let M be a smooth manifold of dimension n. Then its tangent bundle $\mathcal{T}(M)$ is a smooth manifold of dimension $2n$.*

Proof By hypothesis, M be a smooth manifold of dimension n. Consider the projection

$$p : \mathcal{T}(M) \to M, (x, v) \to x.$$

Let (ψ, U) be a chart of M. Then it determines a map

$$T_\psi : p^{-1}(U) \to \psi(U) \times \mathbf{R}^n \subset \mathbf{R}^n \times \mathbf{R}^n, (x, v) \mapsto (\psi(x), d\psi_x(v)).$$

Clearly, T_ψ is a bijective map with its inverse

$$T_\psi^{-1} : \psi(U) \times \mathbf{R}^n \to p^{-1}(U), (a, w) \mapsto (b, d\psi_x^{-1}(w), \quad \text{where } b = \psi^{-1}(a).$$

If (ψ, U) and (ϕ, V) are two compatible charts of M, then the map

$$T_\phi \circ T_\psi^{-1} : \psi(U \cap V) \times \mathbf{R}^n \to \phi(U \cap V) \times \mathbf{R}^n, (a, w) \mapsto T_\phi(b, d\psi_b^{-1}(w)) = (\phi(b), d\phi_b \circ d\psi_b^{-1}(w))$$

$$= (\phi \circ \psi^{-1})(a, d\phi_b \circ d\psi_b^{-1}(w)), \quad \text{where } b = \psi^{-1}(a).$$

Hence, it follows that $T_\phi \circ T_\psi^{-1}$ is a homeomorphism and $\mathcal{T}(M)$ has a unique topology such that

(i) each T_ψ is a homeomorphism;
(ii) $\mathcal{T}(M)$ is second countable and Hausdorff and
(iii) the projection map $p : \mathcal{T}(M) \to M, (x, v) \to x$ is smooth.

Finally, since $T_\phi \circ T_\psi^{-1}$ is a diffeomorphism , the family of charts $\{(p^{-1}(U), T_\psi)\}$ forms a smooth atlas on $\mathcal{T}(M)$. Hence, it follows that $\mathcal{T}(M)$ is a smooth manifold of dimension $2n$. ❑

Example 3.16.4 (*Tangent bundle over S^n*) Let S^n be the n-sphere. For every integer $n \geq 1$, the n-sphere S^n has the tangent bundle $\mathcal{T}(S^n) = (X, p, S^n, \mathbf{R}^n)$, where

(i) the total space X is given by

$$X = \{(x, y) \in \mathbf{R}^{n+1} \times \mathbf{R}^{n+1} : ||x|| = 1, \text{ and } \langle x, y \rangle = 0\};$$

(ii) the projection p is given by

$$p : X \to S^n, (x, y) \mapsto x.$$

It is an n-dimensional real vector bundle. To show its local triviality property, consider an open covering $\{U_i\}$ of S^n, where the open set

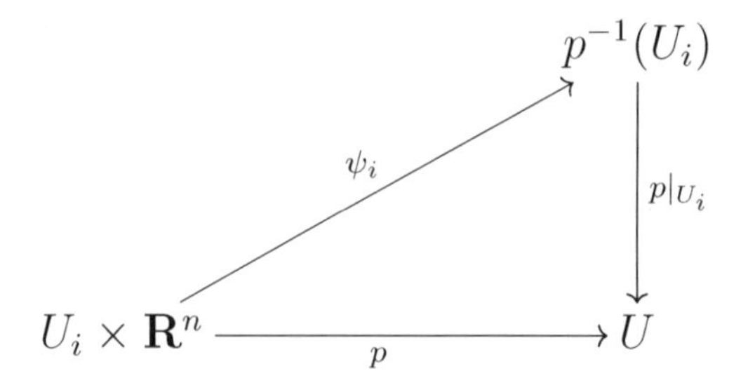

Fig. 3.14 Triangular diagram corresponding to the tangent $\mathcal{T}(S^n)$

$$U_i = \{x = (x_1, x, \ldots, x_{i-1}, x_i, x_{i+1}, \ldots, x_{n+1}) \in \mathbf{R}^{n+1} : ||x|| = 1, x_i \neq 0,\ 1 \leq i \leq n+1\} \subset S^n.$$

Define for each i

$$\psi_i : U_i \times \mathbf{R}^n \to p^{-1}(U_i),\ (x, y) \mapsto (x, f_i(y) - \langle x, f_i(y)\rangle x,$$

where

$$f_i : \mathbf{R}^n \to \mathbf{R}^{n+1},\ (y_1, y_2, \ldots, y_{i-1}, y_i, y_{i+1}, \ldots, y_n) \mapsto (y_1, y_2, \ldots, y_{i-1}, 0, y_i, \ldots, y_n).$$

This implies that ψ_i are linear maps on each fiber (having two components C^+ and C^- corresponding to $x_i > 0$ and $x_i < 0$) such that they are isomorphisms and also diffeomorphisms making the triangular diagram in Fig. 3.14 commutative (i.e., $p \circ \psi_i = p_{U_i}$)

In other words, the tangent bundle $\mathcal{T}(S^n)$ over the n-sphere S^n in the Euclidean $(n+1) - space$ $\mathbf{R}^{n+1}$ is the subbundle $(\mathcal{T}(S^n), p, S^n)$ of the product bundle $(S^n \times \mathbf{R}^{n+1}, p, S^n)$, whose total space $\mathcal{T}(S^n)$ is defined by

$$\mathcal{T}(S^n) = \{(b, x) \in S^n \times \mathbf{R}^{n+1} : \langle b, x\rangle = 0\},$$

and the projection map p is defined by

$$p : \mathcal{T}(S^n) \to S^n, (b, x) \mapsto b.$$

- **(i)** An element of the total space $\mathcal{T}(S^n)$ is said to be a **tangent vector** to S^n at the point $b \in S^n$.
- **(ii)** The fiber $p^{-1}(b) \subset \mathcal{T}(S^n)$ is a vector space of dimension n.
- **(iii)** A cross section of the tangent bundle $\mathcal{T}(\xi)$ over S^n is called a **tangent vector field (or simply vector field)** over S^n.

3.16.2 Normal Bundle Over a Manifold

This subsection defines normal bundle over a manifold M by using the concept of normal space of M at a point $x \in M$ as the orthogonal complement of its tangent space of M at the point x.

Definition 3.16.5 Given an n-dimensional submanifold M of $\mathbf{R}^m$, the normal space $\mathcal{N}_x(M)$ of M, at $x \in M$, is defined by

$$\mathcal{N}_x(M) = \{(x, v) \in M \times \mathbf{R}^m : v \perp T_x(M)\}.$$

Its normal bundle is the vector bundle $(\mathcal{N}(M), p, M)$, where the total space

$$\mathcal{N}(M) = \bigcup_{x \in M} \{\mathcal{N}_x(M)\}$$

(disjoint union) and the projection

$$p : \mathcal{N}(M) \to M, \ (x, v) \to x.$$

Remark 3.16.6 Given an n-dimensional submanifold M of $\mathbf{R}^m$, its normal bundle $(\mathcal{N}(M), p, M)$ is a vector bundle with fiber dimension $m - n$. Moreover,

(i) $\mathcal{N}(M)$ is a manifold of dimension m;
(ii) The projection

$$p : \mathcal{N}(M) \to M, \ (x, v) \to x$$

is a submersion.

Example 3.16.7 (*Normal bundle over* S^n) The normal bundle $\mathcal{N}(S^n)$ over S^n for $n \geq 1$ is the fiber bundle $\xi = (X, q, S^n, \mathbf{R}^1)$, where

$$X = \{(x, y) \in \mathbf{R}^{n+1} \times \mathbf{R}^{n+1} : ||x|| = 1, y = rx, r \in \mathbf{R}\}$$

and $q : X \to S^n, (x, y) \mapsto x$.

(i) The maps

$$\psi : S^n \times \mathbf{R}^1 \to X, (x, r) \mapsto (x, rx)$$

and

(ii)

$$\phi : X \to S^n \times \mathbf{R}^1, (x, y) \mapsto (x, \langle x, y \rangle)$$

are homeomorphisms

such that ψ is a homeomorphism with its inverse ϕ. This asserts that $\xi = N(S^n)$ is an 1-dimensional real trivial bundle.

Definition 3.16.8 The normal bundle $\mathcal{N}(\xi)$ of a bundle ξ over S^n is the subbundle $(\mathcal{N}(S^n), q, S^n)$ of the product bundle $(S^n \times \mathbf{R}^{n+1}, p, S^n)$ whose total space $\mathcal{N}(S^n)$ is defined by

$$\mathcal{N}(S^n) = \{(b, x) \in S^n \times \mathbf{R}^{n+1} : x = tb \text{ for some } t \in \mathbf{R}\}$$

and the projection q is defined by

$$q : \mathcal{N}(S^n) \to S^n, q(b, x) = b.$$

(i) An element of $\mathcal{N}(S^n)$ is called a normal vector to S^n at the point $b \in S^n$.
(ii) The fiber $q^{-1}(b) \subset \mathcal{N}(S^n)$ is a vector space of dimension 1. A cross section of the normal bundle $\mathcal{N}(\xi)$ over S^n is called a **normal vector field** on S^n.

3.16.3 Cotangent Space and Cotangent Bundle of a Smooth Manifold

This subsection communicates the concept of cotangent bundle of a smooth manifold, which is a dual concept of tangent bundle.

Definition 3.16.9 Let M be smooth manifold of dimension n. Then the dual space $T_x(M)^*$ of the vector space $T_x(M)$ at $x \in M$ is called the **cotangent space** of M at x. It consists of all linear maps $L : T_x(M) \to \mathbf{R}$ and is of dimension n. The elements of $T_x(M)^*$ are called **cotangents vectors** at the point $x \in M$.

Definition 3.16.10 Let M be an n-dimensional smooth manifold. Then

$$\mathcal{T}^*(M) = \bigcup_{x \in M} \{T_x(M)^*\}$$

(disjoint union) is called the cotangent bundle of M. It is a vector bundle over M. If (ψ, U) is a chart on U, a natural chart on $\mathcal{T}^*(M)$ is the map

$$\psi_* : \mathcal{T}^*(U) \to \psi(U) \times (\mathbf{R}^n)^* = \psi(U) \times L(\mathbf{R}^n, \mathbf{R}),\ \alpha \mapsto (\psi(x), \alpha\psi_x^{-1}) : \alpha \in M_x^* = L(M_x^*, \mathbf{R})$$

and the projection map

$$p : \mathcal{T}^*(M) \to M,\ T_x(M)^* \mapsto x.$$

Remark 3.16.11 If $f : M \to \mathbf{R}$ is a C^{r+1} map for $1 \leq r \leq \omega$, then for every $x \in M$, the linear map $T_x(f) : T_x(M) \to \mathbf{R}$ is in $(T_x(M))^*$ and it is denoted by $T_x(f) = Df_x \in (T_x(M))^*$.

3.16.4 Vector Fields and Flows on Smooth Manifolds

This section studies vector fields and flows on smooth manifolds. These two concepts are closely related, because every flow on a smooth manifold M of dimension n generates a smooth vector field obtained by assigning to a point $x \in M$, the vector $\alpha(x) \in T_x(M)$ tangent to the curve $\gamma : \mathbf{I} \to M$ and conversely, corresponding to a smooth vector field on M, a coordinate chart in the manifold M can be split into its n-coordinates given by the n functions X_i of n variables $y_1, y_2, \ldots, y_n$:

$$\{X_i(y_1, y_2, \ldots, y_n) : i = 1, 2, \ldots, n.\}$$

Definition 3.16.12 Let M be a smooth manifold. A (smooth) **flow** on M is a smooth map

$$\alpha : \mathbf{R} \times M \to M$$

such that

(i) $\alpha(0, x) = x, \ \ \forall x \in M$ and
(ii) $\alpha(t + s, x) = \alpha(t, \alpha(s, x)) \ \forall x \in M, t, s \in \mathbf{R}$.

Remark 3.16.13 A flow α on M is the same as the action of the topological group (additive) $\mathbf{R}$ on the manifold on M with the condition of smoothness (in addition to conditions (i) and (ii) of Definition 3.16.12).

The next question naturally arises: what is a vector field? Let (ψ, U) be a coordinate chart in a smooth manifold M of dimension n with $\psi = (x_1, x_2, \ldots, x_n)$. Then the map

$$X : U \to T(U), p \mapsto \left[\frac{\partial}{\partial x_i}\right]_p$$

is called a smooth vector field on U. The tangent vectors $\{\frac{\partial}{\partial x_i}]_p\}$ are linearly independent at every point $p \in U$. If X is vector field on U, then X can be represented as

$$X = \Sigma_{i=1}^n X x_i \frac{\partial}{\partial x_i}.$$

Denoting $Xx_i = X_i(or\ a_i)$, the functions X_i are called the component of X for $i = 1, 2, \ldots, n$. This idea leads to define vector field on a smooth manifold in Definition 3.16.14.

Definition 3.16.14 Let M be a smooth manifold of dimension n. A **vector field** on M is a smooth map X such that given local coordinates $x_1, x_2, \ldots, x_n$ near a point $p \in M$,

$$X : M \to \mathcal{T}(M),\ p \mapsto \Sigma_{i=1}^{n} X_i(p)\frac{\partial}{\partial x_i},$$

i.e., the value of X at $p \in M$is a tangent vector $X_p \in T_p(M)$. The map X is said to be a **smooth vector field** if every component function $X_i(p)$ of X is smooth.

Remark 3.16.15 Let $C^{\infty}(M)$ denote the set of all real-valued smooth maps on a smooth manifold M. Since usual addition, product of smooth maps and constant multiples of smooth maps are also smooth, $C^{\infty}(M)$ forms an algebra over **R**. Hence, it follows from Definition 3.16.14 that a smooth vector field X on M is a smooth map

$$X : C^{\infty}(M) \to C^{\infty}(M),\ f \mapsto Xf.$$

such that

(i) $X(rf + sg) = rXf + sXg,\ \forall f, g \in C^{\infty}(M). \forall r, s \in \mathbf{R}$ and
(ii) $X(fg) = f(Xg) + (Xf)g\ \forall f, g \in C^{\infty}(M). \forall r, s \in \mathbf{R}$.

Proposition 3.16.16 *Let $\mathcal{S}(M)$ be the set of all smooth vector fields on M. Then it forms a module over the ring $C^{\infty}(M)$ under the operations*

(i) $(X + Y)f = Xf + Yfm\ \forall X, Y \in \mathcal{S}(M)$ *and* $\forall f \in C^{\infty}(M)$. *and*
(ii) $(fX)g = f(Xg),\ \forall X \in \mathcal{S}(M)$ *and* $\forall f, g \in C^{\infty}(M)$.

Proof It follows from Definition 3.16.14 and Remark 3.16.15. ❑

3.17 Regular Values, Sard's, Brown's and Morse–Sard Theorems

This section extends the concepts of regular and critical values defined in Sect. 3.3.4 for smooth maps between smooth manifolds of same dimensions by introducing the concept of regular values of a smooth map in the general case (for smooth manifolds of different dimensions). It provides useful tools for the study of smooth manifolds from the viewpoint of topology. Sard's Theorem 3.17.20 is an important theorem which asserts that if M is a smooth manifold of dimension m and $\psi : M \to \mathbf{R}^n$ is a smooth map, then the set of critical values of ψ is of measure zero. This section also proves the Brown Theorem as a Corollary 3.17.25 and Sard's theorem in general form or Morse–Sard Theorem 3.17.27 for C^r-maps.

Definition 3.17.1 Let M be a smooth manifold of dimension m, N be a smooth manifold of dimension n and $\psi : M \to N$ be a smooth map. Then a point $x \in M$ is said to

(i) **a critical point** of ψ if its induced map

$$\psi_* = D\psi_x : T_x(M) \to T_{f(x)}(N)$$

has rank less than n;

(ii) **a regular point** of ψ if its induced map

$$\psi_* = D\psi_x : T_x(M) \to T_{f(x)}(N)$$

is onto.

The image of a critical point of ψ in N is called a **critical value**. Otherwise, a point of N which is not a critical value is called a **regular value**.

Remark 3.17.2 Let M be a smooth manifold of dimension m, N be a smooth manifold of dimension n and $f : M \to N$ be a smooth map. Then it follows from Definition 3.17.1 of a regular value in N that a point $y \in N$ is a regular value if

(i) $m \geq n$ asserts that $\psi_* : T_x(M) \to T_{f(x)}(N)$ is onto $\forall\, x \in \psi^{-1}(y)$ and
(ii) $m < n$ asserts that y is not an element of image of ψ.

Hence, it follows that

(i) if the Jacobian matrix $J(f)$ at a point $x \in M$ of the smooth map f has rank $< n$, then $f(x)$ is a critical value of the map f and
(ii) if a point $y \in N$ is a regular value of the smooth map f, it is not a critical value of f.
(iii) By convention, a point of $y \in N$ is a regular value of f, if $y \notin f(M)$. Consequently, if $n > m$, then every point $x \in M$ is a critical point of f and if $n \leq m$ and a point $y \in f(M)$ is regular value of f, then the rank of its Jacobian matrix $J(f)$ is n at every point $x \in f^{-1}(y)$.

Example 3.17.3 Consider the map

$$\psi : \mathbf{R}^n \to \mathbf{R}, x = (x_1, x_2, \ldots, x_n) \mapsto x_1^2 + x_2^2 + \cdots + x_n^2.$$

Then

$$\psi_* = D\psi_x = grad\ \psi(x) = (2x_1, 2x_2, \ldots, 2x_n)$$

asserts that every point $a(\neq 0) \in \mathbf{R}$ is a regular value of ψ. For $a > 0$, its inverse image set

$$\psi^{-1}(a) = \{x \in \mathbf{R}^n : ||x|| = a\}$$

is a smooth manifold of dimension $n - 1$.

Remark 3.17.4 provides an alternative form of critical points, regular points and regular values of a smooth map.

Remark 3.17.4 Let $f : M \to N$ be a smooth map.

(i) A point $x \in M$ is a **critical point** of f if f is not a submersion at x.

(ii) The points of M other than critical points of f are **regular points** of f.
(iii) A point $y \in N$ is a **regular value** of f if $f^{-1}(y)$ contains at least one critical point.
(iv) The points of N other than regular values $y \in N$ of f (including the points of y for which $f^{-1}(y) = \emptyset$) are regular values of f.

Definition 3.17.5 Given $A \subset \mathbf{R}^n$, define the characteristic function

$$\kappa_A : \mathbf{R}^n \to \mathbf{R}, x \mapsto \begin{cases} 1, & \text{if } x \in A \\ 0, & \text{otherwise.} \end{cases}$$

If the function κ_A is integrable, then the number

$$\int_A \kappa_A(x)dx$$

is said to be the **volume Vol** (A), and it is also sometimes denoted by $V(A)$. If in particular, Vol $(A) = 0$, then A is said be a set of volume 0.

Example 3.17.6

(i) If $A \subset \mathbf{R}$, then Vol $(A) =$ length of A.
(ii) If $A \subset \mathbf{R}^2$, then Vol $(A) =$ area of A.
(iii) If $A \subset \mathbf{R}^n$ is an n-cube of length $r > 0$, then Vol $(A) =$ volume of $A = r^n$.

Definition 3.17.7 A subset $A \subset \mathbf{R}^n$ is said to have **measure** 0, if given an $\epsilon > 0$, there is a countable family $\{U_n\}$ of open covering of A such that

$$\Sigma_{n=1} Vol(U_n) < \epsilon.$$

In other words, the set A has measure zero in $\mathbf{R}^n$ if for any $\epsilon > 0$, the set A can be covered by a countable family of n-dimensional cubes such that the sum of their volumes is less than ϵ.

Example 3.17.8 Given a continuous map $f : [a, b] \to \mathbf{R}$, its graph $G_f = \{(x, f(x)\} \subset \mathbf{R}^2$ has measure $zero$.

Remark 3.17.9 A set $A \subset \mathbf{R}^n$ has measure zero iff given an $\epsilon > 0$, there is an open set U such that

$$A \subset U \text{ with } Vol(U) < \epsilon.$$

The concept of measure zero given in Definition 3.17.7 is extended in Definition 3.17.10 to smooth manifolds through local parametrizations. It is used in Sard's Theorem 3.17.20, which gives in a general form of Sard's Theorem 3.17.19

Definition 3.17.10 Let M and N be two smooth manifolds and $f : M \to N$ be a smooth map. A subset $X \subset N$ is said to have **measure zero** in Y, denoted by $\mu(X) = 0$, if for every local parametrization ϕ of N, the preimage $\phi^{-1}(X)$ has measure zero in Euclidean space.

Proposition 3.17.11 *Let $A = A_1 \cup A_2 \cup \cdots$ be a countable union of sets of measure zero, then A is also a set of measure zero.*

Proof By hypothesis,

$$A = \bigcup_{n=1} A_n \ and \ A_n \subset U_n,$$

where U_n is open and $Vol(U_n) = \frac{\epsilon}{2^n}$ for all n. This implies that

$$A \subset U = \bigcup_{n=1} U_n \text{ and } Vol(U) \leq \Sigma_{n=1} Vol(U_n) < \Sigma_{n=1} \frac{\epsilon}{2^n} = \epsilon.$$

Hence, the proposition follows from Definition 3.17.7. ❑

Theorem 3.17.12 proves that the property of being a set of measure zero is invariant under a smooth map. Theorem 3.17.12 was proved by Arthur B. Brown in 1935, which was rediscovered by Dubovickil in 1953 and by Thorn in 1954.

Theorem 3.17.12 *Let the set $A \subset \mathbf{R}^n$ have measure zero. If $f : A \to \mathbf{R}^m$ is an arbitrary smooth map, then $f(A)$ is a set of measure zero in $\mathbf{R}^m$.*

Proof For any point $a \in A$, the smooth map has a smooth extension over a nbd of a in $\mathbf{R}^n$, which is also denoted by the same symbol f. By shrinking this nbd, it is assumed that f is also smooth on a closed n-ball nbd B_a with center at a. For the coordinate functions $u_1, u_2, \ldots, u_n$ in $\mathbf{R}^n$, the partial derivatives $\frac{\partial f_i}{\partial u_j}$ of the components f_i are bounded on the closed ball B_a, which is compact. By using the fundamental theorem of calculus and chain rule, it follows that for some $k > 0$,

$$||f(x) - f(y)|| < k||x - y||, \ \forall x, y \in B_a,$$

known as the **Lipschitz estimate** for the smooth function f.

Given an $\epsilon > 0$, choose a countable covering $\{U_n\}$ of $A \cap B_a$ by n-balls such that

$$\Sigma_{n=1} Vol\ (U_n) < \epsilon.$$

By using the Lipschitz estimate for the smooth function f, it follows that the image $f(B_a \cap U_n) \subset V_n$, where V_n is an n-ball having radius not exceeding k times of the radius of U_n. This implies that $f(B_a \cap U_n) \subset V_n$ is contained in some of the balls of the family $\{V_n\}$ such that its total volume does not exceed the volume

$$\Sigma_{n=1} Vol\ (V_n) < k^n \epsilon.$$

As the total volume can be made as small as we please, it follows that $f(B_a \cap A)$ has measure zero. This implies that $f(A)$ has also measure zero, since it is a union of countably many such sets. ❑

Example 3.17.13 Proposition 3.17.12 fails if we replace the condition of smoothness of f by its continuity condition only. For example, consider the subset $A = \mathbf{I} = [0, 1] \subset \mathbf{R}$. Then the set A has measure zero in $\mathbf{R}^2$. On the other hand, there exists a continuous surjective map

$$f : \mathbf{I} \to \mathbf{I}^2$$

by **Space filling curve theorem** saying that there exists a continuous surjective map

$$f : \mathbf{I} \to \mathbf{I}^2.$$

But its graph $G_f = \{(x, f(x))\} = \mathbf{I} \times \mathbf{I}$ is the unit square which is not a set of measure zero in $\mathbf{R}^2$.

Definition 3.17.14 Let M be an n-dimensional smooth manifold and $A \subset M$. Then the set A is said to have **measure zero** in M if for every coordinate chart $\psi : U \to \mathbf{R}^n$ of M, the set $\psi(A \cap U) \subset \mathbf{R}^n$ has measure zero in $\mathbf{R}^n$.

Proposition 3.17.15 *Let M be an n-dimensional smooth manifold.*

(i) *If $B \subset A \subset M$ and A has measure zero in M, then B has also measure zero in M.*

(ii) *If $\{A_n\}$ is a countable family of subsets of M such that every A_n has measure zero in M, then their set-theoretic union set*

$$\bigcup_{n=1} A_n$$

has also measure zero in M.

Proof It follows from Definition 3.17.14. ❑

Theorem 3.17.16 (Fubini) *Let X be a compact set in $\mathbf{R}^n$ such that every subset $X_t = X \cap (\{t\} \times \mathbf{R}^{n-1})$ has measure zero in the hyperplane $\mathbf{R}^{n-1}$ for each $t \in R$. Then the Lebesque measure $\mu(X) = 0$ in $\mathbf{R}^n$.*

Proof Let $\mathbf{I}^n$ be the n-cube. Suppose that $X \subset \mathbf{I}^n$. Define a function

$$f : \mathbf{I} \to \mathbf{R},\ t \mapsto \mu(X \cap ([0, t] \times \mathbf{I}^{n-1})).$$

Given any positive real number ϵ, there exists by hypothesis an open set $U \subset \mathbf{I}^n$ such that

$$X \cap (t \times \mathbf{I}^{n-1}) \subset t \times U, \text{ where } \quad \text{Vol}(U) < \epsilon.$$

By hypothesis, X is compact. Hence, there exists a $t_0 > 0$ such that

$$X \cap ([t - t_0, t + t_0] \times \mathbf{I}^{n-1}) \subset [t - t_0, t + t_0] \times U.$$

Then for any s sanctifying $0 \leq s < t_0$, the set

$$X \cap ([0, t + s] \times \mathbf{I}^{n-1} \subset (X \cap ([0, t] \times I^{n-1})) \cup ([t, t + s] \times U)$$

can be covered by an open set V with $Vol\ (V) < f(t) + \epsilon s$. Hence, for all s sanctifying $0 \leq s < t_0$, it follows that

$$f(t + s) \leq f(t) + \epsilon s.$$

Proceeding in a similar way, it follows from

$$X \cap ([0, t] \times I^{n-1}) \subset (X \cap ([0, t - s] \times I^{n-1})) \cup ([t - s, s] \times U)$$

that

$$f(t) \leq f(t - s) + \epsilon s, \ \forall s \text{ satisfying } \ 0 \leq s < t_0.$$

It asserts that

$$|\frac{f(t + s) - f(t)}{s}| \leq \epsilon \ \forall s \text{ satisfying } \ |s| < t_0.$$

This implies that f is differentiable at $t \in \mathbf{I}$ with its derivative 0. Since $f(0) = 0$, it also follows that $f(1) = 0$. Hence, the theorem follows from Definition 3.17.7. ❑

Theorem 3.17.17 provides an alternative statement with its proof of the Fubini Theorem 3.17.16

Theorem 3.17.17 (Fubini) *Let A be a closed subset of $\mathbf{R}^n$ such that the subset*

$$A \cap \mathbf{R}_c^{n-m}$$

is a set of measure zero for each $c \in \mathbf{R}^m$, where $V_c = \mathbf{R}_c^{n-m} = \{c\} \times \mathbf{R}^{n-m} \subset \mathbf{R}^n$ is called the ***vertical slice*** *over c. Then A is a set of measure zero.*

Proof By the given condition, the closed subset A of $\mathbf{R}^n$ can be expressed as countable union of compact sets. Hence, without any loss of generality, we may assume that A is compact. We now choose an interval $\mathbf{I} = [a, b]$ such that $A \subset V_{\mathbf{I}} = \mathbf{I} \times \mathbf{I}^{n-1} \subset \mathbf{R}^n$. To prove the theorem, select a covering $A \cap V_c$ for every $c \in \mathbf{I}$ by $(n-1)$-dimensional rectangular solids $R_1(c), \ldots, R_p(c)$ with a total volume $< \epsilon$. Select an interval $K(c)$ in $\mathbf{R}$ such that the rectangular solids $K(c) \times R_i(c)$ cover $A \cap V_{\mathbf{I}}$ (see

Exercise 63). The sets $K(c)'s$ cover the interval $[a, b]$. Use Exercise 62 to replace these sets by a finite family of subintervals K_{I}' having the total length $< 2(b-a)$. Every such K is contained in some interval $K(c_i)$. Hence, the solids $K_{\mathrm{I}}' \times R_i(c_j)$ cover the set A. Since they have the total volume $< 2(b-a)$, the theorem follows by induction on m. ❑

Remark 3.17.18 It has been proved in Proposition 3.3.26 that if M and N are two manifolds in Euclidean spaces such that M is compact and if $f : M \to N$ is a smooth map and $y \in N$ is a regular value of f, then $f^{-1}(y)$ is a finite set (may be empty). In general, the set of critical values of a smooth map $M \to \mathbf{R}^n$ is not finite but it is small in the sense that its Lebesgue measure is zero proved in Theorem 3.17.19. This theorem, known as Sard's theorem, was proved by A. Sard in 1942 based on some earlier work of A. P. Mores. Accordingly, Theorem 3.17.27 is also known as the Morae–Sard Theorem.

Theorem 3.17.19 (Sard's theorem) *Let M be a smooth manifold of dimension m and $f : M \to \mathbf{R}^n$ be a smooth map. The set of critical values of f in $\mathbf{R}^n$ has measure zero.*

Sard's Theorem 3.17.20 gives in a general form from which Sard's Theorem 3.17.19 follows.

Theorem 3.17.20 (Sard's theorem in general form) *Let $f : M \to N$ be a smooth map of manifolds and A be the set of critical points of f in M. Then the set $f(A)$ has measure zero in N.*

Remark 3.17.21 There are several proofs of Sard's theorem. We follow here [Milnor, 1965]. Before giving the proof of Sard's Theorem 3.17.20, we prove Lemma 3.17.22.

Lemma 3.17.22 *Suppose Theorem 3.17.20 is true for every smooth map*

$$f : U \to \mathbf{R}^m,$$

where the set $U \subset \mathbf{R}^n$ is open. Then Theorem 3.17.20 is also true for every smooth map

$$g : U_+ \to \mathbf{R}^m$$

for each open subset $U_+ \subset \mathbf{R}^n_+$.

Proof Suppose $n = dim\ M$ and $m = dim\ N$. Let the set $U_+ \subset \mathbf{R}^n_+$ be open and $f_+ : U_+ \to \mathbf{R}^m$ be a smooth map and A be the set of all critical values of f_+. Then there is an open set $U_+' \subset \mathbf{R}^n$ and a smooth map

$$f_+' : U_+' \to \mathbf{R}^m, \text{ where } U_+ = U_+' \cap \mathbf{R}^n_+ \text{ and } f_+'|_{U_+} = f_+.$$

If A' is the set of all critical points of f_+', then by hypothesis, the set $f_+'(A')$ has measure zero in $\mathbf{R}^m$. Since $f_+(A) = f_+'(A) \subset f_+'(A')$, it follows that $f_+(A)$ is also a set of measure zero in $\mathbf{R}^m$. ❑

We are now in a position to prove Sard's Theorem 3.17.20.

Proof **Sard's Theorem** 3.17.20 is proved by induction on $n = dim M$. If $n = 0$, the proof is trivial. Next suppose that the theorem is true for all manifolds of dimensions less than or equal to $n - 1$. To prove the theorem, it is sufficient to consider by using the second axiom of countability to study the special situation only when

$$f : U \to \mathbf{R}^m : U \subset \mathbf{R}^n_+ \text{ is open and } A \text{ is the set of critical points of } f \text{ in } U.$$

By using Lemma 3.17.22, we suppose that U is an open set in $Int\ \mathbf{R}^n_+$ or in $\mathbf{R}^n$. $B \subset A$ consists of the points of A, where the Jacobian matrix $J(f)$ vanishes. We claim that

(i) the set $f(A)$ has measure zero in $\mathbf{R}^m$;
(ii) the set $f(A - B)$ has measure zero in $\mathbf{R}^m$.

Proof of (i): Let $f_1 : U \to \mathbf{R}$ be the first component of $f = (f_1, f_2, \ldots, f_m)$. If the Jacobian matrix $J(f)$ of f vanishes at some point x, then the Jacobian matrix $J(f_1)$ also vanishes at x. Again if C is the set of points, where the map f_1 vanishes,then C is also the set of critical points of f_1. Hence, $f(B) \subset f_1(C) \times \mathbf{R}^{m-1}$. This shows that if the set $f_1(C))$ has measure zero in $\mathbf{R}$, then the set $f_1(C) \times \mathbf{R}^{m-1}$ and, consequently, the set $f(B)$ have measure zero in $\mathbf{R}^m$ since the set $\mathbf{R}^{m-1}$ has measure zero in $\mathbf{R}^m$. Finally, we consider the case when $m = 1$. Let the set B_i consist of the points of U at which all the partial derivatives of f having $order \leq i$ vanish. Consider a decreasing sequence of closed subsets of U:

$$B = B_1 \supset B_2 \supset \cdots \supset B_n \supset \cdots$$

Then the sets $f(B_i - B_{i+1})$ and $f(B_n)$ have measure zero in $\mathbf{R}^m$ for each $i = 1, 2, \ldots, n$. By second axiom of countability, $(B_i - B_{i+1})$ can be covered by countably many such nbds such that the set $f(B_i - B_{i+1})$ has measure zero in $\mathbf{R}^m$. If $b \notin B_{i+1}$, there exists an ith order derivative of f, say h such that h vanishes on B_i but its Jacobian matrix $J(h)$ is nonzero at b. This implies that h sends a nbd U_b of b by the inverse function theorem diffeomorphically onto an open set V of $\mathbf{R}^n$. Since the Jacobian matrix $J(h)$ vanishes on the set B_i, the set of critical points of $f : U_b \to \mathbf{R}$ is the set $U_b \cap (B_i - B_{i+1})$. Since, h^{-1} is a diffeomorphism, the composite map

$$g = f \circ h^{-1} : V \to \mathbf{R},$$

since by hypothesis, Sard's theorem is true for $n - 1$, it follows by induction that the set

$$g'(\{0\} \times \mathbf{R}^{n-1}) \cap V) = f \circ h^{-1}(\{0\} \times \mathbf{R}^{n-1}) \cap V)$$

has the critical set $h(U_b \cap (B_i - B_{i+1}))$. On the other hand, the set $g = f \circ h^{-1} : V \to \mathbf{R} = \{0\} \times \mathbf{R}^{n-1}$ is also the critical set of the restricted map

$$g' = g|(\{0\} \times \mathbf{R}^{n-1}) \cap V = f \circ h^{-1}(\{0\} \times \mathbf{R}^{n-1}) \cap V) = f(U_b \cap (B_i - B_{i+1}))$$

which has measure zero in $\mathbf{R}^m$.

To prove that the set $f(B_n)$ has measure zero, it is sufficient to show that the set $f(B_n \cap E)$ has measure zero for any n-cube E in U. Let l be the length of each n-cube E and $k > 0$ be an integer. Now, subdivide E into k^n subcubes having length l/k of each subcube and having diameter $\frac{l}{k}\sqrt{n}$. Let $b \in B_n \cap E$ and E_1 be one of the subcubes containing the point b. If $b + h \in E_1$, by Taylor's theorem of order n, there is a constant L independent of k, obtained as a uniform estimate of partial derivatives of f of order $n + 1$ such that

$$|f(b+h) - f(b)| \leq L||h||^{n+1} \leq L\left(\frac{l}{k}\sqrt{n}\right)^{n+1}.$$

This shows that $f(B_n \cap E_1)$ is contained in an interval having length P/k^{n+1}, where P is a constant independent of k. This asserts that $f(B_n \cap E)$ is contained in a union of intervals having total length $\leq Pk^n/k^{n+1} = P/K$. Consequently, $lim_{k\to\infty} B/k = 0 \implies f(B_n \cap E)$ has measure zero in $\mathbf{R}^m$.

Proof of (ii): It says that the set $f(A - B)$ has measure zero in $\mathbf{R}^m$.
Let $b \notin B$. Then there exists a first-order partial derivative of f, say $\frac{\partial f_1}{\partial x_1}$ such that it will not vanish at the point b. Then by applying the inverse transformation theorem as before, it follows that the map

$$h : U \to \mathbf{R}^n, x \mapsto (f_1(x), x_2, \ldots, x_n)$$

maps a nbd U_b of b diffeomorphically onto an open set V of $\mathbf{R}^n$. This implies that the set A_1 of critical points of the composite map

$$\psi = f \circ h : U \to \mathbf{R}^m$$

is the same as the set $h(V \cap A)$ and hence $\psi(A_1) = f(V \cap A)$. Clearly, ψ sends $(t, x_1, x_2, \ldots, x_n) \in U$ into the hyperplane $\{t\} \times \mathbf{R}^{n-1} \subset \mathbf{R}^n$. Consider the restricted map ψ_t of ψ

$$\psi_t : U \cap (\{t\} \times \mathbf{R}^{n-1}) \to \{t\} \times \mathbf{R}^{n-1}.$$

Then a point in $\{t\} \times \mathbf{R}^{n-1}$ is a critical point of ψ_t iff it is a critical point of ψ, because the Jacobian matrix $J(\psi)$ of ψ is of the form

$$\begin{pmatrix} 1 & 0 \\ * & J(\psi_t) \end{pmatrix}.$$

Hence, by induction hypothesis, the set of critical points of ψ_t has measure zero in $\{t\} \times \mathbf{R}^{n-1}$. Finally, using Fubini's Theorem 3.17.17, it follows that the set of critical values of ψ, which is the set $f(V \cap A)$, has measure zero. Let B be the set of points in A where the Jacobian matrix $J(f)$ vanishes. Since both the sets $f(B)$ and $f(A - B)$ have measure zero in $\mathbf{R}^m$, the theorem follows. ❑

Corollary 3.17.23 *Let M and N be two smooth manifolds and $f : M \to N$ be a smooth map.*

(i) *If $dim M < dim N$, then f is not a surjection.*
(ii) *If f is a smooth map having the set of critical points A, then the set $N - f(A)$ is dense in N.*

Proof **(i)** By the given condition, it follows that the critical set of f is M. Hence, if f is a surjection, then the set $N = f(M)$ would have measure zero in N. But it is not tenable.
(ii) It follows from the result that a set having measure zero cannot contain any open set which is nonempty. ❑

Corollary 3.17.24 *Let M and N be two smooth manifolds and $f : M \to N$ be a smooth map and $f_n : M \to N$ be a countable family of smooth maps. Then the set of common regular values of all the members of the family $\{f_n\}$ is dense in N.*

Proof Since every countable union of sets having measure zero has measure zero, the corollary follows immediately. ❑

Corollary 3.17.25 (Brown) *The set of regular values of a smooth map $f : M \to N$ is everywhere dense in N.*

Proof Let M be a smooth manifold of dimension m. Since M has a countable family of open nbds, each of them is diffeomorphic to a open subset of $\mathbf{R}^m$, the corollary follows. ❑

Definition 3.17.26 Let X be a manifold and $x \in M$. Then x is said to be a **critical point** for a C^1 map $M \to N$ if the linear map

$$T_x(f) : T_x(M) \to T_{f(x)}(N)$$

is not surjective. The set of critical points of f is denoted by Σ_f. The set $N - f(\Sigma_f)$ is the **set of regular values of** f.

The above discussion is summarized in a basic result formulated in Theorem 3.17.27.

Theorem 3.17.27 (Sard's theorem in general form or the Morse–Sard Theorem for C^r-Maps) *Let M be a smooth manifold of dimension m, N be a smooth manifold of dimension n and $f : M \to N$ be a C^r-map, where $r > max(0, m - n)$. If Σ_f denotes the set of critical points of f in M, then*

(i) *$f(\Sigma_f)$ has measure zero in N;*
(ii) *the set $N - f(\Sigma_f)$ of regular values of f is dense.*

Remark 3.17.28 For the Morse–Sard Theorem for C^∞ maps, see (Hirsch 1976, p 70).

Sard's theorem has a generalization to manifolds with boundary formulated in Theorem 3.17.29.

Theorem 3.17.29 *Let M be a smooth manifold with boundary ∂M and N be a smooth manifold without boundary. Then for any smooth map $f : M \to N$, almost every point of N is a regular value of both the maps*

$$f : M \to N \text{ and } \partial f : \partial M \to N.$$

Proof Since the derivative of ∂f_x at a point $x \in \partial M$ is the restriction of ∂f_x to the subspace $T_x(\partial M) \subset T_x(M)$, it follows that if ∂f is regular at a point, then f is also regular at that point. This implies that if a point $y \in N$ is not a regular value of both the maps

$$f : M \to N \text{ and } \partial f : \partial M \to N,$$

only when it is a critical value either of

$$f : Int\ M \to N \text{ or } \partial f : \partial M \to N.$$

Since the sets $Int\ M$ and ∂M are both manifolds without boundary, it follows that both sets of critical values are of measure zero. This implies that the complement of the set of common value for the maps f and ∂f has measure zero, because the union of two sets of measure zero is also a set of measure zero. ❑

3.18 Fundamental Theorem of Algebra

This section proves the classical fundamental theorem of algebra. There are several proofs of this theorem. In this chapter, three different proofs are given, of which two are given below based on the concepts of regular values of a smooth map in this section and another proof is given in the proof of Theorem 3.26.6. For other proofs by using the tools of algebraic topology, see **Basic Topology, Volume 3** of the present book series.

Theorem 3.18.1 (Fundamental Theorem of Algebra) *Every polynomial of degree $n \geq 1$ with complex coefficients has a root.*

Proof Let $p(z) = a_0 z^n + a_1 z^{n-1} + \cdots + a_{n-1} z + a_n$ be a complex polynomial with $a_0 \neq 0$ and $n \geq 1$. Then $p(x+iy) = u(x, y) + iv(x, y)$. By using the Cauchy–Riemann equations for the analytic function, $p = u + iv$, it follows that

$$p'(z) = u_x + iv_x = \frac{\partial u}{\partial x} + i\frac{\partial v}{\partial x} = -i\left(\frac{\partial u}{\partial y} + i\frac{\partial v}{\partial y}\right).$$

Then Jacobian determinant $J(u, v, x, y) = u_x^2 + u_y^2 = 0$ iff $p'(z) = 0$.

Proof I: Let $(0, 0, 1)$ be the north pole of the unit sphere $S^2 \subset \mathbf{R}^3$ and the p_+ be the stereographic projection from the north pole

$$p_+ : S^2 - \{(0, 0, 1)\} \to \mathbf{R}^2 \times \{(0, 0, 0)\} \subset \mathbf{R}^3,$$

where any point $x = (x_1, x_2, x_3) \in S^2 - \{(0, 0, 1)\}$ is mapped by p_+ to the point $p_+(x)$ at which the line through the north pole $(0, 0, 1)$ and the point x intersect the plane $\mathbf{R}^2 \times \{(0, 0, 0)\}$, identified with the plane of complex numbers $\mathbf{C}$ (which is the x_1x_2-plane). The polynomial p from $\mathbf{R}^2 \times \{(0, 0, 0)\}$ to itself corresponds to a smooth map

$$f : S^2 \to S^2, x \mapsto \begin{cases} (p_+^{-1} \circ p \circ p_+)(x), & \text{if } x \neq (0, 0, 1) \\ (0, 0, 1) \text{ if } x = (0, 0, 1). \end{cases}$$

f is smooth even in a nbd of $(0, 0, 1)$. It has only a finite number of critical points. Because p is not a local diffeomorphism only at the zeros of the derivative polynomial $p'(z)$. Since p' is not identically zero, there are only finitely many zeros. The set $\mathcal{S}$ of regular values of f is the truncated sphere with finitely many points removed. Hence, the set $\mathcal{S}$ is connected and the locally constant number function $\#p^{-1}(y)$ (the number of points in $p^{-1}(y)$) must be constant on the set $\mathcal{S}$. Since $\#p^{-1}(y)$ fails to be zero everywhere. This implies that it is zero nowhere. Thus, f is an onto map. This proves that the polynomial p must have a root.

Proof II: Let $p(z) = a_0 z^n + a_1 z^{n-1} + \cdots + a_{n-1} z + a_n$ be a complex polynomial with $a_0 \neq 0$ and $n \geq 1$. Then $p(x+iy) = u(x, y) + iv(x, y)$. By using the Cauchy–Riemann equations for the analytic function, $p = u + iv$, it follows that

$$p'(z) = u_x + iv_x = \frac{\partial u}{\partial x} + i\frac{\partial v}{\partial x} = -i\left(\frac{\partial u}{\partial y} + i\frac{\partial v}{\partial y}\right).$$

Then Jacobian determinant $J(u, v, x, y) = u_x^2 + u_y^2 = 0$ iff $p'(z) = 0$. But there exist only a finite number of zeros of p' having this property. This implies that the number of critical points $p : \mathbf{R}^2 \to \mathbf{R}^2$ is only finite. Let $A \subset \mathbf{R}^2$ be this finite set of critical values. The equation

$$|p(z)| = |a_0 z^n + a_1 z^{n-1} + \cdots + a_{n-1} z + a_n| (a_0 \neq 0)$$

asserts that $|p(z)| \to \infty$ as $|z| \to \infty$. This implies that p has a continuous extension over the one-point compactification space of $\mathbf{R}^2$, which is the 2-sphere S^2. (see **Basic Topology, Volume 1** of the present book series.) Clearly, the map $p : \mathbf{C} \to \mathbf{C}$ is a closed map. For any point $w \in \mathbf{C}$, the set $X = p^{-1}(w)$ consists of zeros of the polynomial $p(z) - w$. Then X contains at most n points. If card $X = \# p^{-1}(w) = m$, the number of points in $p^{-1}(w)$ is then say $X = \{z_1, z_2, \ldots, z_m\}$. If w is a regular value, then every z_k has a nbd U_k which is mapped diffeomorphically onto a nbd $V_k \subset \mathbf{R}^2 - A$ of w. Since $\mathbf{R}^2$ is Hausdorff, without loss of generality, we may assume that $U_k' s$ are disjoint. There is an open connected nbd of the point w lying in the open set $V_1 \cap V_2 \cap \cdots \cap V_m - f(\mathbf{R}^2 - (U_1 \cup U_2 \cup \cdots \cup U_m))$. Then p maps $U_k \cap p^{-1} V = V_k$ diffeomorphically onto V. Again, $p(z) \in V$ asserts that

$$z \in p^{-1}(V) \cap (U_1 \cup U_2 \cup \cdots \cup U_m) = W_1 \cup W_2 \cup \cdots \cup \ W_m).$$

Hence, it follows that

$$p^{-1}(V) = W_1 \cup W_2 \cup \cdots \cup \ W_m.$$

This implies that the cardinality card X is locally constant on the set $\mathbf{R}^2 - A$ of regular values. Since the set $\mathbf{R}^2 - X$ is connected, the integer card $X \neq 0$ is constant, otherwise A would be the image of p. Since the image of $p : \mathbf{R}^2 \to \mathbf{R}^2$ is constant. But p is not constant. This asserts that the image p contains the set $(\mathbf{R}^2 - A) \cup A = \mathbf{R}^2$. This proves the theorem.

Proof III: It is given in the proof of Theorem 3.26.6. ❑

3.19 Properties of Smooth Homotopy and Isotopic Embedding

This section studies properties of smooth homotopic maps given in Definition 3.15.21 and proves Isotonic Embedding Theorem 3.19.9. It also introduces the concept of normalized homotopy and proves in Proposition 3.19.6 that continuous homotopy implies smooth homotopy.

3.19.1 Properties of Smooth Homotopy of Smooth Maps

This subsection studies some properties of smooth homotopic maps and proves in Proposition 3.19.6 that if $f, g : M \to N$ are two smooth maps such that they are continuously homotopic, then they are also smoothly homotopic. Recall that

Definition 3.19.1 Two smooth maps $f, g : M \to N$ are called smoothly homotopic if there exists a smooth map

$$F : M \times \mathbf{R} \to N : F(x, 0) = f(x) \text{ and } F(x, 1) = g(x), \ \forall x \in M.$$

If F is just a continuous map, then the smooth maps f and g are called continuously homotopic or simply homotopic (see Definition 3.15.21).

This subsection also introduces the concept of normalized homotopy in Definition 3.19.4, which is closely related to smooth homotopy.

Definition 3.19.2 Let M be a smooth manifold and N be a smooth manifold with a metric d. Given a positive continuous function $\delta : M \to \mathbf{R}$, (i.e., $\delta(x) > 0, \ \forall x \in M$) and two smooth functions $f, g : M \to N$, the function g is said to be a **δ-approximaton** to f if

$$d(f(x), g(x)) < \delta(x), \ \forall x \in M.$$

Example 3.19.3 If n is sufficiently large, then every smooth map $f : M \to \mathbf{R}^n$ can be δ-approximated by an embedding g such that g is homotopic to f by a smooth homotopy

$$F_t : M \to \mathbf{R}^n, x \mapsto (1 - t)f(x) + tg(x)$$

and every homotopy F_t is δ-approximated to f.

Definition 3.19.4 (*Normalized homotopy*) Given a smooth homotopy F between two smooth maps $f, g : M \to N$, there exists a smooth map

$$\tilde{F} : M \times \mathbf{R} \to N, \ (x, t) \mapsto \begin{cases} f(x), & \text{if } t \leq 0 \\ g(x), & \text{if } t \geq 1 \end{cases}$$

formulated by

$$\tilde{F} : M \times \mathbf{R} \to N, \ (x, t) \mapsto F(x, \mathcal{B}(t))$$

using bump function $\mathcal{B} : \mathbf{R} \to \mathbf{R}$ (see Definition 3.14.5 and Example 3.14.6).

The smooth map $\tilde{F} : M \times \mathbf{R} \to N$ is called the normalized homotopy for the given smooth homotopy F. The smooth maps F and $\tilde{F}$ are related by a smooth homotopy

$$H : M \times \mathbf{R} \times \mathbf{R} \to N, \ (x, s, t) \mapsto F(x, (1-s)t + s\mathcal{B}(t)),$$

because $H(x, 0, t) = F(x, t)$ and $H(x, 1, t) = F(x, \mathcal{B}(t)) = \tilde{F}(x, t)$.

Proposition 3.19.5 *Let $\mathcal{S}(M, N)$ be the set of all smooth maps $f : M \to N$. Then the smooth homotopy relation $\mathcal{H}$ is an equivalence relation on $\mathcal{S}(M, N)$.*

Proof The relation $\mathcal{H}$ on $\mathcal{S}(M, N)$ is clearly reflexive and symmetric. To show that the relation $\mathcal{S}(M, N)$ is transitive, take any three smooth maps $f, g, h \in \mathcal{S}(M, N)$ such that f is smooth homotopic to g and g is smooth homotopic to h. Let F be a normalized smooth homotopy between f and g and G be a normalized smooth homotopy between g and h. Define

$$H : M \times \mathbf{R} \to N, \ (x, t) \mapsto \begin{cases} F(x, 3t), & \text{if } t \leq 1/2 \\ G(x, 3t - 2), & \text{if } t \geq 1/2. \end{cases}$$

Then H is well-defined and it is a smooth map, because F and G are smooth maps and $H(x, t) = g(x), \ \forall \, t \in [1/3, 2/3]$. This implies that H is a normalized homotopy between $f, h \in \mathcal{S}(M, N)$ and hence it is proved that the relation $\mathcal{H}$ is also transitive. Consequently, the relation $\mathcal{H}$ is an equivalence relation. ❑

Proposition 3.19.6 *Let $f, g : M \to N$ be two smooth maps. If they are continuously homotopic, then they are also smoothly homotopic.*

Proof Let the smooth maps $f, g : M \to N$ be continuously homotopic and $F : M \times \mathbf{R} \to M$ be a normalized continuous homotopy between them. If $\mathbf{K} = (-\infty, 0] \cup [1, \infty)$, then F is smooth on the closed set $M \times \mathbf{K}$, because $F|_{M \times (-\infty, 0]} = f$ and $F|_{M \times [1, \infty)} = g$. Hence, by using Exercise 27 of Sect. 3.27, there exists a positive continuous function $\delta : M \to \mathbf{R}$ such that

(i) F can be δ-approximated by a smooth map $H : M \times \mathbf{R} \to N$ and
(ii) $F = H|_{M \times K}$. ❑

3.19.2 Isotopic Embedding of Smooth Manifolds

This subsection introduces the concept of isotonic maps and establishes a relation between the concept of embedding of smooth manifolds to the concept of isotonic maps.

Definition 3.19.7 Let M and N be two smooth and manifolds and $f, g : M \to N$ be two embeddings. Then they are said to be **isotopic** if there exists a smooth homotopy $H : M \times \mathbf{R} \to N$ such that

$$H_t : M \to N, x \mapsto H(x, t)$$

is an embedding for every $t \in \mathbf{R}$.

Proposition 3.19.8 *Let M and N be two smooth manifolds and $H_t : M \to N$ be an isotopy. If $f : M_1 \to M$ and $g : N \to N_1$ are two embeddings, then*

$$h = g \circ H_t \circ f : M_1 \to N_1$$

is also is an embedding.

Proof It follows from Definition 3.19.7. ❑

Theorem 3.19.9 relates the two embeddings of smooth manifolds to the corresponding pair of isotonic maps.

Theorem 3.19.9 *Let M be a smooth manifold of dimension n and $f, g : M \to \mathbf{R}^m$ be two embeddings. If $m \geq 2n + 2$, then f and g are isotopic.*

Proof The identity map on $\mathbf{R}^m$ is continuously homotopic to a constant map

$$c : \mathbf{R}^m \to \mathbf{R}^m, x \mapsto x_0 \in \mathbf{R}^m,$$

since $\mathbf{R}^m$ is contractible to a point $x_0 \in \mathbf{R}^m$. Then the given embeddings f and g are continuously homotopic, and hence they are homotopic by a smooth homotopy

$$F : M \times \mathbf{R} \to \mathbf{R}^m.$$

If $m \geq 2n + 2$ is sufficiently large, then the smooth homotopy F can be deformed to an embedding

$$H : M \times \mathbf{R} \to \mathbf{R}^m$$

such that

$$F(x, t) = H(x, t), \ \forall\, x \in M \text{ and } \forall\, t \in (-\infty, 0] \cup [1, \infty).$$

This proves that f and g are isotopic by H. ❑

3.20 Riemann Surfaces

This section studies the Riemann surfaces which form an important class of manifolds for the study of the theory of functions of a complex variable. A connected 1-dimensional complex manifold is also called a Riemann surface. In this subsection, $\mathbf{R}^2$ is regarded as the complex z-plane.

Definition 3.20.1 Let M be a 2-dimensional smooth manifold and $\psi = \{(\psi_i, V_i)\}$ be an atlas on M with the property that if $V_i \cap V_j \neq \emptyset$, then the transition diffeomorphisms

$$\psi_{ji} = \psi_j \circ \psi_i^{-1} : \psi_i(V_i \cap V_j) \to \psi_j(V_i \cap V_j)$$

are complex-valued analytic functions of z in the domains of definition $\psi_i(V_i \cap V_j)$. Then the ordered pair (M, ψ) is called a **Riemann surface**.

Theorem 3.20.2 *The sphere $S^2 = \{w = (x, y, z) \in \mathbf{R}^3 : x^2 + y^2 + z^2 = 1\}$ is a Riemann surface. To show this, consider an analytic structure $V_1 = S^2 - N$, $V_2 = S^2 - S$, where N and S are the north and south points of S^2, respectively.* ***The local coordinates of the point*** *$P = (x, y, z) \in S^2$ in V_1 and V_2 are of the form*

$$w_1 = \frac{x + iy}{1 - z}, \; w_2 = \frac{x - iy}{1 + z},$$

which are obtained by stereographic projections on the equatorial plane from N and Sm respectively. For $P \in V_1 \cap V_2$, it follows that both $w_1 \neq 0$ and $w_2 \neq 0$. Moreover, their product $w_1 w_2 = 1$. This asserts that the transition diffeomorphism $w_1 = 1/w_2$ is an analytic function.

Corollary 3.20.3 *The real projective space $\mathbf{R}P^2$ is surface.*

Proof The proof follows from Theorem 3.20.2, since every point $p \in \mathbf{R}P^2$ has a nbd homeomorphic to an open subset of S^2, which is a surface. ❑

Example 3.20.4 The extended complex plane $\mathbf{C}^* = \mathbf{C} \cup \{\infty\}$ is a complex Riemann surface in the sense that it looks locally like as the complex plane. Let $V_0 = \mathbf{C}$ and $V_1 = \mathbf{C}^* - \{0\}$. Then $\{V_0, V_1\}$ forms an atlas of $\mathbf{C}^*$ such that $V_0 \cap V_1 \neq \emptyset$. Define two homeomorphisms

$$\psi_0 : V_0 \to \mathbf{C}, z \mapsto z, \text{ which is the identity map on } \mathbf{C}$$

and

$$\psi_1 : V_1 \to \mathbf{C}, z \mapsto 1/z.$$

Then $\psi_0(V_0 \cap V_1) = \mathbf{C} - \{0\}$ and the homeomorphism

$$\psi_1 \circ \psi_0^{-1} : \psi_0(V_0 \cap V_1) \to \psi_1(V_0 \cap V_1), z \mapsto 1/z, \; \forall z \in C - \{0\}$$

is analytic.

3.21 Transversability and Transversality Theorem

This section introduces the concept of transversality and studies transverse submanifolds of a smooth manifold. This is a basic concept in differential topology and establishes a good connection between smooth manifolds and smooth maps. The transversality of two smooth manifolds relies on the dimension of the manifold where they are embedded. For example, the two coordinate axes are transverse in $\mathbf{R}^2$ but they are not transverse in $\mathbf{R}^3$. Transversality Theorem 3.21.13 is an important theorem of this section deduced from **Sard's Theorem** 3.17.20 showing that transversality is a generic property.

Definition 3.21.1 Let M and N be two smooth manifolds and $f : M \to N$ be a smooth map. If K is a submanifold of N, then the map f is said to be

(i) **transversal to the submanifold** K **at a point** $x \in f^{-1}(K)$ (*symbolized*, $f \pitchfork_x K$) if the tangent space $T_{f(x)}(N)$ is expressible as

$$df_x(T_x(M)) + T_{f(x)}(K) = T_{f(x)}(N)$$

(LHS is not necessarily the direct sum but it is the set of the sums of the vectors, taken one from each of $df_x(T_x(M))$ and $T_{f(x)}(K)$) at the point $x \in f^{-1}(K)$ and

(ii) **transversal to** K (symbolized, $f \pitchfork_x K$) if either $f^{-1}(K) = \emptyset$ or $f \pitchfork_x K$ holds for all $x \in f^{-1}(K)$.

Example 3.21.2 **(i)** Let $f : \mathbf{R}^1 (= \mathbf{R}) \to \mathbf{R}^2, s \mapsto (0, s)$ be the map and K be the x-axis in $\mathbf{R}^2$. Then $f \pitchfork K$.

(ii) On the other hand, let $g : \mathbf{R}^1 (= \mathbf{R}) \to \mathbf{R}^2, s \mapsto (s, s^2)$ be the map and K be the x-axis in $\mathbf{R}^2$. Then $f \pitchfork K$ is not true in the sense that f is not transversal to K.

Proposition 3.21.3 *Let M and N be two smooth manifolds, K be a submanifold of N and*

$$f : M \to N$$

be a smooth map. The set of points of $f^{-1}(K)$ at which f is transversal to K is an open set in $f^{-1}(K)$.

Proof It follows from Definition 3.21.1. ❑

Throughout our next study, it is assumed that $f^{-1}(K) \neq \emptyset$.

Example 3.21.4 Let $\alpha, \beta : \mathbf{I} \to \mathbf{R}^2$ be two curves in $\mathbf{R}^2$ which intersect at a point. Then they are non-transverse if they are tangent to each other at that point.

Example 3.21.5 In Euclidean space $\mathbf{R}^3$, let M, N, K and f be defined as follows:

(i) M be the plane $x = 0$,
(ii) N be the plane $z = 0$,
(iii) K be the line $y = z = 0$ and
(iv) $f : M \to N, (y, z) \mapsto y$ be the projection.

Then f is transversal to K (i.e., $f \pitchfork K$) but the tangent map

$$df_x = T_x(f) : T_x(M) \to T_{f(x)}(N)$$

is not surjective.

Definition 3.21.6 Let M be a smooth manifold of dimension m and M_1, M_2 be two embedded submanifolds of M of dimensions m_1 and m_2, respectively. Then it is said that M_1 intersects M_2 **transversely** (symbolized, $M_1 \pitchfork M_2$) if

$$T_x(M_1) + T_x(M_2) = T_x(M), \ \ \forall x \in M_1 \cap M_2$$

(the sum in left hand is not direct, it is the set of the sum of vectors, taken one from each of the two subspaces $T_x(M_1)$ and $T_x(M_2)$). This relation is symmetric in M_1 and M_2 and hence $M_1 \pitchfork M_2$ holds iff $M_2 \pitchfork M_1$ holds.

Remark 3.21.7 In the language of linear algebra, the symbol $M_1 \pitchfork M_2$ in M means that for every point $x \in M_1 \cap M_2$, the subspaces $T_x(M_1)$ and $T_x(M_2)$ together span the space $T_x(M)$. If $m_1 + m_2 < m$, then it means that $M_1 \cap M_2 = \emptyset$.

Definition 3.21.8 Let M and N be two smooth manifolds and $f : M \to N$ be a smooth map. The differential of f at the point $x \in M$ along the curve $\alpha : \mathbf{I} \to M$ is the map

$$df = f_* : T_x(M) \to T_{f(x)}(N), \ D_\alpha \mapsto D_{f \circ \alpha},$$

which is well-defined and linear.

Proposition 3.21.9 *Let M and N be two smooth manifolds and $f : M \to N$ be a smooth map. Then $f_* : T_x(M) \to T_{f(x)}(N)$ is represented by a Jacobian matrix.*

Proof Let $x_1, x_2, \ldots, x_m$ be a local coordinate of the point $x \in M$ and $y_1, y_2, \ldots, y_n$ be a local coordinate near the point $f(x) \in N$. Then we can express

$$f(x_1, x_2, \ldots, x_m) = (f_1(x_1, x_2, \ldots, x_m), \ldots, f_n(x_1, x_2, \ldots, x_m)).$$

The linear map $f_* : T_x(M) \to T_{f(x)}(N)$ is represented with respect to the bases $\frac{\partial}{\partial x_i}$ and $\frac{\partial}{\partial y_j}$ by a matrix $(a_{i,j})$ as formulated by

$$f_*\left(\frac{\partial}{\partial x_i}\right) = \Sigma_{i=1}^n \, a_{i,j} \, \frac{\partial}{\partial y_j}$$

and hence it follows that

$$f_*\left(\frac{\partial}{\partial x_i}\right)(y_k) = \Sigma_{i=1}^n \, a_{i,j} \, \frac{\partial}{\partial y_j}(y_k) = a_{k,j}.$$

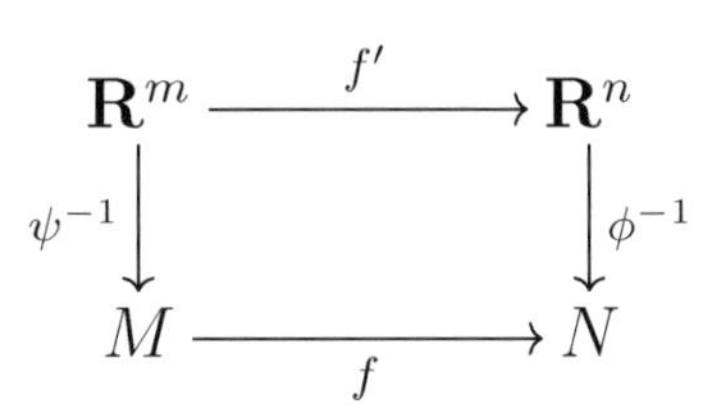

Fig. 3.15 Commutative rectangular diagram in local coordinates

This implies that the linear map f_* is represented by the Jacobian matrix

$$\left(\frac{\partial f_i}{\partial x_j}\right).$$

❑

Theorem 3.21.10 *Let M be a smooth manifold of dimension m and N be a smooth manifold of dimension n. If $f : M \to N$ is a smooth map and if the linear map*

$$f_* : T_x(M) \to T_{f(x)}(N)$$

is a monomorphism at a particular point $x = p \in M$, then there are charts ψ at p and ϕ at $f(p)$ such that the rectangular diagram as shown in Fig. 3.15 is commutative where

$$f' : \mathbf{R}^m \to \mathbf{R}^n,\ (x_1, x_2, \ldots, x_m) \mapsto (x_1, x_2, \ldots, x_m, 0, \ldots, 0)$$

is the standard inclusion. ($f(M)$ is accordingly flat in N in these local coordinates).

Proof Let ψ and ϕ be two arbitrary charts such that the origin in Euclidean space corresponds to p and $f(p)$, respectively. Consequently,

$$f'_* : \mathbf{R}^m = T_0\left(\mathbf{R}^m\right) \to T_0\left(\mathbf{R}^n\right) = \mathbf{R}^n$$

is a monomorphism. Change the coordinates by rotation in $\mathbf{R}^n$ so that

$$Im\ f'_* = \mathbf{R}^m \subset \mathbf{R}^n = \mathbf{R}^m \times \mathbf{R}^{n-m}.$$

Change the coordinates again by a map $\rho : \mathbf{R}^n \to \mathbf{R}^n$ so that the rectangular diagram as shown in Fig. 3.16 is commutative, where $i(x) = (x, 0)$. Then $(\rho \circ i)(x) = f'(x)$, $\forall x \in \mathbf{R}^m$ and hence $\rho(x.y) = f'(x) + y$. Under this choice, ρ_* sends the tangent space of $\mathbf{R}^n \times \{0\}$ onto $\mathbf{R}^n \times \{0\}$ by f'_* and ρ_* sends the tangent space of $\{0\} \times \mathbf{R}^n$ onto $\{0\} \times \mathbf{R}^n$ by the identity map. It proves that ρ_* is an isomorphism and hence it is a diffeomorphism in the nbd of the origin. Again by changing the chart ϕ by $\rho^{-1}\phi$, for the new chart, the theorem also holds. The new f' is given by $i = \rho^{-1} \circ f'$ and $i(x) = (x, 0)$ as required. ❑

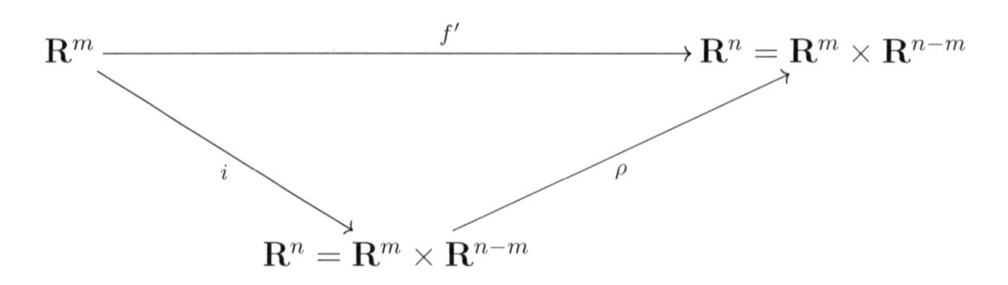

Fig. 3.16 Commutative triangular diagram involving maps f' and ρ

Corollary 3.21.11 asserts that if M is any embedded submanifold of N, then the manifold M has the induced functional structure as subspace of N.

Corollary 3.21.11 *Let $M \subset N$ be an embedded submanifold. Then every smooth real-valued function on M can be extended locally to a smooth function on a nbd in N.*

Theorem 3.21.12 *Let M be a smooth manifold of dimension m and M_1, M_2 be two embedded submanifolds of M of dimensions m_1, m_2, respectively. If $M_1 \pitchfork M_2$ (M_1 intersects M_2 transversely), then*

(i) *$M_1 \cap M_2$ is a submanifold of M of dimension*

$$dim\ (M_1 \cap M_2) = dim\ (M_1) + dim\ (M_2) - dim\ (M);$$

(ii) *$M_1 = \mathbf{R}^{m_1} \times \{0\}$ and $M_2 = \{0\} \times \mathbf{R}^{m_2}$ (locally chosen an appropriate coordinate system).*

Proof By using Theorem 3.21.10, take a chart at the point $p \in M$, in which M_1 is flat and choose a coordinate nbd U of p and also a map $\psi_1 : U \to \mathbf{R}^{m-m_1}$ with 0 as a regular value such that $U \cap M_1 = \psi_1^{-1}(0)$. Similarly, take a map $\psi_2 : U \to \mathbf{R}^{m-m_2}$ with 0 as a regular value such that $U \cap M_2 = \psi_2^{-1}(0)$. Define a map

$$\psi_1 \times \psi_2 : U \to \mathbf{R}^{m-m_1} \times \mathbf{R}^{m-m_2},\ x \mapsto (\psi_1(x), \psi_2(x)).$$

Then $(0, 0) = 0$ is regular value of $\psi_1 \times \psi_2$. Consider its induced linear map

$$(\psi_1 \times \psi_2)_* : T_p(M) \to \mathbf{R}^{m-m_1} \times \mathbf{R}^{m-m_2},\ v \mapsto (\psi_{1*}(v), \psi_{2*}(v)).$$

This asserts that

$$ker\ (\psi_1 \times \psi_2)_* = ker\ (\psi_1)_* \cap ker\ (\psi_2)_* = T_p(M_1) \cap T_p(M_2).$$

This implies that

$$dim\ T_p(M_1) + dim\ T_p(M_2) - dim\ T_p(M) = m_1 + m_2 - m.$$

Consequently, it follows that

$$dim\ (im\ (\psi_1 \times \psi_2)_*) = m - (m_1 + m_2 - m) = 2m - m_1 - m_2 = dim\ (\mathbf{R}^{m-m_1} \times \mathbf{R}^{m-m_2}).$$

This implies that the map $(\psi_1 \times \psi_2)_*$ is onto at p. Hence, it follows that 0 is a regular value for $(\psi_1 \times \psi_2)$ on U and $(\psi_1 \times \psi_2)^{-1}(0) = M_1 \cap M_2$ (locally). It proves the part (i).

For proof of the part (ii), consider the map

$$f : U \to \mathbf{R}^{m-m_2} \times \mathbf{R}^{m_1+m_2-m} \times \mathbf{R}^{m-m_1},\ x \mapsto (\psi_2(x), \phi(x), \psi_1(x)),$$

where ϕ is the projection to $M_1 \cap M_2$ under a coordinate system on U such that $M_1 \cap M_2$ is flat in the sense of Theorem 3.21.10. Hence, it follows that $im\ (f_*)$ contains the middle factor of the tangent space of the above product of Euclidean spaces and $im\ (f_*)$ is

$$im\ (\psi_{2*} \times \psi_{1*}) = \mathbf{R}^{m-m_2} \times \{0\} \times \mathbf{R}^{m-m_1}.$$

This implies that f_* is onto and f is a chart (with possible restriction of its domain). This f_* proves (ii). ❑

Transversality Theorem 3.21.13 is a basic result in differential topology and the key to transversality is the family $\{f_p\}_{p\in P}$ of smooth mappings.

Theorem 3.21.13 (Transversality Theorem) *Let M, N, K and P be smooth manifolds such that only M has boundary, and K is a submanifold of N and let $H : M \times P \to N$ be a smooth map such that both H and ∂H are transverse to K. Suppose for each $p \in P$, the map f_p is defined by*

$$f_p : M \to N,\ x \mapsto H(x, p).$$

Then the family of smooth maps $\{f_p\}_{p\in P}$ is such that for almost all points $p \in P$, both the maps f_p and ∂f_p are transverse to K.

Proof Let $S = H^{-1}(K)$. Then by Exercise 72, it follows that S is smooth manifold with boundary

$$\partial S = S \cap \partial(M \times P).$$

Let $q : M \times P \to P,\ (m, p) \mapsto p$ be the projection map. Then it follows that

(i) if $p \in P$ is a regular value of the restriction $q\ |_S$, then f_p is transverse to K and
(ii) if $p \in P$ is a regular value of the restriction $\partial q\ |\ _{\partial S}$, then ∂f_p is transverse to K.

Since by **Sard's Theorem** 3.17.20, almost every point $p \in P$ is a regular value of both the restricted maps $q\ |_S$ and $\partial q\ |\ _{\partial S}$, then transversality theorem follows. ❑

Example 3.21.14 Let $M = \{(z_1, z_2, z_3) \in \mathbf{C}^3 - \{0\} : z_1^2 + z_2^2 + z_3^2 = 0\}$ and f be the map

$$f : \mathbf{C}^3 - \{0\} \to \mathbf{C} : (z_1, z_2, z_3) \mapsto z_1^2 + z_2^2 + z_3^2.$$

Hence, it follows that 0 is a regular value of f and M is a 4-dimensional manifold. Let $M_1 = S^5 = \{(z_1, z_2, z_3) \in \mathbf{C}^3 : |z_1|^2 + |z_2|^2 + |z_3|^2 = 1\}$. Then $M \pitchfork M_1$ (M intersects M_1 transversely) and $M \cap M_1$ is a 3-dimensional manifold.

3.22 Tubular Nbds and Approximations

This section studies the concept of tubular nbd introduced by Whitney (1936a; 1936c; 1936b). This concept makes a link between vector bundles and the topology of manifolds. If M is a submanifold of a manifold N, then a tubular nbd of M is a nbd of M in N that looks like the normal bundle of M in N. This idea motivates to describe the topology of a nbd of the submanifold. This section proves that every smoothly embedded manifold in $\mathbf{R}^3$ has a nbd that looks like a tube in Euclidean space $\mathbf{R}^3$. This idea is used to prove smooth approximation Theorem 3.22.10 saying that every continuous map can be approximated by a smooth map.

Definition 3.22.1 Let M be a compact smooth manifold of dimension n embedded in $\mathbf{R}^m$. Then the **normal bundle** $\mathcal{N}(M)$ of M in $\mathbf{R}^m$ is defined by

$$\mathcal{N}(M) = \{\langle x, v\rangle \in M \times \mathbf{R}^m : v \perp T_x(M)\}$$

with

$$p : \mathcal{N}(M) \to M, \langle x, v\rangle \mapsto x$$

be projection. It is also denoted by the triple $\xi = (\mathcal{N}(M), p, M)$.

Remark 3.22.2 The notations

$$\theta : \mathcal{N}(M) \to \mathbf{R}^m, \ \langle x, v\rangle \mapsto x + v$$

and

$$\mathcal{N}(M, \epsilon) = \{ \langle x, v\rangle \in \mathcal{N}(M) : ||v|| < \epsilon\}$$

are also used in this section.

Proposition 3.22.3 *Let M be a compact smooth manifold of dimension n. Then every point $x \in M$ has a nbd U such that $p^{-1}(U)$ is homeomorphic to $U \times \mathbf{R}^{m-n}$ with the projection*

$$p : p^{-1}(U) \to U$$

that corresponds to the canonical projection

$$U \times \mathbf{R}^{m-n} \to U.$$

Proof Let $\psi : V \to \mathbf{R}^{m-n} \times \mathbf{R}^n \approx \mathbf{R}^m$ be a chart that makes M flat in the sense $U = V \cap M$ corresponds to the vertical slice $\{0\} \times \mathbf{R}^n$. Let $p_1, p_2, \ldots, p_{m-n}$ be the first $(m-n)$ coordinate projections $\mathbf{R}^m \to \mathbf{R}$. Then each composite map $p_i \circ \psi : V \to \mathbf{R}$ is constant on U. Then the gradient grad $(p_i \circ \psi)$ of $p_i \circ \psi$ at a point of U provides a normal frame to U such that $p^{-1}(U) \approx U \times \mathbf{R}^{m-n}$. Hence, the proposition follows. ❑

Definition 3.22.4 Let $M \subset W$ be a submanifold. A **tubular nbd** of M is a pair (f, ξ) of a vector bundle $\xi = (X, p, M)$ over M and an embedding $f : X \to W$ such that

(i) $f|_M = 1_M$, where M is identified with the zero section of X;
(ii) $f(X)$ is an open nbd of M in W.

Using notations of Definition 3.22.1, one version of **Tubular nbd Theorem** 3.22.5 is proved.

Theorem 3.22.5 (Tubular nbd theorem) *Let M be an n-dimensional compact smooth submanifold of $\mathbf{R}^m$. Then there is an $\epsilon > 0$ such that*

$$\theta : \mathcal{N}(M, \epsilon) \to \mathbf{R}^m$$

is a diffeomorphism onto the nbd $U = \{y \in \mathbf{R}^m : d(M, y) < \epsilon\}$ of M in $\mathbf{R}^m$ with Euclidean distance d under the notation of Remark 3.22.2.

Proof By hypothesis, M is an n-dimensional compact smooth submanifold of $\mathbf{R}^m$. Let $N_x(M)$ be the normal space to $T_x(M)$ in $\mathbf{R}^m$. Since, for fixed x and variable v, the map

$$\theta : \mathcal{N}(M) \to \mathbf{R}^m, \ \langle x, v \rangle \mapsto x + v$$

is a translation, its induced map θ_* is the standard inclusion on $N_x(M) \to \mathbf{R}^m$. Again,

$$\theta_* : T_x(M) \to T_x(\mathbf{R}^m)$$

is the differential of the inclusion $M \hookrightarrow \mathbf{R}^m$. This implies that

$$\theta_* : \mathbf{R}^m = T_x(\mathbf{R}^m) = T_x(M) \times N_x(M) \to \mathbf{R}^m$$

is the identity map on $\mathbf{R}^m$. This implies that for every $x \in M$, the induced homomorphism θ_* is a monomorphism at the point $\langle x, v \rangle$ and hence θ is a diffeomorphism on some nbd of the point $\langle x, v \rangle$. This proves that for every $x \in M$ and $||x||$ small, the homomorphism θ_* is an isomorphism at $\langle x, v \rangle$. By hypothesis, the manifold M is compact. Hence, there exists a $\delta > 0$ such the homomorphism θ_* is an isomorphism at every point of $\mathcal{N}(M, \delta)$. This implies that

$$\theta : \mathcal{N}(M, \delta) \to \mathbf{R}^m$$

is local diffeomorphism. We claim that $\theta : \mathcal{N}(M, \epsilon) \to \mathbf{R}^m$ is injective for some ϵ. Then there exist sequences $0 < \epsilon \leq \delta$. If θ is not injective, then there exist sequences $\langle x_k, v_k \rangle$ and $\langle y_k, w_k \rangle$ in $\mathcal{N}(M, \epsilon)$ for arbitrary $\epsilon > 0$ such that

$$\langle x_k, v_k \rangle \neq \langle y_k, w_k \rangle$$

but it has the properties:

(i) $||v_k|| \to 0$;
(ii) $||w_k|| \to 0$;
(iii) $\theta\langle x_k, v_k \rangle = \theta\langle y_k, w_k \rangle$.

Since by hypothesis, M is a metrizable and compact space, there exists a subsequence such that

$$x_{k_j} \to x \text{ and } y_{k_j} \to y.$$

Hence,

$$\theta(\langle x_k, v_k \rangle) \to \theta(\langle x, 0 \rangle) = x \text{ and } \theta(\langle y_k, w_k \rangle) \to \theta(\langle y, 0 \rangle) = y$$

imply that $x = y$. This asserts that for large k, both the sequences are such that

$$\langle x_k, v_k \rangle \to \langle x, 0 \rangle \text{ and } \langle y_k, w_k \rangle \to \langle y, 0 \rangle.$$

In other words, both the above sequences tend close to the point $\langle x, 0 \rangle$. But it contradicts the fact that θ is locally one-one near the point $\langle x, 0 \rangle$. This contradiction proves that θ is injective on some $\mathcal{N}(M, \epsilon)$. To show that θ is onto, we have to prove that

$$\theta(\mathcal{N}(M, \epsilon)) \subset \{y \in \mathbf{R}^m : d(M, y) < \epsilon\}$$

and

$$\{y \in \mathbf{R}^m : d(M, y) < \epsilon\} \subset \theta(\mathcal{N}(M, \epsilon)).$$

The first inclusion is obvious. For the second inclusion, take any point $y \in \{y \in \mathbf{R}^m : d(M, y) < \epsilon\}$. Then $d(M, y) < \epsilon$. Let $x \in M$ be a point such that $d(y, x) < \epsilon$ (minimum). Then the vector $y - x$ is a normal vector at the point x, and it is a vector having length $< \epsilon$ and hence $y \in \theta(\mathcal{N}(M, \epsilon))$. Consequently, it follows that $\theta(\mathcal{N}(M, \epsilon)) = \mathbf{R}^m$. The above discussion completes the proof of the theorem. ❑

Remark 3.22.6 Using the notation of Tubular Nbd Theorem 3.22.5, consider the map

$$r = p \circ \theta^{-1} : \theta(\mathcal{N}(M, \epsilon)) \to M.$$

Then r is a smooth retraction of the tubular nbd $\theta(\mathcal{N}(M, \epsilon))$ into M. The smooth map r is called the **normal retraction** of the tubular nbd onto M.

Corollary 3.22.7 *Let M be a compact manifold of dimension n and ξ be a nonzero continuous vector field on M that may not be smooth. Then there is a continuous map*

$$f : M \to M$$

such that

(i) *f has no fixed point and*
(ii) *f is homotopic to the identity map $1_M : M \to M$.*

Proof First embed the manifold M in some $\mathbf{R}^m$. Then the given nonzero continuous vector field ξ is a field of tangent vectors to M. Hence, by compactness hypothesis of M, given an $\epsilon > 0$, there is a constant c such that the vector field $c\xi$ contains all vectors of length $< \epsilon$ and hence $x + c\ \xi_x \in \theta(\mathcal{N}(M, \epsilon))$. Let r be the normal retraction of the tubular nbd onto M formulated in Remark 3.22.6. Define

$$f : M \to M, x \mapsto r(x + c\ \xi_x).$$

If $f(x) = x$, then we have $c\ \xi_x \in N_x(M) \cap T_x(M) = \{0\}$, which implies that $\xi_x = 0$. This asserts that f has no fixed points. The last part follows by considering the homotopy

$$H : M \times \mathbf{I} \to M, (x, t) \mapsto r(x + tc\ \xi_x).$$

❑

Remark 3.22.8 A more general version of Theorem 3.22.5 can be proved for smooth submanifolds of any smooth manifold, instead of Euclidean space in almost in the same way using geodesics in some Riemannian structure.

Theorem 3.22.9 approximates arbitrary continuous maps $f : M \to \mathbf{R}^m$.

Theorem 3.22.9 *Let M be a smooth manifold of dimension n and C, D be two closed subsets of M. If $f : M \to \mathbf{R}^m$ is continuous on M and smooth on C with induced structure, then corresponding to an $\epsilon > 0$, there exists a map $\tilde{f} : M \to \mathbf{R}^m$ such that*

(i) *the map $\tilde{f}$ is smooth on $M - D$;*
(ii) *$\tilde{f}(x) = f(x),\ \forall x \in C \cup D$;*
(iii) *$||\tilde{f}(x) - f(x)||, < \epsilon,\ \forall x \in M$ and*

(iv) $\tilde{f} \simeq f$ *rel* $C \cup D$.

Proof Since under the given condition M is metrizable, there exists a metric d on M such that the topology τ of M coincides with τ_d, the topology induced by d on M. For any $x \in M$, let $\epsilon(x)$ be defined by

$$\epsilon(x) = \min\{\epsilon, d(x, D)\},$$

where $d(x, D)$ denotes the distance of D from the point x. By hypothesis, f is smooth on C. Hence, there is a function defined near any point $a \in C$ on M such that it is smooth there and its restriction $f|_C$ is the same as f there. This implies that for every point $x \in M - D$, there exists a nbd V_x of x in $M - D$ and a map

$$F_x : V_x \to \mathbf{R}^m$$

such that

(i) if $x \in C - D$, then F_x is a smooth local extension of $f|_{V_x \cap C}$;
(ii) if $x \notin C \cup D$, then $V_x \cap C = \emptyset$ and if $y \in V_x$, then $F_x(y) = f(x)$, which is constant in y;
(iii) For sufficiently small V_x, if $y \in V_x$, then

$$||f(y) - f(x)|| < \epsilon(x)/2, \ ||F_x(y) - f(x)|| < \epsilon(x)/2 \text{ and } d(x, y) < \epsilon(x)/2.$$

Let $\{U_i\}$ be a local refinement of $\{V_i\}$ having index assignment $i \mapsto x(i)$ and let $\{\lambda_i\}$ be a smooth partition of unity on $M - D$ having *support* $(\lambda_i) \subset U_i$. Then $\lambda_i = 0$ on $C - D$ unless $x(i) \in C - D$. Define

$$\tilde{f} : M \to \mathbf{R}^m, y \mapsto \begin{cases} \Sigma\lambda_i(y)F_{x(i)}(y) \text{ if } y \in M - D \\ f(y), \text{ if } y \in D. \end{cases}$$

Then the map $\tilde{f} : M \to \mathbf{R}^m$ satisfies the required properties as claimed in parts (i)–(iii).

For the last part, construct the homotopy

$$H : M \times \mathbf{R} \to \mathbf{R}^m, \ (x, t) \mapsto tf(x) + (1 - t)\tilde{f}(x).$$

❑

Theorem 3.22.10 gives a sufficient condition for a map between smooth manifolds to be a smooth approximation. For its another form, see Exercise 27 of Sect. 3.27.

Theorem 3.22.10 (Smooth Approximation Theorem) *Let M be a smooth manifold of dimension m and N be a smooth compact metric manifold with a metric d. Given a closed subset $A \subset M$, let $f : M \to N$ be a map such that $f|_A$ is smooth. Then for any $\epsilon > 0$, there is a map*

$$\tilde{f} : M \to N$$

such that

(i) *the map $\tilde{f}$ is smooth;*
(ii) $d(\tilde{f}(x), f(x)) < \epsilon, \ \forall x \in M$;
(iii) $\tilde{f}|_A = f|_A$ *and*
(iv) $\tilde{f} \simeq f$ *rel A.*

Proof First embed the given manifold N in $\mathbf{R}^p$ and use continuity of the inverse of the above embedding. Since by hypothesis, N is compact, it follows by uniform continuity that there exists a $\delta > 0$, such that whenever $||u - v|| < \delta$, then $d(u, v) < \epsilon$. So, without any loss of generality, it is convenient to use the Euclidean metric in $\mathbf{R}^p$ instead of the given metric d on the manifold N. Consider a $\delta/2$-tubular nbd U of N in $\mathbf{R}^p$ (if necessary, take a smaller nbd) and the related normal retraction $r : U \to N$. By using Theorem 3.22.9, approximate f by a smooth map $h : M \to \mathbf{R}^p$ so that $h(M) \subset U$. Consider the map

$$\tilde{f} : M \to N, x \mapsto r(h(x)).$$

Then the map $\tilde{f} : M \to \mathbf{R}^p$ satisfies the required properties as claimed in parts (i)–(iii).

For the last part, construct a homotopy

$$H : M \times \mathbf{R} \to \mathbf{R}^p, \ (x, t) \mapsto r(t\, h(x) + (1 - t) f(x))$$

under which $\tilde{f} \simeq f$ rel A. ❑

Corollary 3.22.11 *Let M be a smooth manifold of dimension m and N be a compact smooth manifold of dimension n.*

(i) *Any continuous map $f : M \to N$ is homotopic to a smooth map and*
(ii) *if a smooth map $g : M \to N$ is homotopic to f, then $f \simeq g$ by a smooth homotopy*

$$H : M \times \mathbf{I} \subset M \times \mathbf{R} \to N.$$

Proof **(i)** Part (i) follows from Smooth Approximation Theorem 3.22.10.
(ii) By hypothesis, the smooth maps f and g are homotopic. Then there exists a homotopy

$$H : M \times \mathbf{I} \to N$$

between f and g. Extend the homotopy H by making it constant on the ends. Then H is a smooth map on the subspace $M \times \{0, 1\}$ and hence Theorem 3.22.10 asserts that there is a smooth map $F : M \times \mathbf{R} \to N$ such that it coincides with H on the subspace $M \times \{0, 1\}$. This proves part (ii). ❑

Definition 3.22.12 Let M be a submanifold of N and $N \subset \mathbf{R}^k$. Then the **normal bundle** $\mathcal{N}(M; N)$ of M in N is defined by

$$\mathcal{N}(M; N) = \{\langle x, v \rangle \in M \times T_x(N) : v \perp T_x(M)\}.$$

Then $\mathcal{N}(M; N)$ is a manifold with the same dimension of M and the map

$$\sigma : \mathcal{N}(M; N) \to M, \ \langle x, v \rangle \mapsto x$$

is a submersion.

Tubular nbd Theorem 3.22.5 is sufficient for most of the cases. However, Theorem 3.22.13 provides another version and its proof is almost similar.

Theorem 3.22.13 (Another Version of Tubular Nbd Theorem) *Let M be a submanifold of N and $N \subset \mathbf{R}^k$. If $\mathcal{N}(M; N)$ denotes the normal bundle of M in N, then there exists a diffeomorphism from an open nbd of M in $\mathcal{N}(M; N)$ onto an open nbd of M in N.*

Remark 3.22.14 Various applications of tubular nbds are given in Exercises of Sect. 3.27.

3.23 Complete Classification of Compact Surfaces

This section conveys the **complete classification of compact surfaces** initiated by A F Möbius (1790–1868) in 1861, published later on, which gives a partial solution of the classification problem of manifolds. A surface is a connected 2-manifold. For example, the 2-sphere is an important example of a surface. The connected sum $S_1\#S_2$ of two surfaces S_1 and S_2 is constructed by cutting a small circular hole in each S_1 and S_2, followed by gluing them together along the boundaries of the holes. The topological type does not depend on the choice of the circular disks in surfaces S_1 and S_2. This idea of connected sum is utilized in the constructions based on gluing two topological spaces described in Sects. 3.23.2, 3.23.3 and 3.23.4 and Exercise 3.27 of Sect. 3.27 for the complete classification of compact surfaces. For this section, the book Massey (1967) is recommended.

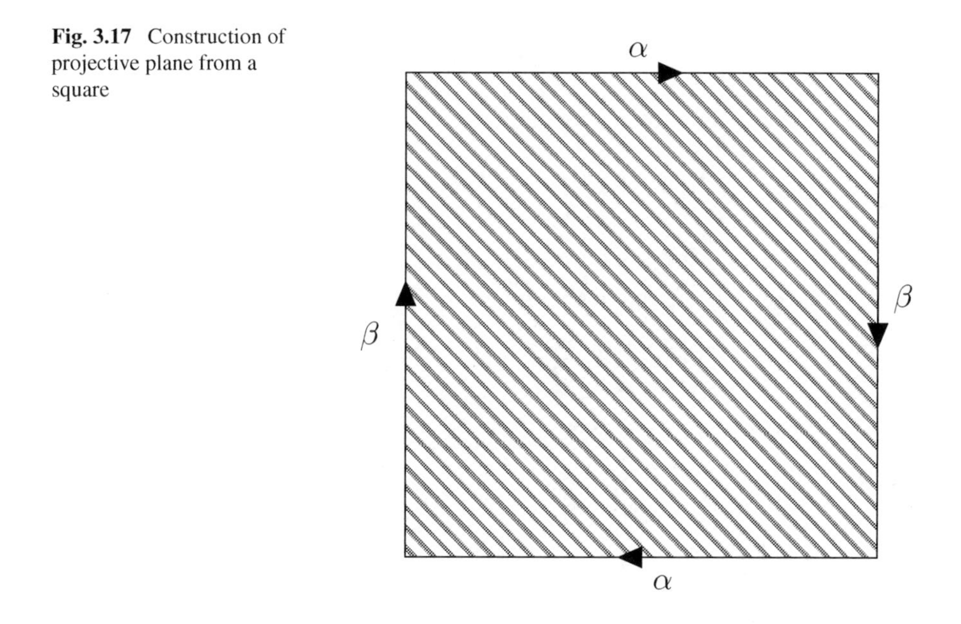

Fig. 3.17 Construction of projective plane from a square

3.23.1 Construction of Projective Plane and Klein Bottle from a Square

Projective plane is constructed from a square as shown in Fig. 3.17.
On the other hand, the Klein bottle is constructed from a square as shown in Fig. 3.18.

3.23.2 Construction of Connected Sum of Tori

This subsection presents a convenient method for construction of connected sum of tori. Recall that the torus T is obtained from a square by identifying its opposite sides. Given two tori T_1 and T_2, each of them is obtained from the square by identifying opposite sides as shown in Fig. 3.19. Then all the four vertices of each square are identified to a single point of the respective torus. To obtain their connected sum $T_1 \# T_2$, cut out a circular hole from each torus as shown in Fig. 3.19 with boundaries of the holes represented by γ_1 and γ_2. Then the complements of the holes in T_1 and T_2 are represented by pentagons as shown in Fig. 3.20, since identification of the indicated edges shows that the two end points of γ_1 and γ_2 are identified. It implies that the octagon is obtained by identifying the segments γ_1 and γ_2 as shown in Fig. 3.23 in which the edges are identified in pairs as indicated in this figure. By this identification, all the eight vertices of the octagon are identified to a single point as shown in Fig. 3.24. This implies that after gluing together, the connected sum $T_1 \# T_2$ is obtained from this octagon by identifying all its eight vertices to a single point in

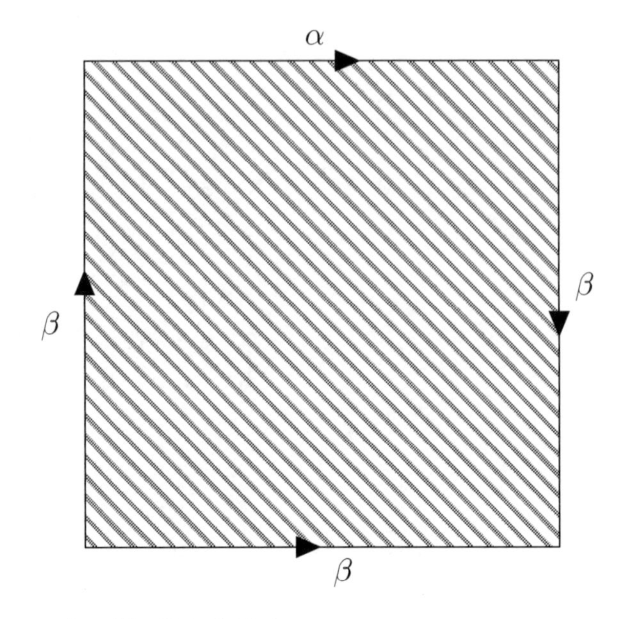

Fig. 3.18 Construction of the Klein Bottle from a square

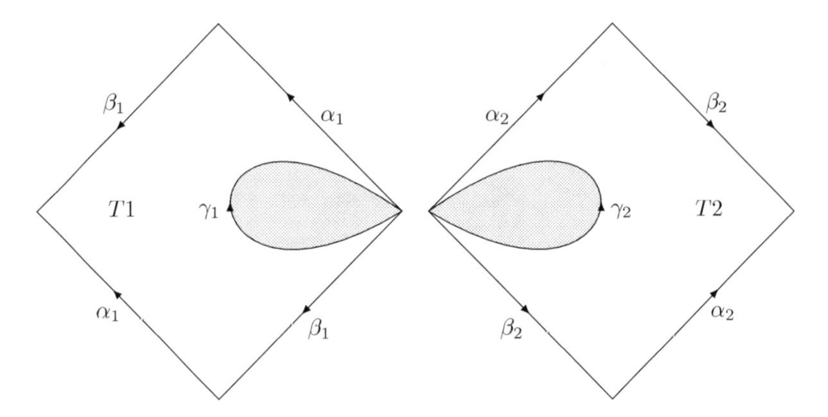

Fig. 3.19 Two disjoint tori T_1 and T_2 with holes γ_1 and γ_2

$T_1\#T_2$. The connected sum $T_1\#T_2\#T_3$ of the three tori is obtained in this way as the quotient space of the 12-gon as shown in Fig. 3.25. By induction on n, the connected sum $T_1\#T_2\#\cdots\#T_n$ of n tori is obtained as the quotient space of $4n$-gon obtained by identifying in pairs as shown above (Figs. 3.21 and 3.22).

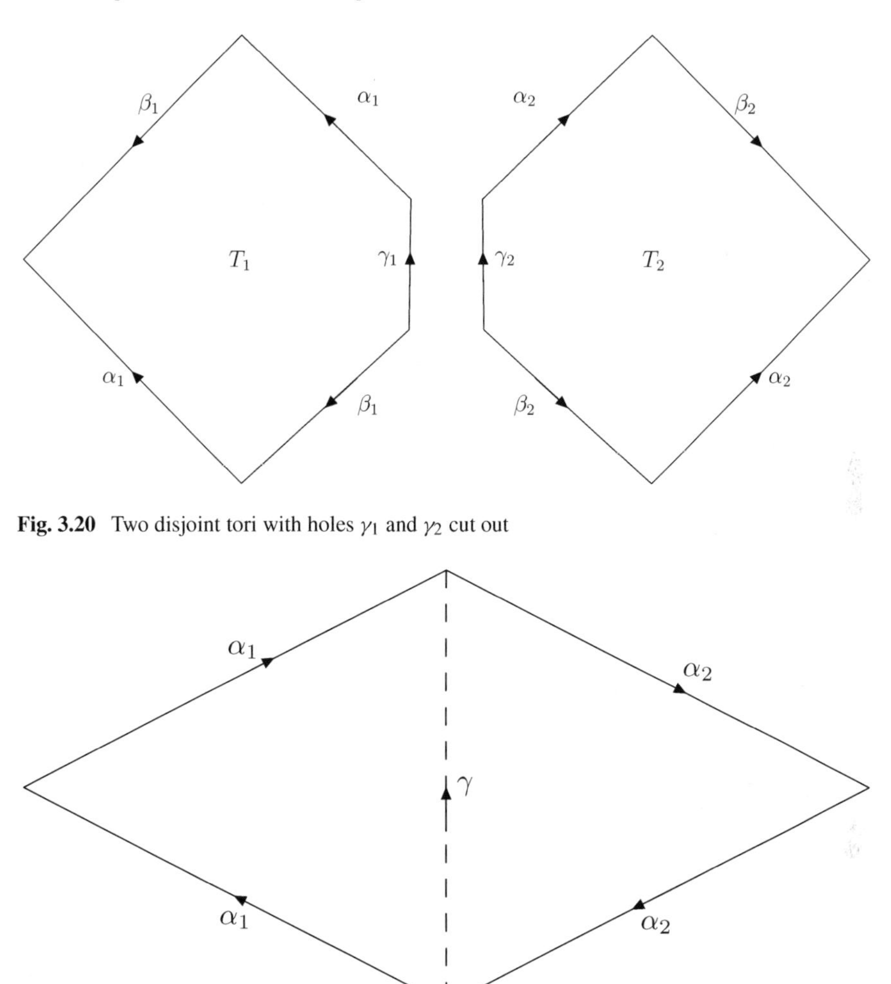

Fig. 3.20 Two disjoint tori with holes γ_1 and γ_2 cut out

Fig. 3.21 Glued together disjoint projective planes with holes γ_1 and γ_2

3.23.3 Construction of Connected Sum of Projective Planes

This subsection conveys a method for construction of connected sum of projective planes. Recall that the projective plane $\mathbf{R}P^2$ is obtained from a circular disk as the quotient space by identifying the diametrically opposite points on its boundary. This boundary circle is divided into two segments by taking a pair of its diametrically opposite points. This asserts that the real projective plane $\mathbf{R}P^2$ is obtained from a 2-gon by identifying its opposite edges as shown in Fig. 3.27. For two disjoint projective planes $\mathbf{R}P_1^2$ and $\mathbf{R}P_2^2$, their **connected sum** $\mathbf{R}P_1^2\#\mathbf{R}P_2^2$ is obtained likewise as the connected sum of two tori as a quotient space of an octagon as described in

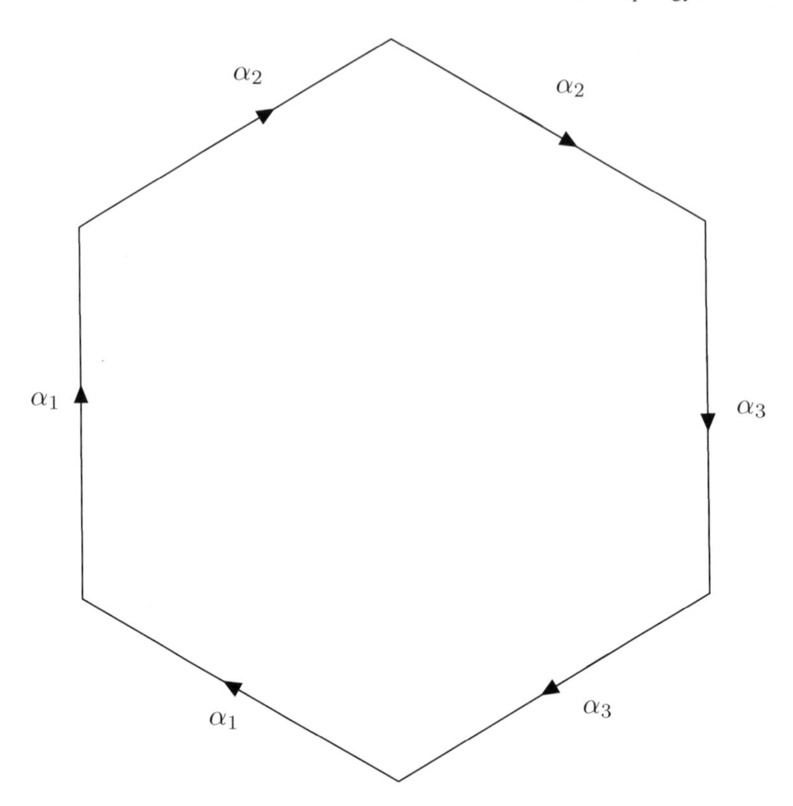

Fig. 3.22 Connected sum of three projective planes constructed by identification in pairs of the sides of a hexagon

Sect. 3.23.2. Repeating this process, the connected sum $\mathbf{R}P_1^2\#\mathbf{R}P_2^2\#\mathbf{R}P_3^2$ of three projective planes is obtained as a quotient space obtained from a hexagon by identifying in pairs its sides as shown in Fig. 3.22. By induction on n, the connected sum $\mathbf{R}P_1^2\#\mathbf{R}P_2^2\#\cdots\#\mathbf{R}P_n^2$ of n projective planes is obtained as the quotient space of a $2n$-gon obtained by identifying in pairs as above and hence by identifying all the vertices of the $2n$-gon to one point in $\mathbf{R}P_1^2\#\mathbf{R}P_2^2\#\cdots\#\mathbf{R}P_n^2$ (Fig. 3.26).

3.23.4 Construction of the Sphere as the Quotient Space of a Polygon

This subsection constructs the sphere as the quotient space of a 2-gon by identifying the sides in pairs as shown in Fig. 3.26.

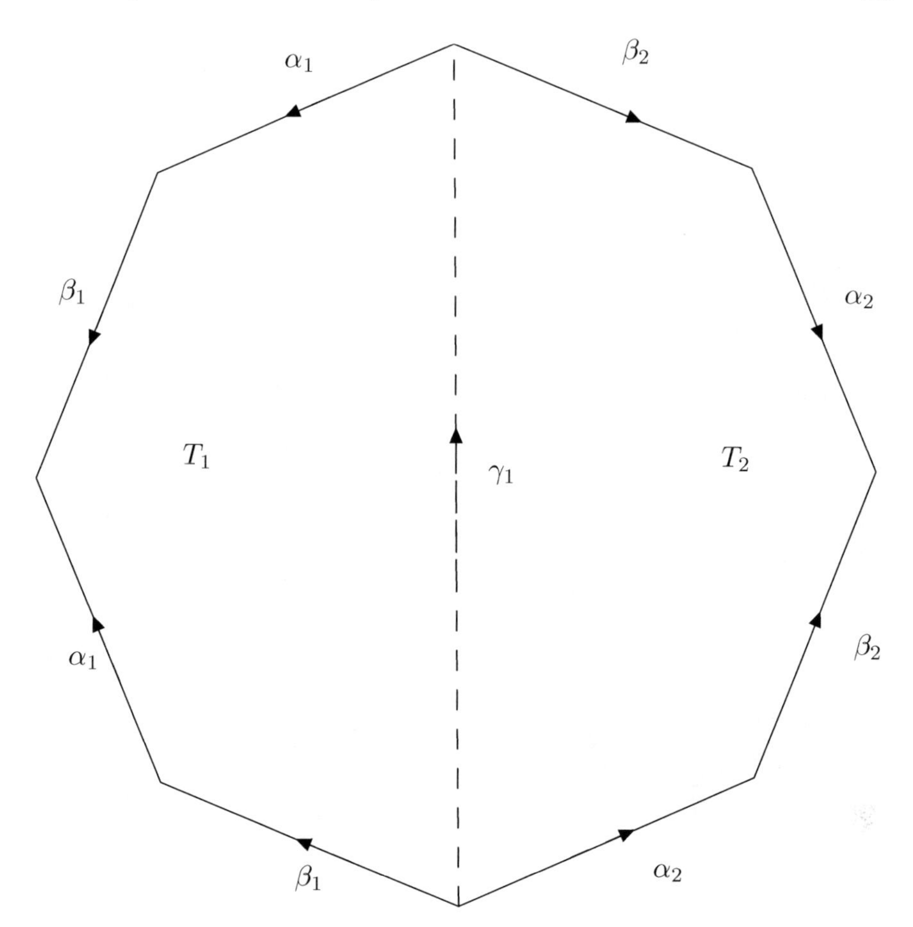

Fig. 3.23 Glued together the tori with holes γ_1 and γ_2

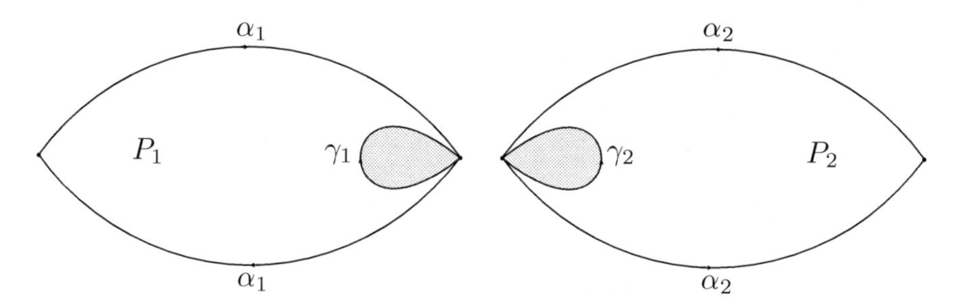

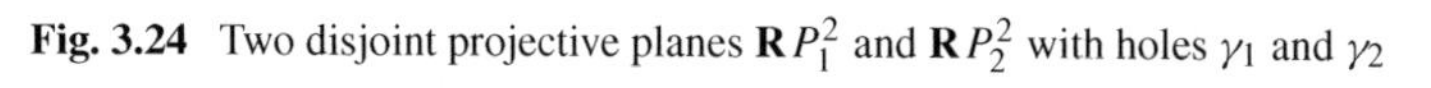

Fig. 3.24 Two disjoint projective planes $\mathbf{R}P_1^2$ and $\mathbf{R}P_2^2$ with holes γ_1 and γ_2

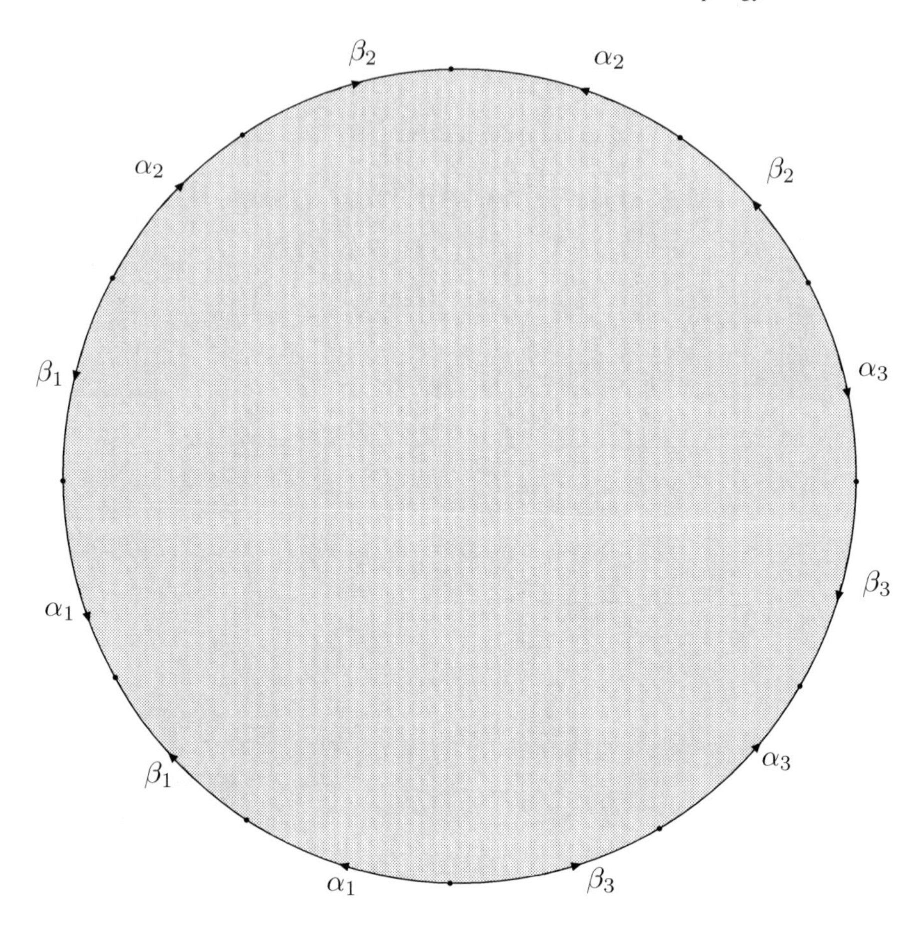

Fig. 3.25 Connected sum of three tori constructed by identifying the edges of a polygon of 12 sides

We can represent the above identifications by the symbols. The identification of

(i) the sphere is represented by the symbol: $\alpha\alpha^{-1}$;
(ii) the connected sum of p tori is represented by the symbol: $\alpha_1\beta_1\alpha_1^{-1}\beta_1^{-1}\alpha_2\beta_2\alpha_2^{-1}\beta_2^{-1}\cdots\alpha_p\beta_p\alpha_p^{-1}\beta_p^{-1}$;
(iii) the connected sum of p projective planes is represented by the symbol: $\alpha_1\alpha_1\alpha_2 a_2 \cdots\alpha_p\alpha_p$.

This above discussion is summarized in more descriptive form in Sect. 3.23.5.

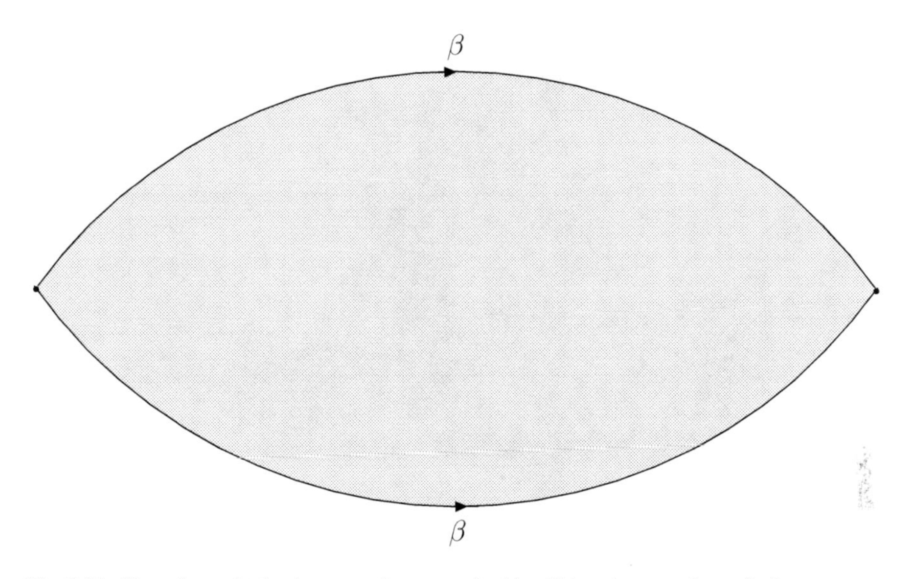

Fig. 3.26 The sphere obtained as a quotient space by identifying the two edges of a 2-gon

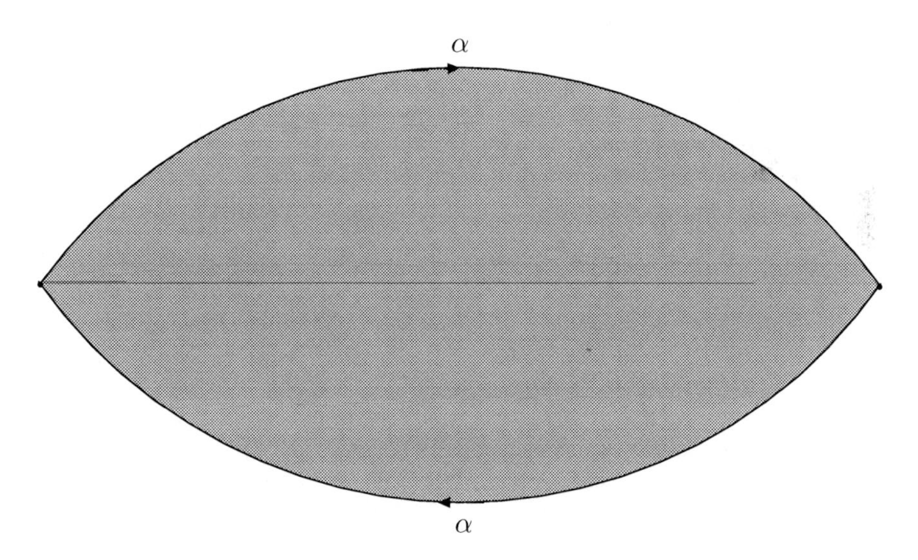

Fig. 3.27 The real projective plane constructed by identifying the opposite two edges of a 2-gon

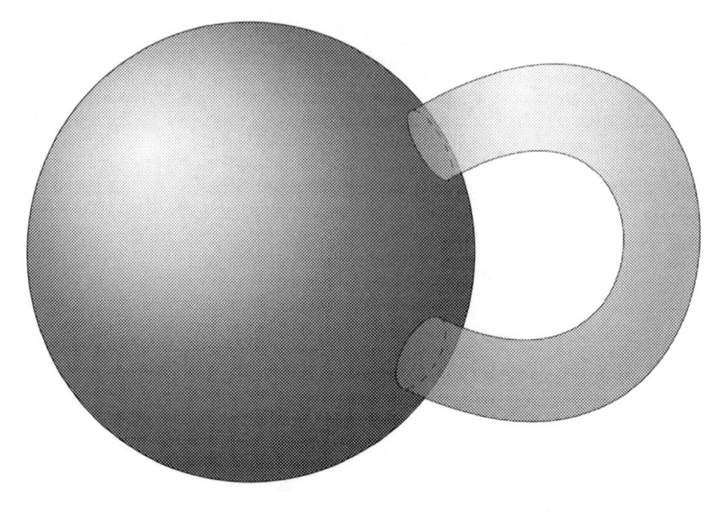

Fig. 3.28 Sphere with one handle

3.23.5 Methods of Constructions

This subsection describes the construction of

(i) a connected sum of a finite number of tori;
(ii) a connected sum of a finite number of projective planes.

(i) **(Adding an handle to a sphere)** Remove two disjoint disks from the 2-sphere S^2 and insert a cylinder by identifying its two end circles with the boundaries of the holes in S^2 as shown in Fig. 3.28. This process is known as adding an handle to the sphere S^2. This process is continued m-times to obtain a sphere with m-handles, called a connected sum of m-tori. The sphere with m-handles is known as the orientable surface of genus m.
(ii) **(Closed surface)** For each integer $m \in \mathbf{N}$, remove m-disjoint disks from the 2-sphere S^2 and replace each of them by a Möbius strip. This process is called a connected sum of m-projective planes. The sphere containing a Möbius strip is nonorientable. For $m = 2$, this process yields the Klein bottle.

We summarize the above discussion together with Exercise 3.27 of Sect. 3.27 in the following basic and important classification theorem of a compact connected surface.

Theorem 3.23.1 (Classification theorem of surfaces) *Any compact connected surface S is either homeomorphic to a sphere or to a connected sum of tori or to a connected sum of projective planes.*

Theorem 3.23.2 gives another form of the complete classification of compact surfaces.

Theorem 3.23.2 (An alternative form of classification theorem of surfaces) *Any compact surface is homeomorphic either*

(i) *to the sphere* S^2,
(ii) *or to the sphere* S^2 *with a finite number of handles glued to the sphere, called a connected sum of tori,*
(iii) *or to the sphere* S^2 *with a finite number of Möbius strips glued to the sphere obtained by a finite number of disks removed and replaced by Möbius strips, called a connected sum of projective planes.*
No two of these surfaces are homeomorphic.

Remark 3.23.3 The assumption of compactness of the surface S in Theorem 3.23.1 asserts that the surface S has a triangulation using only a finite number of triangles and that of connectedness asserts that these triangles can be arranged in sequence $T_1, T_2, \dots, T_p$ in such a way that each triangle in this sequence can be identified along an edge with a previous triangle so that the next triangle can be glued to the geometric figure already constructed.

3.24 Complete Classification of Connected 1-Dimensional Manifolds

This section provides a complete classification of **compact connected** 1-dimensional smooth manifolds in Theorem 3.24.1 and that of a **connected** 1-dimensional smooth manifolds in Theorem 3.24.4 up to diffeomorphism.

Theorem 3.24.1 gives a classification of compact connected 1-dimensional manifolds up to diffeomorphism.

Theorem 3.24.1 *A compact connected* 1*-dimensional smooth manifold is diffeomorphic to either* $[0, 1]$ *or* S^1.

Proof **Proof I**: Let X be a compact connected smooth manifold of dimension one and $\psi : \mathbf{I} \to \psi(\mathbf{I}) \subset X$ be a parametrization in X, and $\mathcal{F} = \{(\mathbf{I}, \psi)\}$ be the family of all pairs $(\mathbf{I}, \psi)$. Then $\mathcal{F}$ can be partially ordered by the relation " $\leq$"

$$(\mathbf{I}, \psi) \leq (\mathbf{K}, \phi) \text{ iff } \mathbf{I} \subset \mathbf{K} \text{ and } \psi = \phi|_{\mathbf{I}}.$$

If the interval $\mathbf{I} = \bigcup_j \mathbf{I}_j$, then every linearly ordered subset (chain) in $\mathcal{F}$

$$(\mathbf{I}_1, \psi_1) \leq (\mathbf{I}_2, \psi_2) \leq \cdots$$

is bounded by an element $(\mathbf{I}, \psi) \in \mathcal{F}$, where ψ is the parametrization formulated by $\psi|_{\mathbf{I}_j} = \psi_i$. Hence, by the Zorn Lemma, $\mathcal{F}$ has a maximal parameterized curve α

in X having parametrization $\phi : \mathbf{I} \to \alpha$. Without any loss of generality (by a change of variable, if needed), the interval $\mathbf{I}$ is exactly one of the intervals

$$[0, 1], (0, 1], [0, 1), (0, 1).$$

Hence, it follows that for any sequence $\{t_n\}$ in $\mathbf{I}$ converging to the point 0 (or 1), the corresponding sequence $\{\phi(t_n)\}$ in the curve α converges to a unique point x_1 (or x_2). Then the closure $\overline{\alpha} = \alpha \cup \{x_1, x_2\}$. The map ϕ has a smooth extension

$$\tilde{\phi} : [0, 1] \to \overline{\alpha} : \tilde{\phi}(0) = x_1 \text{ and } \tilde{\phi}(1) = x_2 \ ; \ \tilde{\phi} = \phi \quad \text{if } x_1, \ x_2 \in \alpha.$$

Consider the two possibilities: either $x_1 = x_2$ or $x_1 \neq x_2$.

Case I: If $x_1 = x_2$ then for $\theta = 2\pi r : 0 \leq r \leq 1$, define a diffeomorphism

$$f : S^1 \to X, (cos\ \theta, sin\ \theta) \mapsto \tilde{\phi}\left(\frac{\theta}{2\pi}\right).$$

Since $f(S^1)$ is compact and open in the connected set X and f is onto, it follows that X is diffeomorphic to S^1.

Case II: If $x_1 \neq x_2$, then the parametrization $\phi : \mathbf{I} \to X$ maps $\mathbf{I}$ onto X and hence it follows that the curve $\alpha = X$; otherwise, every point $x \in X - \alpha$ would have a parameterized curve nbd N_x not intersecting the curve α and α would have a larger parameterized curve contradicting its maximality of α. This implies that α and $\bigcup_{x \in X - \alpha} N_x$ will produce a separation of X. But it contradicts the hypothesis that X is connected. This contradiction proves that X is diffeomorphic to the closed interval $[0, 1]$. ❑

Theorem 3.24.2 gives a classification of connected 1-dimensional smooth manifolds up to diffeomorphism.

Theorem 3.24.2 *Let X be connected smooth manifold of dimension 1.*

(i) *If X is noncompact, then X is diffeomorphic to* $\mathbf{R}$.
(ii) *If X is compact, then X is diffeomorphic to either* $[0, 1]$ *or* S^1.

Proof Let X be a path-connected 1-dimensional manifold. We prove the theorem in two cases depending on additional condition: X is compact or noncompact.

Case I: X is connected but not compact. We prove in this case that X is homeomorphic to $\mathbf{R}$. Let $C = \{x_1, x_2, \ldots\}$ be a countable dense set in X. By hypothesis, as X is connected and locally path-connected, there exists a path α_1 in X joining the points x_1 and x_2 in X such that $C - \alpha_1 \neq \emptyset$, since α_1 is closed in X. Suppose that x_{n_1} be the first point in X outside α_1. This choice is possible, since X is a 1-dimensional connected manifold, any path α from the point x_{n_1} to a point $c \in \alpha_1 - \{x_1, x_2\}$ crosses x_1 or x_2 before entering in $\alpha_1 - \{x_1, x_2\}$.

The homeomorphism $h_1 : \mathbf{I} \to X, t \mapsto \alpha(t)$ is now extended to a homeomorphism $h : \mathbf{R} \to X$.

Let $h_1 : [0, 1] \to \alpha_1$ be a homeomorphism, which is extended as follows. If a path α crosses the point x_1, construct a path $\alpha_2 \subset \alpha$ from the point x_{n_1} to the point x_1, and a homeomorphism h_2 is defined such that

$$h_2 : [-1, 0] \to \alpha_2 : h_2(-1) = x_{n_1},\ h_2(0) = x_1$$

If the path α crosses the point x_2 construct a homeomorphism

$$h_2^* : [1, 2] \to \alpha_2 : h_2^*(1) = x_2,\ h_2^*(2) = x_{n_1}.$$

Hence, $\alpha_1 \cup \alpha_2$ gives a new path in X. Suppose $x_{n_2} \in C - \alpha_1 \cup \alpha_2$ is the first element of C. By induction and using the noncompactness property of X, construct a continuous injective map

$$h : \mathbf{R} \to X : C \subset h(\mathbf{R}).$$

Since by hypothesis, X is a 1-dimensional manifold, the map h is open. Moreover, $h(\mathbf{R}) = X$. Hence, it follows that h is a diffeomorphism.

Case II: Let X be a compact connected smooth manifold.

Proof I: Proceed as in Theorem 3.24.1.

Proof II: We prove in this case that X is homeomorphic to S^1. Let $\mathcal{U} = \{U_1, U_2, \ldots, U_n\}$ be a finite open covering of X such that every open set U_i is homeomorphic to $\mathbf{R}$. Suppose that the open covering $\mathcal{U}$ is minimal in the sense that it is not possible to remove any $U_i \in \mathcal{C}$ covering every point of X. This shows that

$$U_i \cap U_j \neq \emptyset \implies U_i \cap U_j \neq U_i,\ U_j.$$

Let $\psi_i : U_i \to \mathbf{R}$ be a homeomorphism. Since $\mathbf{R}$ is not compact, it follows that $\mathcal{U}$ consists of two or more open sets and the connected components of every open set $U_i \cap U_j \neq \emptyset$ is homeomorphic to an open interval. Let $C_{i,j}$ be any one of such components and $a, b \in \mathbf{R}\ with\ a < b$ such that a be the finite extreme point $\psi_i(C_{i,j})$ and a be the finite extreme point of $\psi_j(C_{i,j})$. If $\psi_i \circ \psi_j^{-1} : \mathbf{R} \to \mathbf{R}$ sends a to b, then $C_{i,j} \cup \{\psi^{-1}(a)\}$ is a connected set in $U_i \cap U_j$ properly containing its component $C_{i,j}$, which is not possible. This shows that for every component $C_{i,j}$, the interval $\psi_i(C_{i,j})$ has one infinite extreme. This asserts that there exist at most two components in every open set $U_i \cap U_j$ and $\psi_j \circ \psi_j^{-1} : \psi_i(C_{i,j}) \to \psi_j(C_{i,j})$ which sends the infinite extremes $-\infty$ and $+\infty$ to the finite extreme points in $\psi_j(C_{i,j})$. This asserts that if $\mathcal{U} = \{U_1, U_2\}$ and the open set $U_1 \cap U_2$ has only one component, then $X = U_1 \cup U_2$ is homeomorphic to $\mathbf{R}$. But it is not possible, since by hypothesis X is compact but $\mathbf{R}$ is not so. This implies that for $n = 2$, the open set $U_1 \cap U_2$ has two components X_1

and X_2, say. Suppose that $\psi_i(X_1) = (-\infty, a_i)$ for $i = 1, 2$ and $\psi_i(X_2) = (b_i, \infty)$ for $i = 1, 2$. Hence,

$$x_i \in X_i \implies \psi_i(x_1) < \psi_i(x_2), \quad \text{for } i = 1, 2.$$

Denote the interval $[\psi_i(x_1), \psi_i(x_2)]$ by the upper-half of S^1 by S^1_+ and the lower half of S^1 by S^1_-. Then it follows that

$$X = \psi_1^{-1}(\mathbf{I}_1) \cup \psi_2^{-1}(\mathbf{I}_2) \text{ and } \psi_1^{-1}(\mathbf{I}_1) \cap \psi_2^{-1}(\mathbf{I}_2) = \{x_1, x_2\}$$

and there exists a homeomorphism $f : X \to S^1$. Finally, use induction on n to prove that there exists a homeomorphism

$$f : X \to S^1.$$

This gives the complete classification of path-connected 1-dimensional smooth manifolds. ❑

Definition 3.24.3 An interval of real numbers is a connected subset of **R**, which is not a point. It may be finite or infinite; it may be open, closed or half-open.

The above discussion is summarized providing a complete classification of smooth, connected 1-dimensional manifolds in Theorem 3.24.4.

Theorem 3.24.4 (Complete Classification of 1-manifolds) *A connected 1-dimensional smooth manifold is diffeomorphic to either some interval of real numbers or to the circle S^1.*

Corollary 3.24.5 *Let ∂X be the boundary of a compact 1-dimensional manifold X with boundary ∂X. Then ∂X consists of an even number of points.*

Proof Since, by hypothesis, X is a compact 1-dimensional manifold X with boundary, it is the disjoint union of finitely many connected components and hence the corollary follows. ❑

3.25 Action of Topological Group on Manifolds and Transformation Group

This section studies action of topological group on manifolds and transformation group of a manifold. The action of a topological group on a topological space defined in Chap. 2 is redefined in Definition 3.25.1. On the other hand, the action of G as a transformation of a manifold M is defined in Definition 3.25.3.

3.25.1 *Topological Transformation Group of a Topological Space*

This subsection recalls the concept of the topological transformation group.

Definition 3.25.1 Let G be a topological group with identity element e and X be a topological space. Then an action of G on X (from left) is a continuous map

$$\mu : G \times X \to X,$$

its image $\mu(g, x)$ is denoted by $g(x)$ or gx such that

(i) $g(hx) = (gh)x. \forall g, h \in G$ and $\forall x \in X$;
(ii) $ex = x, \ \forall x \in X$;
(iii) the map $\psi : X \to X, x \mapsto gx$ is a homeomorphism of X for any $g \in G$.
(iv) If a topological group G acts on a topological space by an action μ, then the pair $(G.\mu)$ is said to be a **topological transformation group** of X.

Example 3.25.2 Let $g \in O(n, \mathbf{R})$ and $G_r(\mathbf{R}^n)$ be the Grassmann manifold given in Definition 3.1.8. If $V \in \mathrm{G_r}(\mathbf{R}^n)$, then $g(V)$ is another r-plane in $\mathbf{R}^n$. Any r-frame is taken into any other by some $g \in O(n, \mathbf{R})$, so the same is also true of r-plane. Hence, the orthogonal group $O(n, \mathbf{R})$ acts transitively on $\mathrm{G_r}(\mathbf{R}^n)$. Since the isotropy group of the standard $\mathbf{R}^r \subset \mathbf{R}^n$ is $O(r, \mathbf{R}) \times O(n-r, \mathbf{R})$, it follows that

$$\mathrm{G_r}(\mathbf{R}^n) \approx O(n, \mathbf{R})/O(r, \mathbf{R}) \times O(n-r, \mathbf{R}).$$

Hence, we get the following homeomorphisms:

$$O(n, \mathbf{R})/O(1, \mathbf{R}) \times O(n-1, \mathbf{R}) \approx \mathbf{R}P^{n-1},$$

$$U(n, \mathbf{C})/U(1, \mathbf{C}) \times U(n-1, \mathbf{C}) \approx \mathbf{C}P^{n-1}$$

and

$$S_p(n, \mathbf{H})/S_p(1, \mathbf{H}) \times S_p(n-1, \mathbf{H}) \approx \mathbf{H}P^{n-1},$$

where $\mathrm{O}(n, \mathbf{R})$, $\mathrm{U}(n, \mathbf{C})$ and $S_n(n, \mathbf{H})$ represent the orthogonal group over $\mathbf{R}$, the unitary group over $\mathbf{C}$ and symplectic group over quaternions $\mathbf{H}$, respectively.

3.25.2 *Topological Transformation Group of a Smooth Manifold*

This subsection considers the action of a topological group on a smooth manifold. The action of the Lie group on a smooth manifold is considered in Chap. 4.

Definition 3.25.3 Let G be a topological group with identity element e and M be a smooth manifold. Then G is said to **act on** M, as a transformation group if there is a smooth function

$$\mu : G \times M \to M,$$

called an **action** of G on M, and its image $\psi(g, x)$ is denoted by gx such that

(i) for any $g \in G$, the map $\psi_g : M \to M,\ x \mapsto \mu(g, x) = gx$ is a diffeomorphism of M;
(ii) $g(hx) = (gh)x. \forall g, h \in G$ and $\forall x \in M$, i.e., $\psi_g \circ \psi_h = \psi_{gh}$;
(iii) $ex = x,\ \forall x \in M$, i.e., ψ_e is the identity map on M.

At a point $x_0 \in M$, the set

$$G_{x_0} = \{g \in G : gx_0 = x_0\}$$

is a subgroup of G, called the **stabilizer group** at the point x_0.

Definition 3.25.4 Let G be a topological group with identity element e and M be a smooth manifold. Then the action of G on M is said to be

(i) **effective,** if $\psi_g(x) = x,\ \forall x \in M$ asserts that $g = e$, i.e., if e is the only element of G with the property that $\psi_g(x) = x,\ \forall x \in M$;
(ii) **free** on M, if e is the only element of G such that $\psi_g(x) = x$ for some $x \in M$;
(iii) **transitive,** if for every pair of elements $x, y \in M$, there is an element $g \in G$ such that $gx = y$.

Example 3.25.5 Let $Sy(n, \mathbf{R})$ be the set of all $n \times n$ real symmetric matrices and is a smooth manifold. The general linear group $GL(n, \mathbf{R})$ acts on $Sy(n, \mathbf{R})$ by action

$$\mu : GL(n, \mathbf{R}) \times Sy(n, \mathbf{R}) \to Sy(n, \mathbf{R}),\ (X, A) \mapsto XA^t X.$$

Example 3.25.6 **(i)** Consider the action of the topological group $G = SO(3, \mathbf{R})$ on $\mathbf{R}^3$. Its stabilizer group G_{x_0} at the point $x_0 = (0, 0, 1) \in \mathbf{R}^3$ is the set of rotations about the z-axis and hence

$$G_{x_0} \approx SO(2, \mathbf{R}).$$

(ii) Since the orthogonal group $O(3, \mathbf{R})$ acts transitively on $\mathbf{R}^3$ with its stabilizer group $G_{x_0} \approx SO(2, \mathbf{R})$ at the point $x_0 = (0, 0, 1) \in \mathbf{R}^3$, it follows that

$$SO(3, \mathbf{R})/SO(2, \mathbf{R}) \approx S^2.$$

(iii) Since the orthogonal (real) group $O(n, \mathbf{R})$ acts transitively on $\mathrm{V}_r(\mathbf{R}^n)$ with isotropy group $O(n - r, \mathbf{R})$ and hence it follows that

$$O(n, \mathbf{R})/O(n - r, \mathbf{R}) \approx \mathrm{V}_r(\mathbf{R}^n).$$

3.25.3 *Properly Discontinuous Action of a Smooth Manifold*

This subsection considers a special action

$$\sigma : G \times M \to M, (g, x) \mapsto gx$$

of a topological group G on a smooth manifold M, known as properly discontinuous action.

Definition 3.25.7 A topological group G with identity e is said to be **discrete** if all points of G are open (equivalently, iff $\{e\}$ is an open set in G).

Definition 3.25.8 Let G be a topological group and M be a smooth manifold. Then G is said to act **discontinuously** (under action σ) if corresponding to any point $x \in M$ and any sequence $\{g_n\}$ of distinct points in G, the sequence $\{g_n x\}$ in M, the sequence $\{g_n x\}$ does not converge to any point in M in the sense that every **orbit space** $orb\ (x) = Gx = \{gx : g \in G\}$ at x is a closed discrete subset of M. On the other hand, G is said to act **properly discontinuously** if

PD(i) for every point $x \in M$, there is an open nbd U_x of x with the property: $U_x g \cap U_x \neq \emptyset$ implies $g = e$;
PD(ii) for every pair of elements $x, y \in X$, $y \notin Gx$, there are open nbds V_x and V_y of x and y, respectively, such that

$$gV_x \cap V_y = \emptyset, \ \forall g \in G,$$

where $gV = \{gv : g \in G\}$.

Remark 3.25.9 Consider the action $\sigma : G \times M \to M$.

(i) The topological space $M/G = M \bmod G$, topologized by the identification map
$$p : M \to X \bmod G, x \mapsto Gx,$$
is the orbit space obtained by the action of σ of G on X.
(ii) Condition **PD**(i) asserts that the action σ is free.
(iii) Condition **PD**(ii) asserts that the orbit space $M/G = M \bmod G$ is a Hausdorff space.

Example 3.25.10 Every action of a finite group on a Hausdorff space is properly discontinuous.

Example 3.25.11 The discrete group $\mathbf{Z}^n$ acts on manifold $\mathbf{R}^n$ by translation

$$\sigma : \mathbf{Z}^n \times \mathbf{R}^n \to \mathbf{R}^n,\ (n_1, n_2, \ldots, n_n, (x_1, x_2, \ldots, x_n) \mapsto (n_1 + x_1, n_2 + x_2, \ldots, n_n + x_n).$$

This action σ is smooth, because the action of a discrete topological group G on a manifold M is smooth iff for every $g \in G$, the map

$$\psi_G : G \to G, x \mapsto gx$$

is smooth.

Remark 3.25.12 Detailed construction of the following quotient spaces is available in Chap. 3 of **Basic Topology, Volume 1** of the present series of books

Example 3.25.13 **(i)** **The cylinder** can be obtained as a quotient space obtained from the square $\mathbf{I} \times \mathbf{I}$ by identifying the point $(0, t)$ to the point $(1, t)$ for all $t \in \mathbf{I}$. Consider the manifold $M = \mathbf{R}^2$. The cylinder is obtained as the quotient manifold $\mathbf{R^2}/\mathbf{Z} = \mathbf{R^2}\ mod\ \mathbf{Z}$ obtained by the properly discontinuous action σ of the group $\mathbf{Z}$ on the manifold $\mathbf{R}^2$

$$\sigma : \mathbf{Z} \times \mathbf{R}^2, (n, (x, y)) \mapsto (x + n, y)).$$

The action σ is properly discontinuous and the group $\mathbf{Z}$ is generated by the single transformation

$$\psi_1 : \mathbf{R}^2 \to \mathbf{R}^2,\ (x, y) \mapsto (x + 1, y).$$

(ii) **The Mӧbius strip (Mӧbius band)** M can be obtained as a quotient space obtained from the square $\mathbf{I} \times \mathbf{I}$ by identifying the point $(0, t)$ to the point $(1, 1 - t)$ for all $t \in \mathbf{I}$. The identification topology on M coincides with the subspace topology induced on M by the usual topology on $\mathbf{R}^3$. Then Mӧbius strip M is embedded in $\mathbf{R}^3$. To show it as the quotient manifold $\mathbf{R}^2/\mathbf{Z}$, consider the group $\mathbf{Z}$ generated by the single transformation

$$\psi : \mathbf{R}^2 \to \mathbf{R}^2, (x, y) \to (x + 1, -y).$$

Since this group acts on $\mathbf{R}^2$ properly discontinuously, it follows that the quotient manifold $\mathbf{R}^2/\mathbf{Z}$ is the Mӧbius band M.

(iii) **The Klein bottle** K can be obtained as a quotient space obtained from the square $\mathbf{I} \times \mathbf{I}$ by identifying the point $(0, t)$ with the point $(1, 1 - t)$and the point $(t, 0)$ with the point $(t, 1)$ for all $t \in \mathbf{I}$. The Klein bottle K is homeomorphic to the quotient space from the topological sum of two Mӧbius strips by identifying the corresponding points of the edges. It is not possible to visualize the Klein bottle K, since K cannot be embedded in $\mathbf{R}^3$. Consider $\mathbf{Z} \times \mathbf{Z}$ as generated by two transformations on $\mathbf{R}^2$:

$$\psi : \mathbf{R}^2 \to \mathbf{R}^2, (x, y) \to (x + 1, -y)$$

and

$$\phi : \mathbf{R}^2 \to \mathbf{R}^2, (x, y) \to (x, y + 1).$$

Since the action of the group $\mathbf{Z} \times \mathbf{Z}$ on $\mathbf{R}^2$ is properly discontinuous, it follows from the above discussion that the quotient manifold $\mathbf{R}^2/\mathbf{Z} \times \mathbf{Z}$ is the Klein bottle K.

(iv) The 2-**torus** T^2 **(or torus)** can be obtained as a quotient space obtained from the square $\mathbf{I} \times \mathbf{I}$ by identifying both pairs of opposite edges via identifying the point $(0, t)$ with the point $(1, t)$ and the point (s, o) with the point $(s, 1)$ for all $t, s \in \mathbf{I}$. To show it as the quotient manifold $\mathbf{R}^2/\mathbf{Z} \times \mathbf{Z}$, consider the action

$$\sigma : (\mathbf{Z} \times \mathbf{Z}) \times (\mathbf{R} \times \mathbf{R}), \ \ ((m, n), (x, y))) \mapsto (x + m, y + n).$$

The action σ is properly discontinuous and the transformation group $\mathbf{Z} \times \mathbf{Z}$ is generated by two transformations

$$\psi_: \mathbf{R}^2 \to \mathbf{R}^2, \ (x, y) \mapsto (x + 1, y)$$

and

$$\phi : \ \mathbf{R}^2 \to \mathbf{R}^2, (x, y) \to (x, y + 1).$$

Since the group $\mathbf{Z} \times \mathbf{Z}$ acts on $\mathbf{R}^2$ properly discontinuously, it follows from the above discussion that the quotient manifold $\mathbf{R}^2/\mathbf{Z} \times \mathbf{Z}$ is the 2-totus T^2.

3.26 Applications of Morse–Sard's Theorem and the Brouwer Fixed-Point Theorem

This section proves the Brower fixed-point theorem from the viewpoint of differential topology and without using tools of algebraic topology. The same results are proved in **Basic Topology, Volume 3** of the present book series by using algebraic topology.

The usual technique of studying maps in differential topology consists of

(i) approximating by a smooth map,
(ii) finding a regular value and then
(iii) exploiting the topology of the inverse image of the regular value.

On the other hand, the technique of an extension

(i) does not use a regular value,
(ii) instead, uses a submanifold to which the map is transverse.

To obtain transversality, further approximation theorems are used.

Theorem 3.26.1 *Let ∂M be the boundary of a compact manifold M with boundary. Then there exists no retraction onto ∂M in the sense that there exists no smooth map*

$$f : M \to \partial M \text{ such that } \partial f : \partial M \to \partial M$$

is the identity map.

Proof Let M be any compact manifold of dimension n. We prove the theorem by the method of contradiction.

If possible, there exists a retraction $f : M \to \partial M$. Since f is a smooth map, it follows by Sard's Theorem 3.17.19 that there exists a point $x \in \partial M$ which is a regular value of the map f. This asserts that $f^{-1}(x)$ is a submanifold U of M with boundary $\partial U = U \cap \partial M$. The codimension of $f^{-1}(x) = U$ in M equals to the codimension of $\{x\}$ in ∂M, which is $n - 1$. This shows that U is 1-dimensional. Since U is closed, it is compact. Thus, $f^{-1}(x) = U$ is 1-dimensional and closed. Since, by hypothesis, $\partial f : \partial M \to \partial M$ is the identity map, U consists of only single point x, and hence $U = \{x\}$ contradicts Corollary 3.24.5. This proves the non-existence of any smooth retraction f onto ∂M. ❑

Remark 3.26.2 Theorem 3.26.1 is also valid without smoothness condition of M but it needs algebraic topology for its proof. Lemma 3.26.3 conveys an interplay between maps and manifolds. It is proved by using the result that every compact 1-dimensional manifold has an even number of boundary components (see Corollary 3.24.5). This shows that a simple topology of a 1-dimensional manifold facilitates a proof of highly nontrivial results on maps.

Theorem 3.26.3 *There is no retraction $f : \mathbf{D}^{n+1} \to S^n$ in the sense that there exists no continuous map $f : \mathbf{D}^{n+1} \to S^n$ such that*

$$f|_{S^n} = 1_{S^n} \text{ (identity map)}.$$

Proof We prove the theorem by the method of contradiction.

Proof I: If possible, let $f : \mathbf{D}^{n+1} \to S^n$ be a continuous retraction. Consider the continuous map

$$h : \mathbf{D}^{n+1} \to \mathbf{D}^{n+1},\ x \mapsto \begin{cases} 2x, \ if & 0 \leq ||x|| \leq 1/2 \\ x/||x||, \ if & 1/2 \leq ||x|| \leq 1. \end{cases}$$

The continuity of h follows by using pasting lemma. Again, consider the composite map

$$g = f \circ h : \mathbf{D}^{n+1} \to S^n.$$

Then g is also a retraction and it is also smooth on a closed subset C of S^n. Then the map g can be approximated by a smooth map

$$\psi : \mathbf{D}^{n+1} \to S^n \text{ such that } g|C = \psi|C.$$

This implies that ψ is a smooth retraction, which contradicts Theorem 3.26.1.

Proof II: If possible, $f : \mathbf{D}^{n+1} \to S^n$ has retraction in the sense that there exists a continuous map $f : S^n \to: S^n$ which is the identity map on S^n. Then there exists a new retraction $r : \mathbf{D}^{n+1} \to S^n$, which is C^∞ on a nbd of S^n in $\mathbf{D}^{n+1}$ such as

$$r : \mathbf{D}^{n+1} \to S^n, x \mapsto \begin{cases} f(x/||x||), \text{ if } 1/2 \leq ||x|| \leq 1 \\ f(x), \text{ if } 0 \leq ||x|| \leq 1/2. \end{cases}$$

Let $g : \mathbf{D}^{n+1} \to S^n$ be a C^∞ map that approximates r and agrees with r on a nbd of S^n. Then g is also a C^∞ retraction. Consider the regular value $y \in S^n$ of g (existence of y is ensured by Sard's theorem).

(i) $g^{-1}(y)$ is a compact 1-dimensional submanifold of $U \subset \mathbf{D}^{n+1}$ and
(ii) $\partial U = U \cap S^n$.

This shows that y is a boundary point of U and the component of U which contains the point y is diffeomorphic to a closed interval. Hence, it has another boundary point $z \in S^n$ such that $z \neq y$. This implies that $g(z) = z$, which contradicts that $z \in g^{-1}(y)$. This contradiction proves that there is no retraction $f : \mathbf{D}^{n+1} \to S^n$. ❑

Theorem 3.26.4 (The Brouwer Fixed-point Theorem for dimension n) *Every continuous map* $f : \mathbf{D^n} \to \mathbf{D^n}$ *has a fixed point.*

Proof If possible, let $f : \mathbf{D^n} \to \mathbf{D^n}$ have no fixed point. Then $f(x) \neq x, \ \forall x \in \mathbf{D^n}$. Then this f determines the directed line segment from $f(x)$ to x and let $g(x)$ be the point where this directed line segment hits the boundary S^{n-1} of $\mathbf{D^n}$. Consider $g : \mathbf{D}^n \to S^{n-1}$ as a map, which is the identity map on S^{n-1}. We claim that g is continuous. For any point x lying on the line segment joining the points $f(x)$ and $g(x)$,

$$g(x) = tx + (1-t)f(x), \ \forall t \geq 1.$$

This shows that g is continuous if t is a continuous function of x. Since $||g(x)|| = 1, \ \forall x \in \mathbf{D}^n$, by taking the dot product of vectors of both sides of the above formula, we have a quadratic equation in t :

$$t^2||x - f(x)||^2 + 2tf(x) \cdot [x - f(x)] + ||f(x)||^2 - 1 = 0.$$

This quadratic equation asserts that it has a unique positive root expressed in terms of continuous functions of x. Hence, the continuity of g contradicts Lemma 3.26.3.

❑

Theorem 3.26.5 gives another form of Theorem 3.26.4, replacing continuity of f by its smoothness condition.

Theorem 3.26.5 (The Brouwer Fixed-point Theorem of another form) *Every smooth map $f : \mathbf{D^n} \to \mathbf{D^n}$ has a fixed point.*

Proof Suppose that there exists a smooth map $f : \mathbf{D^n} \to \mathbf{D^n}$ which has a fixed point having no fixed point. Proceed as in Theorem 3.26.4, construct a retraction $g : \mathbf{D}^n \to S^{n-1}$ and prove that g is smooth function of x by showing that t depends smoothly on x.

❑

Theorem 3.26.6 is an important theorem from which fundamental theorem of algebra Theorem 3.18.1 follows as a corollary. Two other proofs are given in the proof of Theorem 3.18.1.

Theorem 3.26.6 *Let $p(z)$ and $q(z)$ be two complex polynomials and $1S^2$ be the Reimann sphere, which is the one-point compactification of $\mathbf{C}$ by ∞, i.e., $S^2 = \mathbf{C} \cup \infty$. The rational function $p(z)/q(z)$has a smooth extension*

$$f : S^2 \to S^2.$$

Then the map f is either constant or it is surjective.

Proof The key result used to prove this theorem is that $z \in S^2$ is a regular point of $f : S^2 \to S^2$ iff its complex derivative $f'(x) \neq 0$ and hence in this case the determinant of the real derivative $Df_z : \mathbf{R}^2 \to \mathbf{R}^2$ is positive. If f is not constant, then its complex derivative f' is not identically zero. Hence, there exists a regular point $z \in S^2$ of f. It follows from inverse function theorem that there exists an open set U ofS^2 about the point z that contains only regular points such that $f(U)$ is open in S^2. Let $v \in f(U)$ be a regular value of f. Then $f^{-1}(v) \neq \emptyset$. Thus, every point in $f^{-1}(v)$ is of positive type. It asserts that $deg\ f > 0$ and hence the map f is surjective. ❑

Theorem 3.26.7 (Fundamental Theorem of Algebra) *Every polynomial of degree $n \geq 1$ with complex coefficients has a root.*

Proof It follows as a direct consequence of Theorem 3.26.6 or see proof of Theorem 3.18.1. ❑

Theorem 3.26.8 (Frobenius theorem) *Every real square matrix A with non-negative entries has a real non-negative eigenvalue.*

Proof Without any loss of generality, it is assumed that the matrix A is nonsingular; otherwise, 0 will be an eigenvalue. Let A_L denote the associated linear map of $\mathbf{R}^n$. Consider the map

$$A_L : S^{n-1} \to S^{n-1},\ x \mapsto Ax/|Ax|.$$

Let

$$Q_1 = \{x = (x_1, x_2, \ldots, x_n) \in S^{n-1} : x_i \geq 0,\ \forall i = i, 2, \ldots, n\}$$

denote the first quadrant. Then A_L maps Q_1 into itself and Q_1 is homeomorphic to $\mathbf{D}^{n-1}$. This asserts that there exists a continuous bijective map

$$f : Q_1 \to \mathbf{D}^{n-1}$$

with a continuous inverse. Consequently, the map

$$\psi_f : \mathbf{D}^{n-1} \to \mathbf{D}^{n-1},\ x \mapsto A_L(x)/|A_L(x)|$$

is a continuous map and hence it has a fixed point by the Brouwer fixed-point Theorem 3.26.4. If x_0 is this fixed point, then

$$A_L(x_0)/|A_L(x_0)| = x_0 \implies A_L(x_0) = ||A_L(x_0)||x_0.$$

This proves that the matrix A has a positive eigenvalue. ❑

3.26.1 More Applications

Proposition 3.26.9 *Let* $\mathbf{B_0}(r)$ *denote the open ball in* $\mathbf{R}^n$ *with center at* $\mathbf{0} = (0, 0, \ldots, 0)$ *and radius* r. *Then*

(i) $\mathbf{B_0}(r)$ *is diffeomorphic to* $\mathbf{R}^n$ *and*
(ii) *every point of an* n*-dimensional manifold has an open nbd diffeomorphic to* $\mathbf{R}^n$.

Proof Consider the two maps

$$f : \mathbf{B_0}(r) \to \mathbf{R}^n,\ x \to \frac{rx}{[r^2 + ||x||^2]^{\frac{1}{2}}}$$

and

$$g : \mathbf{R}^n \to \mathbf{B_0}(r),\ y \to \frac{ry}{[r^2 + ||y||^2]^{\frac{1}{2}}}.$$

Then f and g are inverse maps of each other and they are also smooth maps. This shows that $\mathbf{B_0}(r)$ is diffeomorphic to $\mathbf{R}^n$.

For the second part, for every $x \in X$, consider a local parametrization $\psi : U \to X$ at x and the open ball $\mathbf{B} = B_{\psi^{-1}(x)}(r)$ such that $\mathbf{B} \subset U$ (by taking its radius $r > 0$ sufficiently small). Then $\psi|U$ is also a local parametrization at x. For a diffeomorphism $\phi : \mathbf{R}^n \to \mathbf{B}$, the composite map

$$\psi|\mathbf{B} \circ \phi : \mathbf{R}^n \to X$$

is a local parametrization at x. ❑

Proposition 3.26.10 (Linear coordinates) *Let V be an m-dimensional vector subspace of $\mathbf{R}^n$. Then*

(i) *V is a manifold diffeomorphic to $\mathbf{R}^m$;*
(ii) *all linear transformations on V are smooth;*
(iii) *if $f : \mathbf{R}^m \to V$ is a linear isomorphism, then its corresponding coordinate functions (known as linear coordinates) are all linear functionals on V.*

Proof Consider a linear transformation $T : \mathbf{R}^n \to \mathbf{R}^m$. Since $dT_x = T, \ \forall x \in \mathbf{R}^n$, it follows that T is a smooth map. Moreover, every linear map f on V is smooth, because f has a linear extension over $\mathbf{R}^n$. Let $\psi : V \to \mathbf{R}^n$ be a linear isomorphism with respect to a chosen basis $\mathbf{B_V}$ of V. Then ψ has a linear extension $\overline{\psi} : \mathbf{R}^n \to \mathbf{R}^m$ over $\mathbf{R}^n$. Hence, ψ is smooth and its inverse $\psi^{-1} : \mathbf{R}^m$ is linear and smooth. This proves that V is a manifold diffeomorphic to $\mathbf{R}^m$. To prove the last part, let $p_i : \mathbf{R}^n \to \mathbf{R}, \ x = (x_1, x_2, \ldots, x_i, \ldots, x_n) \mapsto x_i$ be the projection map and $\psi = (\psi_1, \psi_2, \ldots, \psi_i, \ldots, \psi_n)$ be the coordinate functions of ψ. Then $p_i \circ \psi = \psi_i, \ \forall i = 1, 2, \ldots, n$, where each

$$\psi_i : V \to \mathbf{R}, \ x = (x_1, x_2, \ldots, x_i, \ldots, x_n) \mapsto x_i$$

is a linear functional. ❑

Proposition 3.26.11 *The union of two coordinate axes in the Euclidean plane is not a manifold.*

Proof Consider $M = \{x = 0\} \cup \{y = 0\}$, the union of two coordinate axes in the Euclidean plane $\mathbf{R}^2$. If its origin $(0, 0)$ is deleted from M, then we have four disconnected components; on the other hand, if any point is deleted from $\mathbf{R}$, then we have only two disconnected components. As connectedness is a topological property, this connectivity argument proves the proposition. ❑

Proposition 3.26.12 *The map*

$$f : \mathbf{R} \to \mathbf{R}, \ x \mapsto \begin{cases} e^{\frac{1}{x^2}}, \ if \, x > 0, \\ 0, \ if \, x \leq 0 \end{cases}.$$

is smooth.

Proof Since for all $x > 0$, the nth derivative $f^n(x)$ is a rational polynomial such that

$$lim_{x \to 0} f^n(x) = 0, \ \forall n = 0, 1, 2, \ldots,$$

it follows that the map f is smooth. ❑

Proposition 3.26.13 *Let $f : \mathbf{R} \to \mathbf{R}$ be a local diffeomorphism. Then*

(i) *Imf is an open interval;*

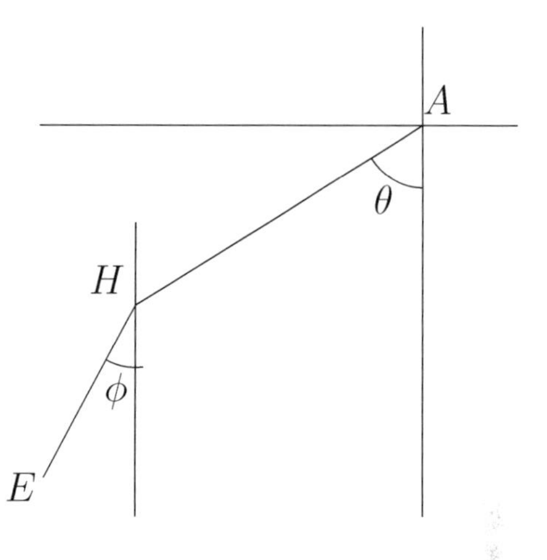

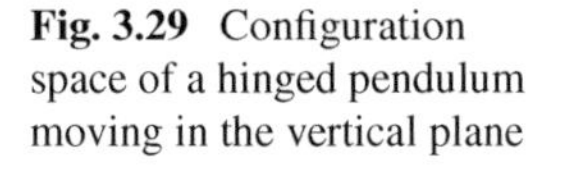
Fig. 3.29 Configuration space of a hinged pendulum moving in the vertical plane

(ii) $\psi : \mathbf{R} \to Imf,\ x \mapsto f(x)$ *is a diffeomorphism.*

Proof Since $\mathbf{R}$ is connected, it follows that Imf is an interval. Again, since by hypothesis, f is a local diffeomorphism, it follows that f is an open map such that $f'(x) \neq 0$. Thus, f is an injective local diffeomorphism such that its image is an open interval and hence ψ is a diffeomorphism, because every local homeomorphism is open and every bijective local diffeomorphism is a diffeomorphism. ❑

3.26.2 Configuration Space of a Hinged Pendulum

This subsection presents an idea of a configuration space to locate the set of positions of a moving system. It is an important example of smooth manifolds and is used in mechanics, physics and some other applied sciences. Proposition 3.26.14 gives an interesting example of a configuration space.

Proposition 3.26.14 *The* ***configuration space*** *of a hinged pendulum swinging in the vertical plane is the* 2*-torus* T^2.

Proof Let the pendulum be attached at the point A, hinged at the point H and end at the point E as shown in Fig. 3.29. Every position of the system can be described by the direction of the rod AH together with the direction of the rod HE or by the pair of angles θ and ϕ, which vary independently such that $0 \leq \theta < 2\pi,\ 0 \leq \alpha < 2\pi$. It asserts that the configuration space of this system is the product space $S^1 \times S^1$, which is topologically equivalent to the 2-torus T^2. ❑

3.26.3 Solution Set of a System of Equations

This subsection studies the solution set of a system of equations from the viewpoint of the manifold and proves by using the concept of solution set that the n-sphere S^n is an n-dimensional submanifold in $\mathbf{R}^{n+1}$ of C^∞-class.

Consider a given system of m equations in n variables ($n \geq m$) given by

$$f_1(x_1, x_2, \ldots, x_n) = 0,$$
$$f_2(x_1, x_2, \ldots, x_n) = 0,$$
$$\ldots$$
$$f_m(x_1, x_2, \ldots, x_n) = 0,$$

where $f_1, f_2, \ldots, f_m : \mathbf{R}^n \to \mathbf{R}$ are functions of C^r-class, for any $r \geq 1$. The family $\mathcal{F} = \{f_1, f_2, \ldots, f_m\}$ of functions determine a C^r-function

$$f : \mathbf{R}^n \to \mathbf{R}^m,\ x \mapsto (f_1(x), f_2(x), \ldots, f_m(x)),$$

which gives the solution set of the given system of equations described in Definition 3.26.15.

Definition 3.26.15 The inverse image $f^{-1}\{0\} = M$ of

$$f : \mathbf{R}^n \to \mathbf{R}^m,\ x \mapsto (f_1(x), f_2(x), \ldots, f_m(x)),$$

is called the **solution set of the given system of equations** determined by the family $\mathcal{F}$.

Theorem 3.26.16 is now proved by applying the concept of solution set.

Theorem 3.26.16 *S^n is an n-dimensional submanifold in $\mathbf{R}^{n+1}$ of C^∞-class.*

Proof Consider the function

$$f : \mathbf{R}^n \to \mathbf{R},\ (x_1, x_2, \ldots, x_n) \mapsto x_1^2 + x_2^2 + \cdots + x_{n+1}^2 - 1.$$

Then S^n is the solution set of the equation

$$x_1^2 + x_2^2 + \cdots + x_{n+1}^2 - 1 = 0.$$

The rank of the Jacobian matrix $\frac{\partial f}{\partial x}|_x$ of f is 1, $\forall\, x \in S^n$. Then it follows by Exercise 20 of Sect. 3.27 that S^n is an n-dimensional submanifold in $\mathbf{R}^{n+1}$ of C^∞-class. ❑

For the study of the solution set of a given system of equations from the viewpoint of manifold, see Exercise 20 of Sect. 3.27.

3.27 Exercises

1. Let M be an n-dimensional C^r-manifold and U be an open set of M. Show that U is also a C^r-manifold.
2. Show that every connected manifold M is arcwise connected in the sense that given any two points $a, b \in M$ there is a smooth curve $\alpha : [0, 1] = \mathbf{I} \to M$ such that $\alpha(0) = a,\ \alpha(1) = b$.
3. Show that there is a smooth function $f : \mathbf{R} \to \mathbf{R}$ such that

   (i) $f(x) > 0,\ \forall x \in (-1, 1)$;
   (ii) $f(x) = 0,\ \forall x\ satisfing\ |x| > 1$.

[Consider $f(x) = e^{-1/(x-1)^2} e^{-1/(x+1)^2},\ \forall x\ satisfying\ |x| < 1;\ and\ f(x) = 0,$ $otherwise$.

4. Given smooth manifolds M, M', N, N', let $f : M \to M'$ and $g : N \to N'$ be two maps and their the product map be defined by

$$f \times g : M \times N \to M' \times N',\ (x, y) \mapsto (f(x), g(x)).$$

   Show that

   (i) if f and g are both smooth, then $f \times g$ is also ;
   (ii) if f and g are both diffeomorphisms, then $f \times g$ is also so.

5. Given smooth manifolds M, N, show that the projection maps

   (i) $p_1 : M \times N \to M,\ (x, y) \to x$ and
   (ii) $p_2 : M \times N \to M,\ (x, y) \to y$

   are both smooth maps.
6. Show that the map $f : (0, 2\pi) \to \mathbf{R}^2,\ x \mapsto (sinx, sin2x)$

   (i) is continuous and bijective but f^{-1} is not continuous;
   (ii) is a C^∞-embedding.

7. A smooth function $f : \mathbf{R}^n \to \mathbf{R}^n$ is said be **locally diffeomorphism,** if every point $x \in \mathbf{R}^n$ has a nbd U such that $f|_U$ is a diffeomorphism. Show that the function

$$f : \mathbf{R}^2 \to \mathbf{R}^2,\ (x, y) \mapsto (e^x cos\ y,\ e^x sin\ y)$$

   is locally diffeomorphism but f is not a diffeomorphism.
8. Let $U \subset \mathbf{R}^n$ be an open set and $f : \mathbf{R}^n \to \mathbf{R}^n$ be a differentiable function on U such that f has a differentiable inverse. Show that the Jacobian determinant of f is nonzero, $\forall x \in U$. (For its partial converse, see inverse function theorem).
9. **(Inverse function theorem)** Let $U \subset \mathbf{R}^n$ be an open set and $f : \mathbf{R}^n \to \mathbf{R}^n$ be a differentiable function on U such that the Jacobian determinant of f is nonzero

at some point $x \in U$. Show that there exists a nbd U_x of x and a nbd $U_{f(x)}$ of $f(x)$ such that

(i) f maps U_x homeomorphically onto $U_{f(x)}$;
(ii) the inverse of f on $U_{f(x)}$ is also a differentiable function onto U_x.

10. A Hausdorff topological space (X, τ) is said to be a **paracompact-manifold** if it is paracompact and there is an integer n such that every point of X has an open nbd homeomorphic to an open subset of the Euclidean n-space $\mathbf{R}^n$. Prove that

(i) every paracompact-manifold (X, τ) is metrizable;
(ii) every Hausdorff topological (X, τ) is a manifold iff it is a paracompact-manifold having only countably many components.

11. **(Paracompactness property of manifold)** Show that

(i) every second countable, locally compact Hausdorff space is paracompact;
(ii) every topological manifold with or without boundary is paracompact.

12. Show that the paraboloid in $\mathbf{R}^3$ defined by $x^2 + y^2 - z^2 = r$ is a manifold for all $r > 0$ but $x^2 + y^2 - z^2 = 0$ is not a manifold.
[Hint: Use the connectivity property to show that $(0, 0, 0)$ has no nbd diffeomorphic $\mathbf{R}^2$.]

13. Let M be a smooth manifold and A be a closed subset of M. If $f : A \to \mathbf{R}$ is a smooth map in the induced structure, show that

(i) f has a locally smooth extension at each point $x \in A$ to a smooth function on some nbd U_x of x in M and
(ii) there exists a smooth function $\overline{f} : M \to \mathbf{R}$, extending f.

14. Let M be a smooth manifold. Show that the diagonal

$$\Delta(M) = \{(x, x)\} \subset M \times M$$

is a smooth manifold diffeomorphic to M.

15. Given smooth manifolds M and N, let $f : M \to N$ be a smooth map and $G_f = \{(x, f(x)) : x \in M\} \subset M \times N$ be its graph. Show that

(i) G_f is a manifold and
(ii) the map

$$\psi : M \to G_f : x \mapsto (x, f(x))$$

is a diffeomorphism.

16. Show that the subspaces $\mathbf{R}_1^n$ and $\mathbf{R}_2^n$ of the Euclidean space $\mathbf{R}^n$ defined by

(i)

$$\mathbf{R}_1^n = \{x \in \mathbf{R}^n : ||x|| \leq 1\}$$

and

(ii)

$$\mathbf{R}^n_2 = \{x \in \mathbf{R}^n : ||x|| \geq 1\}$$

are both n-dimensional C^∞-manifolds with S^{n-1} as their common boundary.

17. Consider the 3-sphere S^3 concising of the set of unit quaternions

$$S^3 = \{q = x + iy + jz + kw : x, y, z, w \in \mathbf{R} \text{ and } x^2 + y^2 + z^2 + w^2 = 1\}.$$

Define the set

$$U = \{x + iy + jz + kw \in S^3 : x > 0\} \subset S^3$$

and the map

$$\psi : U \to \mathbf{R}^3, \; x + iy + jz + kw \mapsto (y, z, w).$$

Consider ψ as a chart. Define for each $q \in S^3$, a chart $\psi_q(u) = \psi(q^{-1}u)$. Prove that the set of all charts constitutes an atlas for a smooth structure of dimension 3 on the 3-sphere S^3.

18. Let M be a smooth manifold and A, U be two subsets of M such that

(i) $A \subset U \subset M$;
(ii) A is compact and U is open.

Show that there is a smooth function

$$f : M \to [0, \infty) \text{ such that } f(x) > 0, \; \forall x \in A \text{ with support } f \subset U.$$

19. Let M be a differentiable manifold of dimension m and N be a differentiable manifold of dimension n ($\geq m$). Given a differentiable map $f : M \to N$, show that under suitable local coordinates $(x_1, x_2, \ldots, x_m)$ about a point $x \in M$ and $(y_1, y_2, \ldots, y_n)$ about the point $f(x) \in N$, the map f can be expressed locally as follows:

(i) $y_k = x_k, \; \forall k = 1, 2, \ldots, m$;
(ii) $y_k = f_k(x_1, x_2, \ldots, x_m), \; \forall k = m + 1, m + 2, \ldots, n$, where f_k are differentiable functions.

20. Consider a system (A) of m equations in n variables $(n \geq m)$
$f_1(x_1, x_2, \ldots, x_n) = 0$,
$f_2(x_1, x_2, \ldots, x_n) = 0$,
$\ldots$
$f_m(x_1, x_2, \ldots, x_n) = 0$,
where $f_1, f_2, \ldots, f_m : \mathbf{R}^n \to \mathbf{R}$ are functions of C^r-class for every $r \geq 1$.

Let M be the nonempty solution set of of the above system (A) of equations and $f : \mathbf{R}^n \to \mathbf{R}^m$ $(n \geq m)$ be a C^r-function determined by the family of functions $\{f_1, f_2, \ldots, f_n\}$. If for every point $x \in M$, the rank of Jacobian matrix $\frac{\partial f}{\partial x}|_x$ of f is m, show that $M = f^{-1}(0)$ is an $(n-m)$-dimensional submanifold of class C^r in $\mathbf{R}^n$.
[Hint: See Sect. 3.26.3.]

21. Let S_1 and S_2 be two compact connected 2-manifolds and $S_1 \# S_2$ be their connected sum. Let $\mathcal{S}$ be the set of compact connected 2-manifolds. Show that

(i) $S_1 \# S_2$ is also a compact connected manifold, i.e., the operation # is closed on $\mathcal{S}$;
(ii) the operation # is on $\mathcal{S}$ is commutative and associative with the 2-sphere as a unit (up to homeomorphism);
(iii) the set $\mathcal{S}$ endowed with the operation # forms a commutative semigroup with identity element;
(iv) any compact connected 2-manifold is homeomorphic either to the sphere S^2 or the torus $T = S^1 \times S^1$ or the real projective plane $\mathbf{R}P^2$ or to a connected sum of several copies of T and $\mathbf{R}P^2$;
(v) the connected sums $\mathbf{R}P^2 \# T$ and $\mathbf{R}P^2 \# \mathbf{R}P^2 \# \mathbf{R}P^2$ are equal (up to homeomorphism) (which asserts that while expressing a compact connected 2-manifold as a connected sum, there is no need of combination of tori and projective planes).
[Hint: See Sect. 3.23 or Massey (1967).]

22. Let $X(x, y) = -y\, \partial/\partial x + x\, \partial/\partial y$ be a vector field in the Euclidean plane $\mathbf{R}^2$. Show that

(i) the action

$$\psi : \mathbf{R} \times \mathbf{R}^2 \to \mathbf{R}^2,\ (t, (x, y)) \mapsto (xcost - ysint,\ xsint + ycost)$$

is a flow generated by the vector field X satisfying the following properties:
(ii) the flow through (x, y) is the circle having the center at the origin;
(iii) if $\psi_t \equiv \psi(t, -)$, then $\psi_t = \psi_{2n\pi+t}$;
(iv) one-parameter group $\{\psi_t = \psi(t, -)\}$ is isomorphic to $SO(2, \mathbf{R})$ or to the circle group $S^1 \cong U(1, \mathbf{R})$.

23. Let M be a compact manifold of dimension n and $f : \mathbf{R}^n \to M$ be a smooth map and $y \in M$ be a regular value of f. If $K = f^{-1}(y)$ is compact, show that

(i) there is an open nbd U of K lying in a tubular nbd of K with normal retraction $r : U \to K$ and
(ii) an open nbd $V \approx \mathbf{R}^n$ of y in M such that

$$r \times f : U \times \mathbf{R}^n \to K\mathbf{V}$$

is a diffeomorphism.

24. Let G be a finite group acting smoothly on a smooth manifold M. If σ is this action and $x_0 \in M$ is a fixed point under σ, show that there a nbd U of x_0 in which there is a local coordinate system such that σ is linear with respect to local coordinate system.

25. Let $X(x, y) = y\ \partial/\partial x + x\ \partial/\partial y$ be a vector field in the Euclidean plane $\mathbf{R}^2$. Find the flow generated by the vector field $X(x, y)$.

26. **(Existence of Finite Partition of Unity)**

(i) Let (X, τ) be a normal space and $\mathcal{C} = \{V_1, V_2, \ldots, V_n\}$ be an open covering of X. Show that there exists a partition of unity subordinate to the covering $\mathcal{C}$.

(ii) Let M be a C^r-manifold of dimension n for $r = 1, 2, \ldots, \infty$ and $\mathcal{C} = \{U_i\}$ be an open covering of M. Show that there exists a partition of unity of C^r-class subordinate to $\mathcal{C}$.

27. **(Smooth Approximation Theorem)** Let M and N be two smooth manifolds and $f, g : M \to N$ be two smooth maps. If d is a metric on N and $\delta : M \to \mathbf{R}$ is a positive-valued real function, then g is said to be a δ-approximation to f, if

$$d(f(x), g(x)) < \delta(x),\ \ \forall x \in M.v$$

Let A be a closed subset of M. If $h : M \to N$ be a continuous map such that it is smooth on A, show that there exist

(i) a positive continuous function $\delta : M \to \mathbf{R}$ and

(ii) a smooth map $k : M \to N$ such that k agrees with f on A and k is a δ-approximation to f.

28. **(Alternative form of the Smooth Urysohn Lemma)** Let M be a smooth manifold and P, Q be two disjoint smooth closed subsets of M. Show that there is a smooth function $f : M \to \mathbf{R}$
such that

$$f(x) = \begin{cases} 0, & \text{for all } x \in P \\ 1, & \text{for all } x \in Q \end{cases}$$

and

$$0 \leq f(x) \leq 1 \text{ for all } x \in M.$$

[Hint: Use partition of unity.]

29. Let M be a C^r-manifold for $r = 1, 2, \ldots, \infty$ and $K \subset V \subset M$, where K is compact and V is open. Show that there exists a C^r-function f such that

$$f : M \to [0, 1], x \mapsto \begin{cases} 0, & \text{if } x \in M - V, \\ 1, & \text{if } x \in V. \end{cases}$$

30. Let M and N be two smooth manifolds and $f : M \to N$ be a proper injective immersion. Show that f is an embedding.
[Hint: Use the result that every injective immersion that is a homeomorphism onto its image is an embedding.]

31. Let M be a smooth manifold of dimension n and $f : M \to \mathbf{R}^m$ be an immersion with $m > 2n$ and $\alpha : M \to \mathbf{R}$ be a positive continuous function. Show that

(i) the immersion $f : M \to \mathbf{R}^m$ that is injective can be α-approximated by an injective immersion $h : M \to \mathbf{R}^m$;

(ii) if the immersion f is injective on an open nbd U of a closed subset $A \subset M$, then the map h can be so chosen such that

$$h(x) = f(x), \ \forall\, x \in U.$$

32. **(Whiteney Embedding Theorem)** Let M be a compact n-dimensional smooth manifold. Prove that

(i) there exists a smooth embedding

$$f : M \to \mathbf{R}^{2n+1};$$

(ii) M can be embedded as a submanifold in the Euclidean space $\mathbf{R}^{2n+1}$ and also as a closed subspace of $\mathbf{R}^{2n+1}$.

[Hint: See Theorem 3.14.20.]

33. Let M be an n-dimensional topological manifold. Show that the following statements are equivalent.

(i) M is a paracompact space;
(ii) M is a second countable space;
(iii) M is a Lindelöf space;
(iv) M is a metrizable space.

34. Let X be a smooth manifold and $\{U_a : a \in \mathbf{A}\}$ be open covering of X. Then there is a partition of unity $\{f_b : b \in \mathbf{B}\}$ subordinate to this covering with the property that the functions f_b are all smooth.
[Hint: Assume that the covering is locally finite and each member is a subset of the domain of coordinate chart and has compact closure. Then there is a open covering $\{V_a : a \in \mathbf{A}\}$ such that $\overline{V_a} \subset U_a, \ \forall\, a \in \mathbf{A}$.]

35. Show that

(i) the general linear group $\mathrm{GL}(n, \mathbf{C})$ is an open complex submanifold of $\mathbf{C}^{n^2}$;
(ii) $dim_{\mathbf{C}}\, GL(n, \mathbf{C}) = n^2$.

36. Let M and N be two manifolds with ∂M and ∂N be their boundaries. Show that

(i)

$$\partial(M \times N) = (\partial M \times N) \cup (M \times \partial N);$$

(ii) if $\psi : \partial M \to \partial N$ is a homeomorphism and $M \cup_\psi N$ is obtained by gluing the manifolds M and N along their boundaries by identifying $x \in \partial M$ with $\psi(x) \in \partial N$, then $M \cup_\psi N$ is also a manifold with boundary.

37. Let $\psi : M \to N$ be a local homeomorphism in the sense that ψ is continuous and every point of M has an open nbd V such that $\psi(V)$ is open in N and the restriction $\psi|_V : V \to \psi(V)$ is a homeomorphism. Show that M is a manifold iff N is a manifold.

38. Let M and N be two smooth manifolds and $f : M \to N$ be a smooth map. Show that f induces a smooth map

$$df : T(M) \to T(N),\ (x, v) \mapsto (f(x), df_x(v)).$$

39. Show that the family of all n-dimensional C^r-manifolds ($n \geq 1$) and their C^r-mappings form a category, whose equivalences (in the sense of category theory) are C^r-diffeomorphisms of manifolds.

40. Let M be an n-dimensional manifold with boundary ∂M. Show that ∂M is a manifold of dimension $(n-1)$ without boundary.

41. Let M be a smooth manifold with boundary and $f : M \to \mathbf{R}$ be a smooth function having regular value 0. Show that

(i) the subset $S = \{x \in M : f(x) \geq 0\} \subset M$ is a manifold with boundary;
(ii) its boundary ∂S is given by $\partial S = f^{-1}(0)$.

42. Let M be a smooth manifold and $\psi : \mathbf{I} \to C$ and $\phi : \mathbf{I} \to D$ be two parameterizations in M. If $C \cap D$ is connected, show that $C \cup D$ is a parameterized curve.

43. Let M be a noncompact connected manifold with $dim\ M = 1$. Show that it is diffeomorphic to an interval.
[Hint: See Sect. 3.24.]

44. Let M be a manifold of dimension one and N be a subset of M diffeomorphic to an open interval in $\mathbf{R}$. Show that the $\overline{M} - N$ consists of at most two points.

45. Let $\psi : \mathbf{R}^m \to \mathbf{R}^n$ be a linear map. Identifying $T_x(\mathbf{R}^p)$ with $\mathbf{R}^p$ by using $\partial/\partial x_i$ as the ith standard basis vector, prove that the induced linear transformation ψ_* coincides with ψ.

46. Let the curve $\psi : \mathbf{R} \to \mathbf{R}^n$ be an embedding. Show that the concept $\psi_* \left(\frac{d}{dt}\right)$ and the classical concept of tangent vector to the curve $\psi : \mathbf{R} \to \mathbf{R}^n$ coincides under the identification of the tangent space with the Euclidean space $\mathbf{R}^n$.

47. Let $\alpha : \mathbf{R} \to \mathbf{R}^2$ be a smooth curve and A be the set of all points $r \in \mathbf{R}$ such that the circle C_r with center at the origin and radius r is tangent to the curve α at some point. Show that in $\mathbf{R}$ the set A has an empty interior.

48. Let X be the union of two coordinate axes in $\mathbf{R}^{n+1}$. Show that X is not a manifold. [Hint: Any nbd of the origin in X can be decomposed into at least four connected components but it is not true if the origin is deleted.]

49. Let M and N be two smooth manifolds such that $dim\ M < dim\ N$. If $f : X \to Y$ is a smooth map, show that $f(M) \neq N$.
[Hint: Use Sard's theorem.]

50. Let $f : U(n, \mathbf{C}) \to S^{2n-1}$ be the map which assigns to every unitary matrix in $U(n, \mathbf{C})$ its first column.

(i) Show that all points of f are regular.
(ii) Determine the inverse image $f^{-1}(z)$.

[Hint: Use the rank of the Jacobian matrix $J(f)$ in local coordinates.]

51. Show that the Klein bottle $\mathbf{K}$ is a nonorientable 2-dimensional manifold. [Hint: Proceed representing $\mathbf{K}$ as a square identified its opposite sides and transferring the basis consisting of the tangent vectors along the midline.]

52. Let $S^n \subset \mathbf{R}^{n+1}$ be the n-sphere with its north pole $N = (0, 0, \ldots, 1)$. Show that the stereographic projection

$$f : S^n - N \to \mathbf{R}^n$$

is a diffeomorphism.

53. **(i)** Let $S^n \subset \mathbf{R}^{n+1}$ be the n-sphere with its north pole $N = (0, 0, \ldots, 1)$. Show that the stereographic projection $f : S^n - N \to \mathbf{R}^n$ is a diffeomorphism.
(ii) Let f be the stereographic projection of a sphere from its north pole onto a tangent plane at its south pole. Show that f is a diffeomorphism at all points of the sphere excepting its north pole.

[Hint: Use local coordinates and determine the rank of the Jacobian matrix $J(f)$ in local coordinates.]

54. Prove the following statements:

(i) The real projective line $\mathbf{R}P^1$ is diffeomorphic to the circle S^1.
(ii) The complex projective line $\mathbf{C}P^1$ is diffeomorphic to the sphere S^2.
(iii) There exists a natural diffeomorphism between the Grassmann manifolds.

$$\psi : G_k(\mathbf{R}^n) \to G_{n-k}(\mathbf{R}^n) : .$$

55. Show that the action of a discrete topological group G on a smooth manifold M is smooth iff for every $g \in G$, the map

$$\psi_G : G \to G, x \mapsto gx$$

is smooth.

56. Let G be a topological group, M be a smooth manifold and σ be an action of G on M. Prove the following statements:

(i) If the action σ is properly discontinuous, then the orbit space $M\ mod\ G$ has the structure of a quotient manifold having the same dimension as M such that the projection

$$p : M \to M\ mod\ G, x \mapsto Gx$$

is a submersion.

(ii) Given any transformation group G of M and a diffeomorphism f of G with $f \notin G$, the set

$$S_f = \{fgf^{-1} : g \in G\}$$

is a transformation group of M such that if the orbit space $M\ mod\ G$ has the structure of a quotient manifold, then the orbit space $M\ mod\ S_f$ has a differentiable structure diffeomorphic to $M\ mod\ G$.

57. Show that

(i) The extended complex plane $\mathbf{C}^* = \mathbf{C} \cup \{\infty\}$ is a complex manifold, called the **Riemann surface** in the sense that it looks locally like the complex plane.

(ii) The topological space $\mathbf{C}^n$ is a complex manifold of complex dimension n with the unique chart (V, ψ), where $V = \mathbf{C}^n$ and $\psi : V \to V$ is the identity map.

(iii) The complex projective space $\mathbf{C}P^n$ is an n-dimensional complex manifold with $(n + 1)$ charts (V_i, ψ_i), formulated by

$$V_i = \{(z_1, z_2, \ldots, z_{n+1}) \in \mathbf{C}P^n : z_i \neq 0\}$$

and

$$\psi_i : V_i \to \mathbf{C}^n, (z_1, z_2, \ldots, z_{n+1}) \mapsto \left(\frac{z_1}{z_i}, \frac{z_2}{z_i}, \ldots, \frac{z_{i-1}}{z_i}, \frac{z_{i+1}}{z_i} \ldots, \frac{z_{n+1}}{z_i}\right) \in \mathbf{C}^n.$$

58. Show that

(i) The complex projective $\mathbf{C}P^n$ is orientable as a real manifold of dimension $2n$.

(ii) The real projective $\mathbf{R}P^{2n}$ is not orientable.

(iii) The real analytic structure of $\mathbf{C}P^n$ is different from that of $\mathbf{R}P^{2n}$.

59. Let $\mathcal{X}^n$ be the solution set of the system of equations:

$$f_i(x_1, x_2, \ldots, x_m) = 0, \quad \text{for } i = 1, 2, \ldots, m - n$$

such that the rank $\{\partial f_i/\partial x_j\} = m - n$. Show that $\mathcal{X}^n$ is a submanifold. [Hint: Use the implicit function theorem.]

60. Let M be a smooth manifold of dimension m and N be a smooth manifold of dimension n and $f : M \to N$ be a smooth map. If $m \geq n$ and $y \in N$ is a regular value of f, show that

(i) the subset $f^{-1}(y)$ of M is a smooth manifold with dimension $m - n$;
(ii) using (i), the unit sphere S^m is a smooth manifold of dimension m.

[Hint: Consider

$$f : \mathbf{R}^{m+1} \to \mathbf{R},\ x = (x_1, x_2, \ldots, x_{m+1}) \mapsto x_1^2 + x_2^2 + \cdots + x_{m+1}^2.$$

Any positive $y \in \mathbf{R}$ is a regular value of f and the $f^{-1}(1)$ is the smooth manifold S^m of dimension m.]

61. Let M and N be smooth manifolds and $f : M \to N$ be a smooth map. Show that

(i) the null space of

$$df_x : T_x(M) \to T_{f(x)}(N)$$

is the same as the tangent space $T_x(M') \subset T_x(M)$ of the submanifold $M' = f^{-1}(y)$;
(ii) df_x maps the orthogonal complement of $T_x(M')$ isomorphically onto $T_{f(x)}(N)$.

62. Let $U_1, U_2, \ldots, U_n$ be a covering of the closed interval $[a, b]$ in $\mathbf{R}^1 = \mathbf{R}$. Show that there exists another covering $V_1, V_2, \ldots, V_m$ of $[a, b]$ such that

(i) each V_j is contained in some U_i and
(ii) $\Sigma_{j=1}^m$ length $(V_j) < 2(b - a)$.

63. Let $\mathbf{I}_1 \subset \mathbf{R}$, $V_{\mathbf{I}_1} = \mathbf{I}_1 \times \mathbf{R}^{n-1} \subset \mathbf{R}^n$ and A be a compact subset of $\mathbf{R}^n$. If for some $c \in \mathbf{R}_1$, the set $A \cap V_c$ is contained in an open set U of V_c, show that there exists a suitably small interval $\mathbf{I}_1$ about the point c in $\mathbf{R}$ such that

$$A \cap V_{\mathbf{I}_1} \subset \mathbf{I}_1 \times U.$$

[Hint: If the statement is not true, there exists a sequence of points $\{x_n, c_n\}$ in A such that $c_n \to c$ and $x_n \notin U$. To obtain a contradiction, replace this sequence by a convergent sequence.]

64. **(Mini-Sard)** Let $U \subset \mathbf{R}^n$ be open and $f : U \to \mathbf{R}^m$ be a smooth map. If $m \geq n$, show that $f(U)$ has measure zero in $\mathbf{R}^m$.

65. Let $U \subset \mathbf{R}^n$ be open and $f : U \to \mathbf{R}^m$ be a smooth map. If C is the set of critical points of f, show that $f(C)$ has measure zero in $\mathbf{R}^m$.

66. Show that every compact 1-dimensional manifold with boundary is diffeomorphic to either $[0, 1]$ or S^1.

67. Let $f : \mathbf{R}^3 - \{(0,0,0)\} \to \mathbf{R}$, $(x, y, z) \mapsto [2 - (x^2 + y^2)^{1/2}]^2 + z^2$. Show that 1 is a regular value of f and identify the manifold $M = f^{-1}(1)$. Further show that

(i) if S is the surface $S = \{(x, y, z) \in \mathbf{R}^3 : x^2 + y^2 = 4\}$, then the statement $M \pitchfork S$ (M is transverse to S) is true;

(ii) if S is the surface $S = \{(x, y, z) \in \mathbf{R}^3 : x^2 + y^2 = 1\}$, then the statement $M \pitchfork S$ is not true;

(iii) if P is the plane $P = \{(x, y, z) \in \mathbf{R}^3 : x = 1\}$, then the statement $M \pitchfork P$ is not true.

Identify the manifolds $M \cap S$ and $M \cap P$.

68. The **codimension** of an arbitrary submanifold M_1 of M in the manifold M, denoted by $codim\ M_1$, is defined by the formula

$$codim\ M_1 = dim\ M - dim\ M_1.$$

Let $f : M \to N$ be a smooth map and $N_1 \subset N$ be a submanifold of N. If $f \pitchfork N_1$ (f is transversal to the submanifold N_1), then show that

(i) the preimage $f^{-1}(N_1)$ is a submanifold of M and

(ii) the codimension of $f^{-1}(N_1)$ in M = the codimension of $f^{-1}(N_1)$ in N.

69. Let M and X be both submanifolds of a manifold N such that they are transversal ($M \pitchfork X$). Show that

(i) $M \cap X$ is also a submanifold of N;

(ii) the codim of $M \cap X$ is given by

$$codim\ (M \cap X) = codim\ M + codim\ X.$$

70. Let M and N be two smooth manifolds and $f : M \to N$ be a smooth map. If f is transverse to a submanifold K of N, where none of M, N and K has boundary, show that

(i) $f^{-1}(K)$ is a submanifold of M with codimension in M same as codimension of K in N;

(ii) if $dim\ M = codim\ K$, then the submanifold $f^{-1}(K)$ consists of isolated points only.

Hence, prove that the intersection $N_1 \cap N_2$ of two transverse manifolds N_1 and N_2 of N, where none of N, N_1 and N_2 has boundary, is a submanifold such that

$$codim\ (N_1 \cap N_2) = codim\ N_1 + codim\ N_2.$$

71. Let $f : X \to \mathbf{R}$ be a smooth map with a regular value at 0. Show that $f^{-1}[0, \infty)$ is a manifold with boundary $f^{-1}(0)$.

72. Let M be a smooth manifold with boundary ∂M and $f : M \to N$ be a smooth map and N be a manifold without boundary, K be submanifold of N without boundary and $f : M \to N$ be a smooth map. If both f and $\partial f = f|_{\partial M}$ are transverse to K, show that

(i) $f^{-1}(K)$ is a submanifold of M with boundary

$$\partial(f^{-1}(K)) = f^{-1}(K) \cap \partial M;$$

(ii) the codimension of $f^{-1}(K)$ in M = the codimension of K in N.

73. **(Transversality of a transverse map)** Let $f : M \to N$ and $g : N \to P$ be smooth maps and K be a submanifold of P such that $g \pitchfork_x K$. Show that

(i) $f \pitchfork g^{-1}(K) \Leftrightarrow (g \circ f) \pitchfork K$ and
(ii) if $M = N$ and if f is a diffeomorphism, then $(g \circ f) \pitchfork K$.

74. Given any smooth manifold M, let N be submanifold of M such that N is a closed subset of M. If $\xi = (X, p, M)$ is a vector bundle over M, show that every smooth section s of the restricted bundle $\xi|N$ can be extended to a smooth section of ξ.
[Hint: Consider the smooth section s as a map with values in a vector space and an open nbd U of for every point $p \in N$ and a section $\tilde{s}$ of $\xi|U$ such that $\tilde{s} = s$ on $U \cap N$. Use Smoothing theorem (see Exercise 27) and partition of unity subordinate to an open covering of M consisting of such open nbds $\{U_i\}$ in N together with the open set $M - N$.]

75. Let $\xi = (X, p, M)$ and $\xi' = (X', p', M)$ be two vector bundles of same dimension over a manifold M and N be a closed submanifold of M. Show that every isomorphism

$$\psi : \xi|N \to \xi'|N$$

has an extension to an isomorphism

$$\tilde{\psi} : \xi|U \to \xi'|U$$

over an open nbd U of N.

76. Show that the Lens spaces $L(p, q)$ and $L(p, q')$ are diffeomorphic iff either

(i) $q + q' \equiv 0 \ (mod\ p)$ or
(ii) $qq' \equiv 0 \ (mod\ p)$.

Multiple Choice Exercises

Identify the correct alternative (s) (there may be more than one) from the following list of exercises:

1. **(i)** The Euclidean plane $\mathbf{R}^2$ is a manifold of dimension 2.
   (ii) The standard sphere $S^2 \subset \mathbf{R}^3$ is a manifold of dimension 3.
   (iii) The real projective space $\mathbf{RP}^3$ is a manifold of dimension 3.
2. Let $V_r(\mathbf{R}^n)$ denote the Stiefel manifold of (orthogonal) r-frames in $\mathbf{R}^n$ and $G_r(\mathbf{R}^n)$ denote the Grassmann manifold of r-planes of $\mathbf{R}^n$ through the origin. Then

   (i) $V_1(\mathbf{R}^n) = S^n$;
   (ii) $G_1(\mathbf{R}^n) = \mathbf{RP}^n$;
   (iii) the orthogonal group $O(n, \mathbf{R})$ acts transitively on $G_r(\mathbf{R}^n)$.

3. Let $\mathbf{R}^n$ be the Euclidean n-space.

   (i) Let X be the circle at which the standard sphere S^2 in $\mathbf{R}^3$ intersects the plane $z = 0$. Then X is a submanifold of S^2.
   (ii) The Euclidean subspace $X = \{x = (x_1, x_2, x_3, x_4, x_5) \in \mathbf{R}^5 : x_4 = x_5 = 0\} \subset \mathbf{R}^5$ is a submanifold of $\mathbf{R}^5$.
   (iii) Let $X = \{M \in M(n, \mathbf{R}) : M\ is\ symmetric\} \subset M(n, \mathbf{R})$. It is not a submanifold of $M(n, \mathbf{R})$.

4. The following plane curves are smooth manifolds:

   (i) a triangle;
   (ii) a circle;
   (iii) two triangles with only one common vertex.

5. The following manifolds are orientable:

   (i) the n-torus T^n;
   (ii) the n-sphere S^n;
   (iii) the real projective space $\mathbf{R}P^n$.

6. **(i)** The space X of nonzero real numbers (with usual topology) is a 1-dimensional Lie group under the usual multiplication of real numbers.
   (ii) The space X of nonzero complex numbers (with usual topology) is a 1-dimensional (real) Lie group under usual multiplication of complex numbers.
   (iii) The orthogonal group $O(2, \mathbf{R})$ is a 4-dimensional Lie group under usual multiplication of matrices.
   (iv) The real projective plane $\mathbf{RP}^2$ is a surface.
   (i) The Euclidean plane $\mathbf{R}^2$ is a manifold of dimension 2.
   (ii) The standard sphere $S^2 \subset \mathbf{R}^3$ is a manifold of dimension 3.
   (iii) The real projective space $\mathbf{RP}^3$ is a manifold of dimension 3.

7. Let $V_r(\mathbf{R}^n)$ denote the Stiefel manifold of (orthogonal) r-frames in $\mathbf{R}^n$ and $G_r(\mathbf{R}^n)$ denote the Grassmann manifold of r-planes of $\mathbf{R}^n$ through the origin. Then

(i) $V_1(\mathbf{R}^n) \approx S^n$;
(ii) $G_1(\mathbf{R}^n) \approx \mathbf{RP}^n$;
(iii) the orthogonal group $O(n, \mathbf{R})$ acts transitively on $G_r(\mathbf{R}^n)$.

8. Let $\mathbf{R}^n$ be the Euclidean n-space.

(i) Let X be the circle at which the standard sphere S^2 in $\mathbf{R}^3$ intersects the plane $z = 0$. Then X is a submanifold of S^2.
(ii) The Euclidean subspace $X = \{x = (x_1, x_2, x_3, x_4, x_5) \in \mathbf{R}^5 : x_4 = x_5 = 0\} \subset \mathbf{R}^5$ is a submanifold of $\mathbf{R}^5$.
(iii) Let $X = \{M \in M(n, \mathbf{R}) : M \ is \ symmetric\} \subset M(n, \mathbf{R})$. It is not a submanifold of $M(n, \mathbf{R})$.

9. **(i)** The space X of nonzero real numbers (with usual topology) is a 1-dimensional Lie group under usual multiplication of real numbers.
(ii) The space X of nonzero complex numbers (with usual topology) is a 1-dimensional (real) Lie group under usual multiplication of complex numbers.
(iii) The orthogonal group $O(2, \mathbf{R})$ is a 4-dimensional Lie group under usual multiplication of matrices.
(iv) The real projective plane $\mathbf{RP}^2$ is a surface.

References

Adhikari, M.R.: Basic Algebraic Topology and its Applications. Springer, New Delhi (2016)
Adhikari, M.R.: Basic Topology, Volume 3: Algebraic Topology and Topology of Fiber Bundles. Springer, India (2022)
Adhikari, M.R., Adhikari, A.: Groups. Rings and Modules with Applications. Universities Press, Hyderabad (2003)
Adhikari, M.R., Adhikari, A.: Textbook of Linear Algebra: An Introduction to Modern Algebra. Allied Publishers, New Delhi (2006)
Adhikari,M.R., Adhikari, A.: Basic Modern Algebra with Applications. Springer, New Delhi, New York, Heidelberg (2014)
Adhikari, A., Adhikari, M.R.: Basic Topology, Volume 1: Metric Spaces and General Topology. Springer, India (2022)
Apostol, T.M.: Mathematical- Analysis. Addison-Wesley, Boston (1957)
Bredon, G.E.: Topology and Geometry. Springer, New York (1993)
Brickell, F., Clark, R.S.: Differentiable Manifolds. Van Nostrand Reinhold, London (1970)
Dieudonné, J.: A History of Algebraic and Differential Topology, 1900–1960. Modern Birkhäuser, Basel (1989)
Donaldson, S.K.: Self-dual connections and the topology of smooth 4-manifolds. Bull Am Math Soc **8**(1), 81–83 (1983)
Guillemin, V., Pollack, A.: Differential Topology. Prentice-Hall, Englewook Cliffs (1974)

Hirsch, M.W.: Differential Topology. Springer, New York (1976)
Lie, S.: Theorie der Transformations gruppen. Math Ann **16**, 441–528 (1880)
Mandelbaum, R.: Four-dimensional topology: An introduction. Bull Am Math Soc **2**(1), 1–159 (1980)
Massey, W.S.: Algebraic Topology: An Introduction. Harcourt, Brace and World, New York (1967)
Milnor, J.: Construction of Universal bundles I, II. Ann Math **63**(2), 272–284 (1956), 63(2), 430–436 (1956)
Milnor, J.: Morse Theory. Notes by M. Spivak and R. Wells. Princeton University Press, New Jersey (1963)
Milnor, J., Weaver, D.W.: Topology from the Differentiable Viewpoit. University Press of Virginia, Charlottesville (1969)
Moise, E.E.: Geometric Topology in Dimensions 2 and 3. Springer-Verlag, New York (1977)
Mukherjee, A.: Differential Topology. Hindustan Book Agency, New Delhi (2015)
Nakahara, M.: Geometry. Topology and Physics. Institute of Physics Publishing, Taylor and Francis, Bristol (2003)
Singer, I.M., Thorpe, J.A.: Lecture Notes on Elementary Topology and Geometry. Springer, New York (1967)
Wallace, A.: Differential Topology. First Steps. W.A. Benjamin, New York (1968)
Whitney, H.: A Function not constant as a connected set of critical points. Duke Math J **1**, 514–517 (1935)
Whitney, H.: Analytic extensions of differentiable functions defined in closed sets. Trans Am Math Soc **36**, 63–89 (1936)
Whitney, H.: Differentiable manifolds. Ann Math **37**(3), 645–680 (1936)
Whitney, H.: The imbedding of manifolds in families of analytic manifolds. Ann Math **37**(4), 865–878 (1936)
Whitney, H.: The self-intersections of a smooth manifold in 2n-space. Ann Math **45**, 220–246 (1944)

Chapter 4
Lie Groups and Lie Algebras

This chapter conveys the basic principles governing the theory of Lie groups. A Lie group is a combination of three simultaneously different **structures** such as abstract group structure together with topological and manifold structures which are interrelated with each other by smooth functions. Lie groups consist of two most important special families: a family of differentiable manifolds and a family of topological groups. Their important examples include classical matrix groups $GL(n, \mathbf{R})$, $O(n, \mathbf{R})$, $U(n, \mathbf{C})$, $SL(n, \mathbf{R})$, $SL(n, \mathbf{C})$ and their Hermitian analogues. Almost all important groups of geometry and analysis are Lie groups. The theory of Lie groups studies topological groups, differentiable manifolds and Lie algebra. It establishes a relationship between a Lie group and its associated Lie algebra, especially, its Lie algebra of left-invariant vector fields, correspondence between subgroups of a Lie group and subalgebras of the associated Lie algebra and also correspondence between homomorphisms of Lie groups and homomorphisms of the associated Lie algebras. In this way, this theory provides a key link between Lie groups and Lie algebra. This link facilitates a study of Lie theory.

Many problems of Lie groups are solved by transferring the corresponding problems in their associated Lie algebra. For example, a connected Lie group G is abelian iff its Lie algebra is abelian. A fundamental theorem in Lie group theory says that there exists a one-to-one correspondence between the connected Lie subgroups of a Lie group G and the Lie subalgebras of its Lie algebra $\mathcal{G}$. Moreover, there exists a special finite-dimensional Lie algebra associated with every Lie group. Lie groups, named after Sophus Lie (1842–1899), arose in mathematics through the study of continuous transformations, and such groups constitute manifolds in a natural way. Lie groups form a versatile class of differentiable manifolds with group structure compatible with the smooth structure and provide many important examples of topological groups. More precisely, a Lie group is a topological group G whose underlying space $|G|$ is a manifold having analytic group operations.

A. Adhikari and M. R. Adhikari, *Basic Topology 2*,
https://doi.org/10.1007/978-981-16-6577-6_4

Historically, S. Lie laid the foundation of the theory of continuous transformation groups and developed his theory of continuous transformations to investigate differential equations in his landmarking paper published in **Math. Ann., Vol 16, 1880** (Lie 1880), which provides important tools for the study of different areas in contemporary mathematics and modern theoretical physics.

Hilbert's Fifth Problem and Lie Groups

David Hilbert (1862–1943) in his address at International Congress of Mathematicians, 1900 (ICM 1900) posed 23 mathematical problems, which stimulated a monumental amount of research in the twentieth century. His **fifth problem says whether each locally Euclidean topological group admits a Lie group structure.** To solve this problem during the first half of the twentieth century, many mathematicians made significant work on locally compact groups (see Chap. 2). Mathematicians such as Gleason, Iwasawa, Montgomery, Yamabe and Zippin gave a positive answer to this fifth problem determining the structure of almost connected locally compact groups. The family of almost connected groups is versatile, since this family includes every compact group and every connected locally compact groups. The second half of the twentieth century witnessed more work on the structure and representation theory of locally compact groups. The study of a Lie group G is based on the behavior of its differentiable coordinates. On the other hand, its classical theory studies local Lie groups.

For this chapter, the books (Adhikari 2016, 2022, Adhikari and Adhikari 2003, 2006, 2014, 2022, Baker and Andrew 2002, Bredon 1983 and Brickell and Clark 1970, Chevelly 1957, Hall Brian and Hall Amiya 2015, Hirsch 1976, Lie 1880, Moise 1977, Nakahara 2003, Pontragin 1939, Postnikov 1986, Singer and Thorpe 1967, Sorani 1969, Warner 1983) and some others are referred to in Bibliography.

4.1 Lie Group: Introductory Concepts

This section starts with Lie groups which are continuous transformation groups. Lie groups occupy a vast territory of topological groups carrying a differentiable structure and play a key role in the study of topology, geometry and physics. S. Lie developed his theory of continuous transformations with an eye to investigating differential equations. The basic ideas of his theory appeared in his paper (Lie 1880) published in " Math. Ann., Vol 16, 1880". A Lie group is a topological group possessing the structure of a smooth manifold on which the group operations are smooth functions. Its mathematical formulation is given in Definition 4.1.1.

Definition 4.1.1 A topological group G with identity element e is called a **real Lie group** if

(i) G is a real differentiable manifold;
(ii) the group multiplication

$$m : G \times G \to G, \ (x, y) \mapsto xy$$

and the group inversion

$$v : G \to G, \ x \mapsto x^{-1}$$

are both differentiable.

Definition 4.1.2 A topological group G is called a **complex Lie group** if

(i) G is a complex manifold;
(ii) the group multiplication

$$m : G \times G \to G.(x, y) \mapsto xy$$

and the group inversion

$$v : G \to G, \ x \mapsto x^{-1}$$

are both holomorphic.

Definition 4.1.3 The **dimension of a Lie group** is defined to be its dimension as a manifold. Its alternative formulation is given in Definition 4.5.13.

Example 4.1.4 Examples of Lie groups are plenty.

(i) The real line **R** is a Lie group under usual addition of real numbers.
(ii) $\mathbf{R}^+ = \{x \in \mathbf{R} : x > 0\}$ is a Lie group under usual multiplication of real numbers.
(iii) $\mathbf{R}^2$ is a Lie group under pointwise addition given by

$$(x_1, y_1) + (x_2, y_2) = (x_1 + x_2, \ y_1 + y_2).$$

(iv) Every finite-dimensional vector space V over **R** is a Lie group, since the map

$$f : V \times V \to V, (x, y) \mapsto x + y$$

is linear and hence differentiable.

Similarly, the map

$$g : V \to V, x \to -x$$

is smooth. In particular, the additive group $\mathbf{R}^n$ with its standard structure as a differentiable manifold is a Lie group.
(v) Examples of **Classical Lie groups of matrices** are given in Sect. 4.1.2.

Example 4.1.5 Every discrete topological group G is a Lie group of dimension 0. Because the definition of a Lie group does not require that G is to be connected, thus, any finite group is a 0-dimensional Lie group.

Example 4.1.6 All topological groups are not Lie groups. There are some topological groups which fail to be Lie groups. For example,

(i) Let V be an infinite-dimensional vector space. Then $(V, +)$ is not a Lie group, because it is not a finite-dimensional manifold.
(ii) The additive group $\mathbf{Q}$ of rational numbers with topology inherited from the real number space $\mathbf{R}$ is a topological group but it is not a Lie group, because it is a countable space having no discrete topology.
(iii) The additive group of the p-adic numbers is not a Lie group, because it is not a totally disconnected group and its underlying space is not a real manifold.

Remark 4.1.7 A Lie group is not necessarily connected in the sense that its underlying manifold is not connected.

Definition 4.1.8 A connected 1-dimensional complex manifold is also called a **Riemann surface**.

Example 4.1.9 The three sphere S^3 **is a Lie group**. To show it, consider the 3-sphere S^3 consisting of the set of unit quaternions

$$S^3 = \{q = x + iy + jz + kw : x, y, z, w \in \mathbf{R} \; and \; x^2 + y^2 + z^2 + w^2 = 1\}.$$

It is a group under usual multiplication of quaternions. Define

$$U = \{x + iy + jz + kw \in S^3 : x > 0\} \subset S^3$$

and

$$\psi : S^3 \to \mathbf{R}^3, \; x + iy + jz + kw \mapsto (y, z, w).$$

Consider ψ as a chart and for each $q \in S^3$, define a chart $\psi_q(u) = \psi(q^{-1}u)$. Then the set of all charts constitutes an atlas for a smooth structure of dimension 3 on the 3-sphere S^3 and S^3 forms a Lie group.

Remark 4.1.10 An abstract subgroup and a submanifold H of a Lie group G, which is a Lie group with its induced smooth structure, is called a Lie subgroup.

Definition 4.1.11 Let G be a Lie group. If H is an abstract subgroup and also a submanifold of G, then H is said to be a **Lie subgroup** of G.

Remark 4.1.12 Since every Lie group G is a topological group, every Lie subgroup of G is necessarily a closed subgroup of G. It is also a smooth submanifold. Its detailed study is available in Sect. 4.5.

Definition 4.1.13 Let G and H be two Lie groups and $f : G \to H$ be an (abstract) group homomorphism. Then

(i) f is said to be a **Lie group homomorphism** if f is smooth;

(ii) f is said to be a **Lie group isomorphism** if f is a group isomorphism and f, f^{-1} are both smooth.

Remark 4.1.14 Let G and H be two Lie groups and $f : G \to H$ be a Lie group homomorphism. A Lie subgroup arises as $Im\ f$ is also an embedding. If H is a closed abstract subset of a Lie group G, then H is a Lie subgroup of G (see (Chevelly 1957)). Every compact Lie group can be embedded in the Lie group $GL(n, \mathbf{C})$ by the celebrated Peter–Wyel theorem (proof is beyond the scope of the book). Exercise 50 of Sect. 4.11 says that every closed subgroup of a Lie group is an embedded Lie subgroup of $GL(n, \mathbf{R})$.

Definition 4.1.15 (*Product of Lie groups*) Let G and H be two Lie groups. A Lie group structure on the set $G \times H$ is defined by taking the manifold structure and the group structure as products. Then $G \times H$ is called the product of Lie groups G and H.

Example 4.1.16 The torus $T^2 = S^1 \times S^1$ is a Lie group under a standard structure.

4.1.1 Topology of a Lie Group Induced by Its Differential Structure

Chapter 3 studies the topology of a smooth manifold induced by its C^∞-structure. Since a Lie group is a differential manifold, a similar study can be made for a Lie group. However, this subsection studies some of them. Moreover, a Lie group endowed with this induced topology is a topological group.

Theorem 4.1.17 *If G is a connected Lie group and $X = |G|$ is its underlying manifold, then X admits a countable basis for its topology.*

Proof Let G be a connected Lie group of dimension n with identity e and $X = |G|$ be its underlying manifold. Consider a coordinate nbd U of e such that $U = U^{-1}$ (symmetric nbd). Take a countable subset S of U such that S is dense in U. This choice is possible, because U is homeomorphic to an open subset of $\mathbf{R}^n$. Consider the set

$$A = S \cup S^{-1} = A^{-1} \subset U.$$

Then A is dense in U. Let $K = \langle S \rangle$ be the group generated by S and hence K consists of all the finite products of points of A. This implies that K is countable. Claim that the family $\{kU : k \in K\}$ of coordinate domains forms a covering of the Lie group G. To show it, take any $g \in G$. Then

$$g = g_1 g_2 \cdots g_m : g_i \in U,\ i = 1, 2, \ldots m,$$

because G is connected and G is generated by U. Since the group multiplication in G is continuous, there exists a nbd U_i of the point g_i such that

$$U_1U_2\cdots U_m \subset gU.$$

Since A is dense in U, there exists a point $a_i \in A$ such that $a_i \in U_i$. Then the product

$$k = a_1a_2\cdots a_m \in K \quad \text{and} \quad k \in gU.$$

This implies that $g \in kU^{-1} = kU$. This proves the theorem. □

4.1.2 Examples of Classical Lie Groups of Matrices and Others

This subsection gives some examples of classical Lie groups of matrices and others.

Example 4.1.18 **(i)** Every finite-dimensional vector space V over $\mathbf{R}$ is a Lie group, since the map

$$f : V \times V \to V, (x, y) \mapsto x + y$$

is linear and hence differentiable. Similarly, the map

$$g : V \to V, x \to -x$$

is smooth.

(ii) In particular, the additive group $\mathbf{R}^n$ with its standard structure as a differentiable manifold is a Lie group.

(iii) The group $\mathbf{R}^*$ (set of nonzero real numbers) is a 1-dimensional Lie group with 1 as its identity element under usual multiplication of real numbers.

(iv) The group $\mathbf{C}^*$ (set of nonzero complex numbers) is a 2-dimensional Lie group under usual multiplication of complex numbers.

(v) The special orthogonal group $SO(2, \mathbf{R})$ is the group of all 2×2 real orthogonal matrices M having determinant $det M = 1$. It is a 1-dimensional Lie group, because the group $O(2, \mathbf{R})$ is homeomorphic to the unit circle $x^2 + y^2 = 1 \subset \mathbf{R}^2$ by the homeomorphism f defined by

$$(x, y) \mapsto \begin{pmatrix} x & y \\ -y & x \end{pmatrix}.$$

Alternatively,

$$f : SO(2, \mathbf{R}) \to S^1, \begin{pmatrix} \cos\theta & -\sin\theta \\ \sin\theta & \cos\theta \end{pmatrix} \mapsto e^{i\theta}.$$

Moreover, if the group $SO(2, \mathbf{R})$ is given the smooth structure, then f is a diffeomorphism. Hence, it follows that the group $SO(2, \mathbf{R})$ is a 1-dimensional Lie group.

(vi) The general linear group $GL(n, \mathbf{R})$ of all invertible $n \times n$ real matrices is a Lie group. It is neither compact nor connected, because $GL(n, \mathbf{R})$ is a group under usual multiplication of matrices.

Moreover,

$$GL(n, \mathbf{R}) = \{M \in M(n, \mathbf{R}) : det M \neq 0\}$$

is an open subset, because the det function is continuous. Hence, it follows that $GL(n, \mathbf{R})$ is a manifold of dimension n^2. Again, the matrix multiplication

$$m : M(n, \mathbf{R}) \times M(n, \mathbf{R}) \to M(n, \mathbf{R})$$

involving products and sum of the entries is a differentiable function. Hence, its restriction to $GL(n, \mathbf{R}) \times GL(n, \mathbf{R})$ is also differentiable. Again, the inversion map

$$v : G(n, \mathbf{R}) \to GL(n, \mathbf{R}), M \mapsto M^{-1}$$

is differentiable, since, by Cramer's rule, the entries of M^{-1} involve rational functions in the entries of M with the determinant in the denominator. Consequently,$GL(n, \mathbf{R})$ is a Lie group. But it not compact, because it is the inverse image of the nonzero real numbers of the determinant function

$$det : M(n, \mathbf{R}) \to \mathbf{R}, M \mapsto det\ M,$$

which is continuous, as it is a polynomial of matrix coefficient. Again, since $\{0\}$ is closed in $\mathbf{R}$ and det function is continuous, it follows that $det^{-1}\{0\}$ is closed in $M(n, \mathbf{R})$ and hence its complement $GL(n, \mathbf{R})$ is open in $M(n, \mathbf{R})$. This implies that $GL(n, \mathbf{R})$ is not compact. $GL(n, \mathbf{R})$ is also not connected, because the matrices with positive and negative determinants are both open sets in $GL(n, \mathbf{R})$ such that they produce a partition of $GL(n, \mathbf{R})$.

(vii) The group $GL(n, \mathbf{C})$ is the group of all nonsingular $n \times n$-complex matrices. It represents the set of all nonsingular linear transformations $T : \mathbf{C}^n \to \mathbf{C}^n$. The unitary group $U(n, \mathbf{C})$, special linear group $SL(n, \mathbf{C})$ and special unitary group $SU(n, \mathbf{C})$ defined by

$$SL(n, \mathbf{C}) = \{M \in GL(n, \mathbf{C}) : det M = 1\},$$

$$U(n, \mathbf{C}) = \{M \in GL(n, \mathbf{C}) : MM^* = I\},$$

where M^* is the Hermitian conjugate, and

$$SU(n, \mathbf{C}) = U(n, \mathbf{C}) \cap SL(n, \mathbf{C})$$

are important examples of Lie groups. The Lie group $SL(n, \mathbf{C})$ of dimension $2(n^{2-1})$ is not compact, since it is a closed subset of $GL(n, \mathbf{C})$ but $SU(n, \mathbf{C})$ is a compact subgroup of $GL(n, \mathbf{C})$.

(viii) Special linear group $SL(n, \mathbf{R}) = \{M \in GL(n, \mathbf{R}) : detM = 1\}$ is a connected real Lie group of dimension $n^2 - 1$. It is not compact and is a hypersurface of $GL(n, \mathbf{R})$.

(ix) Orthogonal group $O(n, \mathbf{R}) = \{M \in GL(n, \mathbf{R}) : MM^t = I\}$ is a compact real Lie group of dimension $\frac{n^2-n}{2}$. It is not connected, because it contains matrices having determinants 1 and -1.

(ix) The circle with center at the origin 0 and radius 1 in the complex plane $\mathbf{C}$ is a Lie group with complex multiplication.

Remark 4.1.19 The classical Lie groups $\mathrm{GL}(n, \mathbf{R})$, $\mathrm{SL}(n, \mathbf{R})$, $\mathrm{O}(n, \mathbf{R})$, $\mathrm{SO}(n, \mathbf{R})$ are all manifolds, because for each point x of any one of these groups, there exists an open neighborhood homeomorphic to a Euclidean space. All of them are real Lie groups, called classical Lie groups. For example, the topological group $\mathrm{GL}(n, \mathbf{R})$ is a real Lie group of dimension n^2, $\mathrm{SL}(n, \mathbf{R})$ is a real Lie group of dimension $n^2 - 1$, $\mathrm{O}(n, \mathbf{R})$ is a real Lie group of dimension $n(n-1)/2$ and $\mathrm{SO}(n, \mathbf{R})$ is a real Lie group of dimension $n(n-1)/2$. Their complex and quaternionic analogues are also Lie groups.

Example 4.1.20 (*Some other Examples of Lie groups*)

(i) Every finite-dimensional vector space V over $\mathbf{R}$ is a Lie group, since the map $f : V \times V \to V, (x, y) \mapsto x + y$ is linear and hence differentiable. Similarly, the map $g : V \to V, x \to -x$ is smooth. In particular, the additive group $\mathbf{R}^n$ with its standard structure as a differentiable manifold is a Lie group. Moreover, if dim $V = n$, then the automorphism group $\mathcal{A}ut(V)$ of all automorphisms of V, which is a group of all nonsingular linear operators of V is isomorphic to the group $GL(n, R)$ is a Lie group. An analogous result holds for every finite-dimensional vector space V over $\mathbf{C}$.

(ii) The group $\mathbf{R}^*$ (set of nonzero real numbers) is a 1-dimensional Lie group with 1 as its identity element under usual multiplication of real numbers.

(iii) The group $\mathbf{C}^*$ (set of nonzero complex numbers) is a 2-dimensional Lie group under usual multiplication of complex numbers.

(iv) The circle with center at the origin 0 and radius 1 in the complex plane $\mathbf{C}$ is a Lie group with complex multiplication.

4.1.3 Matrix Lie Groups

This subsection considers special subgroups G of the complex linear group $GL(n, \mathbf{C})$ that are closed in the sense of Definition 4.1.22 to introduce the concept of matrix Lie groups. **All subgroups of $GL(n, \mathbf{C})$ are not matrix Lie groups** (see Example 4.1.24). More precisely, a subgroup G of $GL(n, \mathbf{C})$ is called a matrix Lie group if a smoothness is endowed on G to make it a Lie group and the embedding $i : G \to GL(n, \mathbf{C})$ is smooth and hence it is a homomorphism of Lie groups. This subsection makes an approach to define a matrix Lie group by using the concept of convergence of sequences of complex matrices.

Definition 4.1.21 Let $M(n, \mathbf{C})$ be the topological space identified with $\mathbf{C}^{n^2}$ and the standard notion of convergence be used in $\mathbf{C}^{n^2}$. Then a sequence $\{A_k\}$ of complex matrices in $M(n, \mathbf{C})$ is said to converge to a matrix $A \in M(n, \mathbf{C})$ if every entry of A_k converges to the corresponding entry of A as $k \to \infty$.

Consider the closed subgroups of the linear group $GL(n, \mathbf{C})$. Such subgroups are called matrix Lie groups. They are formulated in Definition 4.1.22 in another way.

Definition 4.1.22 A **matrix Lie group** G is a subgroup of $GL(n, \mathbf{C})$ such that if $\{A_k\}$ is any sequence of complex matrices in G, and the sequence $\{A_k\}$ converges to some matrix A in $M(n, \mathbf{C})$, then either $A \in G$ or A is not invertible.

Remark 4.1.23 The condition prescribed in Definition 4.1.22 on G asserts that G is a closed subset of the group $GL(n, \mathbf{C})$ (not necessarily closed in $M(n, \mathbf{C})$). This implies that this definition is equivalent to the assertion that a matrix Lie group is a closed subgroup of $GL(n, \mathbf{C})$. So, while considering topological properties of a matrix Lie group G, it is sufficient to consider the subspace topology of G inherited as a subset of $M(n, \mathbf{C})$ identified with the space $\mathbf{C}^{n^2}$.

Example 4.1.24 All the subgroups of $GL(n, \mathbf{C})$ are not matrix Lie groups. Most of the subgroups of $GL(n, \mathbf{C})$ are matrix Lie groups but there exist subgroups of $GL(n, \mathbf{C})$ which are not closed and hence they fail to be matrix Lie groups. For example, consider the subgroup G of $GL(n, \mathbf{C})$ consisting of all $n \times n$ invertible matrices with rational entries. This subgroup is not closed, because there exists a sequence of invertible matrices with rational entries that converges to a matrix in $A \in GL(n, \mathbf{C})$ such that A has some irrational entries. This is possible, because every nonsingular matrix with real entries is the limit of some sequence of nonsingular matrices with rational entries.

Example 4.1.25 Another example of a group of matrices is a non-matrix Lie group. Given any irrational number α, define

$$G = \left\{ \begin{pmatrix} e^{it} & 0 \\ 0 & e^{i\alpha t} \end{pmatrix} : t \in \mathbf{R} \right\} \subset GL(2, \ \mathbf{C}).$$

Then G is a subgroup of the group $GL(2, \ \mathbf{C})$ with its closure

$$\overline{G} = \left\{ \begin{pmatrix} e^{i\phi} & 0 \\ 0 & e^{i\theta} \end{pmatrix} : \phi, \ \theta \in \mathbf{R} \right\}$$

in the space $M(2, \mathbf{C})$. The group G is called an **irrational line in torus.** Since $G \neq \overline{G}$, it follows that the group G is not a closed subset of the group $GL(2, \mathbf{C})$ and hence G is not a matrix group.

Definition 4.1.26 formulates the Heisenberg group. This group has many applications. It is used to study the Heisenberg–Weyl communication relations in physics. Some properties of the Heisenberg group are given in Exercises 34 and 35 of Sect. 4.11.

Definition 4.1.26 **(Heisenberg group)** Let G be the set of all 3×3 matrices with real entries of the form

$$A = \begin{pmatrix} 1 & x & y \\ 0 & 1 & z \\ 0 & 0 & 1 \end{pmatrix}.$$

Then $G \subset GL(3, \mathbf{R})$ forms a subgroup of the group $GL(3, \mathbf{R})$ having the inverse of the matrix A

$$A^{-1} = \begin{pmatrix} 1 & -x & xz - y \\ 0 & 1 & -z \\ 0 & 0 & 1 \end{pmatrix}.$$

This group G is called the Heisenberg group.

4.2 Topological Properties of Matrix Lie Groups

This section studies some topological properties such as the compactness, connectedness and simple connectedness properties of matrix Lie groups. Here, every matrix Lie group is considered as a subset of the topological space $M(n, \mathbf{C})$ identified with Euclidean space $\mathbf{R}^{2n^2}$ endowed with standard topology.

4.2.1 Compactness Property of Matrix Lie Groups

This subsection discusses the compactness property of matrix Lie groups.

Definition 4.2.1 Let $G \subset GL(n, \mathbf{C})$ be a matrix Lie group. Then the group G is said to be a **compact matrix Lie group** if it is a compact subset of the topological space $M(n, \mathbf{C})$.

Proposition 4.2.2 characterizes the compactness of matrix Lie groups with the help of the classical Heine–Borel theorem.

Proposition 4.2.2 *Let $G \subset GL(n, \mathbf{C})$ be a matrix Lie group. Then G is compact iff*

(i) *whenever the matrix $A_k \in G$ and $A_k \to A$, then the matrix $A \in G$ and*
(ii) *there exists a constant $\delta > 0$ such that $|A_{i,j}| < \delta, \ \forall\ 1 \leq i, j \leq n$.*

Proof Using Definition 4.1.22, the proof of the proposition follows from the Heine–Borel theorem which characterizes that a subset of the Euclidean space $\mathbf{R}^{2n^2}$ is compact iff it is a closed and bounded subset of $\mathbf{R}^{2n^2}$. □

Example 4.2.3 (*Compact matrix Lie groups*) All the following matrix Lie groups G are compact, because each of these groups G is closed in $M(n, \mathbf{C})$ and each matrix $A = (A_{i,j}) \in G$ is such that $|A_{i,j}| \leq 1$ (since the columns of A are unit vectors).

(i) The orthogonal group $G = O(n, \mathbf{R})$ is a compact subgroup of the general linear group $GL(n, \mathbf{R})$ for any $n \geq 2$.
(ii) The special orthogonal group $G = SO(n, \mathbf{R}) \subset GL(n, \mathbf{R})$ is compact for any $n \geq 2$.
(iii) The unitary group $G = U(n, \mathbf{C}) \subset GL(n, \mathbf{C})$ is compact.
(iv) The special unitary group $G = SU(n, \mathbf{C}) = \{M \in U(n, \mathbf{C}) : det\ M = 1\} \subset GL(n, \mathbf{C})$ is compact for any $n \geq 2$.
(v) The symplectic group

$$G = Sp(n, \mathbf{H}) = \{A \in M(n, \ \mathbf{H}) : A^*A = I\},$$

where A^* is the quaternionic conjugate transpose of A and the conjugation is in the sense of reversal of all three imaginary components, is compact for all $n \geq 2$.

Example 4.2.4 (*Noncompact matrix Lie groups*) The special (real) linear group $SL(n, \mathbf{R}) = \{M \in GL(n, \mathbf{R}) : det M = 1\}$ is not compact and is a hypersurface of $GL(n, \mathbf{R})$, because $SL(n, \mathbf{R})$ is unbounded except in the trivial case for $n = 1$. For example, for $k \neq 0$, the matrix

$$A_k = \begin{pmatrix} k & 0 & 0 \\ 0 & 1/k & 0 \\ 0 & 0 & 1 \end{pmatrix} \in SL(3, \mathbf{R})$$

but it not bounded. In general, for any integer $m \geq 2$, the diagonal matrix D_m with diagonal entries $m, \frac{1}{m}, 1, \ldots, 1$ is such that its determinant $|D_m| = 1$ and hence $D_m \in SL(3, \mathbf{R})$ but its every diagonal element is not bounded.

4.2.2 *Path Connectedness and Connectedness Properties of Matrix Lie Groups*

This subsection discusses the path connectedness and connectedness properties of a matrix Lie group. It is well known that every path-connected space is connected but its converse is not necessarily true (see **Chap. 5, Basic Topology, Volume 1** of the present series of books). Hence, to show that a matrix Lie group is connected, it is sufficient to show that each matrix A can be connected to the identity matrix $I \in G$ by a continuous path $\alpha : \mathbf{I} \to G$. This idea is formulated in Definition 4.2.6.

Definition 4.2.5 A matrix Lie group G is said to be a **connected matrix Lie group** if its underlying manifold is connected.

Definition 4.2.6 Let $G \subset GL(n, \mathbf{C})$ be a matrix Lie group. Then the group G is said to be **connected** if given any two matrices $A, B \in G$, there exists a continuous path

$$\alpha : \mathbf{I} \to G : \alpha(0) = A \text{ and } \alpha(1) = B.$$

The **identity component** of G, denoted by G_e, is the set

$$G_e = \{A \in G : \exists \text{ a continuous path } \alpha : \mathbf{I} \to G \text{ with } \alpha(0) = I \text{ and } \alpha(1) = A\}.$$

Example 4.2.7 All the following matrix Lie groups are path-connected, and hence all of them are connected.

(i) The special orthogonal group $SO(n, \mathbf{R}) = \{M \in O(n, \mathbf{R}) : det\ M = 1\}$ is connected, and hence $SO(n, \mathbf{R})$ is the component of $O(n, \mathbf{R})$ containing its identity element.
(ii) The matrix Lie group $Sp(n, \mathbf{C})$ is connected for every $n \geq 1$.
(iii) The symplectic group $Sp(n, \mathbf{H})$ is connected for every $n \geq 1$.

Definition 4.2.8 A connected nonabelian Lie group is said to be **simple** if it has no nontrivial connected normal subgroups.

Example 4.2.9 The only connected Lie group with dimension one is the real line space $\mathbf{R}$ (with the group operation being addition). The circle group S^1 of complex numbers with absolute value one (with the group operation being multiplication) is an important Lie group. The group S^1 is often denoted as $U(1, \mathbf{C})$, the group of 1×1 unitary matrices.

4.2.3 Simply Connectedness Property Matrix Lie Groups

This subsection discusses the simply connectedness property of a matrix Lie group.

Definition 4.2.10 Let $G \subset GL(n, \mathbf{C})$ be a matrix Lie group. Then the group G is said to be **simply connected**

(i) if G is connected and
(ii) each loop (closed path) in G can be continuously deformed to a point in G.

Remark 4.2.11 If a matrix Lie group G is connected, then its simple connectedness property implies that for every continuous path

$$\alpha : \mathbf{I} \to G : \alpha(0) = \alpha(1),$$

there exists a continuous function

$$H : \mathbf{I} \times \mathbf{I} \to G$$

such that

(i) $H(s, 0) = H(s, 1), \forall s, t \in \mathbf{I}$,
(ii) $H(0, t) = \alpha(t), \forall t \in \mathbf{I}$ and
(iii) $H(1, t) = H(1, 0) \forall t \in \mathbf{I}$.
It is considered that α is a loop in G and H as a family of loops. Then condition (i) asserts that every value of the parameter s gives a loop in G, condition (ii) implies that if $s = 0$, then the loop $H(0, t)$ is the specified loop $\alpha(t)$ and condition (iii) says that if $s = 1$, then the loop $H(1, t)$ is deformed to a point in G. The concept of simply connectedness is used in the study of Lie algebra.

Example 4.2.12 Example of a simply connected matrix Lie group: The special unitary group $SU(2, \mathbf{C})$ is topologically the same as the group S^3, the group of unit quaternions, which is simply connected (see Chap. 2, **Basic Topology, Volume 3** of the present series of books). This implies that $SU(2, \mathbf{C})$ is simply connected.

4.3 Lie Group Action on Differentiable Manifolds

This section studies the concept of Lie group action on differentiable manifolds and its basic properties. It relates the concepts of translation and diffeomorphism on Lie groups with the help of Lie group action in Sect. 4.3.1.

4.3.1 Translation and Diffeomorphism on Lie Groups

This subsection establishes relations between the concepts of translation and diffeomorphism on Lie groups with the help of Lie group action.

Definition 4.3.1 Let G be a Lie group and $g \in G$ be an arbitrary element. Then the function

$$L_g : G \to G, x \mapsto gx$$

is said to be a **left translation**.

A **right translation** is defined in a similar way:

$$R_g : G \to G, \ x \mapsto xg.$$

Definition 4.3.2 Let G and H be two Lie groups. Then they are smooth manifolds, A map

$$f : G \to H$$

is called a **diffeomorphism** if f is bijective and both f and f^{-1} are smooth maps.

Proposition 4.3.3 *Let G be a Lie group. For any element $g \in G$,*

(i) *the left translation* $L_g : G \to G, x \mapsto gx$ *is differentiable and it is also a diffeomorphism and*
(ii) *the right translation* $R_g : G \to G, x \mapsto xg$ *is differentiable and it is also a diffeomorphism.*

Proof **(i)** Since the group operation on the Lie group G is differentiable, L_g is differentiable. Again since its inverse L_{g-1} is also differentiable, it follows that L_g is a diffeomorphism.
(ii) can be proved following the similar argument as in (i).

□

Proposition 4.3.4 *Let G be a Lie group with its identity element e. Then the function*

$$f : G \to G, x \mapsto x^{-1}$$

is a diffeomorphism.

Proof It is sufficient to show that f is differentiable, since it is bijective and is its own inverse. First, show that it is differentiable at its identity element e, because for any $x \in G$, the function

$$f = R_{x^{-1}} \circ f \circ L_{x^{-1}} : G \to G$$

and hence by Proposition 4.3.3 the differentiability of f at the point x follows from the differentiability of f at the point e. □

Remark 4.3.5 Let G and H be two Lie groups. Them $f : G \to H$ is a Lie group homomorphism iff

$$f \circ L_g = L_{f(g)} \circ f.$$

4.3.2 *Actions of Lie Groups*

Definition 4.3.6 Let M be a differentiable manifold and G be a Lie group with identity e. Then G **is said to act as a transformation group on** M **(differentiably)** from the left if there is a C^∞-map (differentiable or smooth map)

$$\psi : G \times M \to M, \ (g, x) \ \mapsto \psi(g, x) = gx$$

such that

(i) $e(x) = x$ for all $x \in M$;
(ii) $(gk)(x) = g(k(x))$, for all $x \in M$, and $g, k \in G$.

Then M is also called a **left** G**-manifold.** The element gx is also written as $g(x)$. If ψ is continuous, then the action is said to be a continuous action.

In particular, an action $\mathbf{R} \times M \to M$ of the group $(\mathbf{R}, +)$ on a manifold M is said to be a **flow.**

A right G**-manifold** is defined in an analogous way.

Remark 4.3.7 There is a one-to-one correspondence between the left and right G-manifold structures on M. So it is sufficient to study only one of them according to the situation.

Definition 4.3.8 Let G be Lie group and M be a left G-manifold under an action: $\psi : G \times M \to M$.

(i) Two given elements x, y in M are called G**-equivalent**, if $y = g(x)(i.e., y = gx)$ for some $g \in G$.
(ii) The relation of being G-equivalent is an equivalence relation on M and the corresponding quotient space $M mod\ G$ endowed with quotient topology induced from X (i.e., the largest topology such that the projection map $p : M \to M mod\, G, x \mapsto G(x)$ is smooth, where $G(x) = Gx = \{gx : g \in G\}$ is called the **orbit of** x, the equivalence class of x for every $x \in M$.
(iii) For an element $x \in G$, the subgroup G_x of G defined by $G_x = \{g \in G : g(x) = gx = x\}$ is called the **stabilizer or isotropy group at** x of the corresponding Lie group action ψ.

Example 4.3.9 (*Action of Lie group* $O(n, \mathbf{R})$ *on* $\mathbf{R}^n$) Consider the action of the Lie group $O(n, \mathbf{R})$ on $\mathbf{R}^n$

$$\psi : O(n, \mathbf{R}) \times \mathbf{R}^n \to \mathbf{R}^n, (A, x) \to Ax.$$

Then $Orb(x)$ is the sphere S^{n-1} of radius $||x||$. This implies that this action provides the orbits that divide $\mathbf{R}^n$ into a family of disjoint spheres parameterized by the radius. This defines an equivalence relation $\sim$ on $\mathbf{R}^n$. The equivalence class $[x]$ is the orbit $Orb(x)$. The orbit space $\mathbf{R}^n$ modulo $O(n, \mathbf{R})$ is the half-line space $[0, \infty)$, because each equivalence class is parameterized by the radius.

Example 4.3.10 (*Action of* $O(n+1, \mathbf{R})$ *on* $\mathbf{R}P^n$) Consider the action of the Lie group $O(n+1, \mathbf{R})$ on $\mathbf{R}P^n$ by using the usual action of $O(n+1, \mathbf{R})$ on $\mathbf{R}^{n+1}$. The real projective space $\mathbf{R}^{n+1}$ can be constructed from $\mathbf{R}^{n+1}$ by an equivalence relation $\sim$: if $x, y \in \mathbf{R}^{n+1}$ and $g \in O(n+1, \mathbf{R})$, then $x \sim y$, (in the sense that $y = tx$ for some $t \in \mathbf{R} - \{0\}$) and $gx \sim gy$ (in the sense that $gy = tgx$ for some $t \in \mathbf{R} - \{0\}$). This implies that the action

$$\psi : O(n+1, \mathbf{R}) \times \mathbf{R}^{n+1} \to \mathbf{R}^{n+1}, \ (A, x) \to Ax$$

induces a natural action

$$\psi_* : O(n+1, \mathbf{R}) \times \mathbf{R}P^n \to \mathbf{R}P^n, \ (A, [x]) \to [Ax].$$

The action ψ_* is transitive. Let $p \in \mathbf{R}P^n$ be corresponding to the point $x = (1, 0, 0, \ldots, 0) \in \mathbf{R}^{n+1}$ in the sense that $p = [x]$, then the isotropy group at the point p is

$$G_p = \mathbf{Z}_2 \times O(n, \mathbf{R}).$$

Hence, it follows that

$$O(n+1)/(\mathbf{Z}_2 \times O(n, \mathbf{R})) \cong S^n/\mathbf{Z}_2 \cong \mathbf{R}P^n.$$

Example 4.3.11 The action of the Lie group $G(n, \mathbf{R})$ (or any one of its subgroups) on the Euclidean n-space $\mathbf{R}^n$ is defined under usual matrix action on a vector in the sense that

$$\psi : G(n, \mathbf{R}) \times \mathbf{R}^n \to \mathbf{R}^n, \ (A, \ x) \mapsto Ax.$$

This implies that the Lie group $GL(n, \mathbf{R})$ (or any one of its subgroups) acts on $\mathbf{R}^n$ from the left under the usual multiplication of matrices.

Example 4.3.12 **(i)** The Lie group $O(n, \mathbf{R})$ acts on the sphere $S^{n-1} \subset \mathbf{R}^n$.
(ii) The Lie group $U(n, \mathbf{C})$ acts on the sphere $S^{2n-1} \subset \mathbf{C}^n$.
(iii) A periodic flow on a smooth manifold M is an action of $SO(2, \mathbf{R})$ on M.

Proposition 4.3.13 **(i)** *Let M be a real differentiable manifold and G be real Lie group acting on M. Then for every element $g \in G$, the map*

$$\psi_g : M \to M, x \mapsto gx$$

is a diffeomorphism.
(ii) *Let M be a complex manifold and G be a complex Lie group. Then for every element $g \in G$, the map*

$$\psi_g : M \to M, x \mapsto gx$$

is biholomorphic (i.e., ψ_g is a homeomorphism such that ψ_g and ψ_g^{-1} are both holomorphic).

Proof The proposition follows from the concept of Lie group action of G on M. Because for every $g \in G$, the map ψ_g is differentiable and

$$\psi_g \circ \psi_g^{-1} = \psi_e = 1_M = \psi_g^{-1} \circ \psi_g.$$

□

Definition 4.3.14 Let ***diffeo***(M) be the set of all diffeomorphisms ψ_g defined in Proposition 4.3.13. It is called **the group of all diffeomorphisms of** M under the usual composition of mappings.

Proposition 4.3.15 *Let M be a real differentiable manifold and G be a real Lie group acting on M by an action $\sigma : G \times M \to M, (g, x) \mapsto gx$. Then σ induces a group homomorphism*

$$f : G \to \textit{\textbf{diffeo}}(M), \ g \mapsto \psi_g.$$

Proof Consider the map

$$f : G \to \textit{\textbf{diffeo}}(M), \ g \mapsto \psi_g.$$

It is well-defined by Proposition 4.3.13. Moreover for any $g, k \in G$,

$$f(gk) = \psi_{gk} = \psi_{gk} = \psi_g \circ \psi_k = f(g) \circ f(k)$$

proves that f is a group homomorphism. □

Proposition 4.3.16 *Let M be a real differentiable manifold and G be real Lie group acting on M from right by an action $\sigma : M \times G \to M, (x, g) \mapsto xg$. Then σ induces an anti-homomorphism*

$$f : G \to \textit{\textbf{diffeo}}(M), \ g \mapsto \psi_g,$$

in the sense that

$$f(gk) = f(k) \circ f(g), \ \forall \, g, k \in G,$$

where

$$\psi_g : M \to M, \ x \mapsto xg.$$

Proof For any $g, k \in G$,

$$f(gk) = \psi_{gk} = f(k) \circ f(g)$$

proves that f is an anti-homomorphism. □

Definition 4.3.17 Let G be a Lie group with identity element e acting on a manifold M. Then the action

$$\psi : G \times M \to M, \ (g, \ x) \mapsto \psi(g, x) = gx$$

is said to be

(i) **free** if for every element $g \neq e \in G$, there is no fixed element in G in the sense that

$$\psi(g, \ x) = x \implies g = e;$$

(ii) **transitive** if for every pair of elements $x, y \in M$, there exists an element $g \in G$ such that

$$\psi(g, x) = y;$$

(iii) **effective** if the identity element e is the unique element which determines the trivial action in the sense that

$$\psi(g, x) = x, \ \forall x \in M \implies g = e.$$

Example 4.3.18 (*Free and transitive action*)

(i) Let G be a Lie group. Given $g \in G$, consider the left (right) translation

$$L_g : G \to G, x \mapsto L_g(x) = gx \ (R_g(x) = xg)$$

of a Lie group G. Then each of them is a free and transitive action.

(ii) Consider the action of the Lie group $O(n, \mathbf{R})$ on S^{n-1}

$$\psi : O(n, \mathbf{R}) \times S^{n-1} \to S^{n-1}, (A, x) \to Ax.$$

This action is transitive.

(iii) For more transitive actions, see Exercise 36 of Section 4.11.

Example 4.3.19 (*A non-transitive action*) Consider the action of the Lie group $O(n, \mathbf{R})$ on $\mathbf{R}^n$

$$\psi : O(n, \mathbf{R}) \times \mathbf{R}^n \to \mathbf{R}^n, \ (A, x) \mapsto Ax,$$

because, for $x, y \in \mathbf{R}^n$ with $||x|| \neq ||y||$, there exists no orthogonal matrix $A \in O(n, \mathbf{R})$ such that $Ax = y$.

Definition 4.3.20 (*Representation of a Lie group*) Let G be a Lie group and V be a vector space over $F = \mathbf{R}$ or $\mathbf{C}$. If $Hom(V, V) = GL(V)$ denotes the Lie group of nonsingular linear transformations of V onto itself under usual composition of mappings, then a linear representation of G is a group homomorphism

$$\psi : G \to GL(V)$$

such that if V is finite-dimensional, then the map

$$\sigma : G \times V \to V, \ (g, v) \mapsto \psi(g)(v)$$

is smooth and the map ψ is a homomorphism of Lie groups. If V is finite-dimensional, then the degree of the representation is the **dimension** of V. Sometimes, V is called a representation of G.

Remark 4.3.21 Results on the representation of compact Lie groups are available in Exercise 33 of Section 4.11.

Proposition 4.3.22 *Let a Lie group G act on a differentiable manifold M as a Lie transformation group. If H is any Lie subgroup of G, then H also acts on M as a Lie transformation group.*

Proof Suppose that G acts on M as a Lie transformation group with an action

$$\psi : G \times M \to M, (g, x) \mapsto gx.$$

Let $i : H \hookrightarrow G$ be the natural inclusion map. Consider the map

$$\psi_H = \psi \circ (i \times 1_d) : H \times M \to G \times M,$$

where

$$(i \times 1_d) : H \times M \to G \times M, (h, x) \mapsto (h, x).$$

It asserts that H also acts on M as a Lie transformation group. □

Corollary 4.3.23 *The orthogonal group $O(n, \mathbf{R})$ acts on $\mathbf{R}^n$ as a Lie transformation group.*

Proof Since Lie group $GL(n, \mathbf{R})$ acts on the Euclidean n-space $\mathbf{R}^n$ as a Lie transformation group under the action

$$\psi : GL(n, \mathbf{R}) \times \mathbf{R}^n, \ (A, \ x) \mapsto Ax,$$

here the vector x is considered as a column matrix. The action has exactly two orbits which are $\{0\}$ and $\mathbf{R}^n - \{0\}$. Hence, the corollary is proved by using Proposition 4.3.22. □

Definition 4.3.24 Let a Lie group G act on a differentiable manifold M as a Lie transformation group with a group action

$$\psi : G \times M \to M.$$

Then a subset $A \subset M$ is said to be **invariant under** G if

$$\psi(G \times A) \subset A.$$

Proposition 4.3.25 *Let a Lie group G act on a differentiable manifold M as a Lie transformation group with Lie transformation group action ψ. If N is a regular submanifold of M, then G also acts on N as a Lie transformation group.*

Proof Let N be a regular submanifold of M with inclusion

$$i : N \hookrightarrow M$$

and natural injection.

$$j = (1_d \times i) : G \times N \hookrightarrow G \times M.$$

By hypothesis, the action

$$\psi : G \times M \to M, (g, x) \to gx$$

is smooth. Consider the map

$$\psi_N = \psi \circ (1_d \times i) : G \times N \to N$$

where the map

$$(1_d \times i) : G \times N \to G \times M$$

is smooth. It asserts that the map $\psi_N = \psi \circ (1_d \times i)$ is differentiable and hence G also acts on N as a Lie transformation group. □

Definition 4.3.26 Let G be a transformation group on M and $\sim$ be an equivalence relation on M. Then G is said to preserve the relation $\sim$ if $x \sim y$ and $g \in G$, then $gx \sim gy$.

Proposition 4.3.27 *Let G be a Lie transformation group with an action ψ on M and $\sim$ be an equivalence relation preserved by G on M. Then G acts in a natural way on the quotient manifold $M/\sim$ as a Lie transformation group.*

Proof By hypothesis, $\psi : G \times M \to M$ is a differentiable function. Let $p : M \to M/\sim$ be the natural projection. Then

$$p \circ \psi : G \times M \to M/\sim$$

is also a differentiable function and $G \times (M/\sim)$ is a quotient manifold of $G \times M$. Clearly, $p \circ \psi$ is an invariant of the given equivalence relation $\sim$ on $G \times M$, and hence it projects to the differentiable function

$$\psi_* : G \times (M/\sim) \to M/\sim, \ (g, p(x)) \mapsto p(gx).$$

This asserts that G acts in a natural way on the quotient manifold $M/\sim$ as a Lie transformation group. □

Example 4.3.28 If G is a Lie group and H is a Lie subgroup of G, then G acts on the quotient manifold G/H in a natural way as a Lie transformation group. Since G

acts on itself and the subgroup H defines an equivalence relation $\sim$ on G given by $g \sim k$ iff $g = kh$ for some $h \in H$, then the group action is defined as

$$\psi : G \times G/H \to G/H, \ (g, kH) \mapsto (gk)H.$$

It asserts that G acts on the quotient manifold G/H as a Lie transformation group.

Theorem 4.3.29 *Let G be a matrix Lie group with identity component G_0 of G. Then G_0 is a closed normal subgroup of G.*

Proof By hypothesis, G is a matrix Lie group with identity element as the identity matrix I and G_0 is the identity component of G containing I. Then it is a closed subset of G, and hence G_0 is a matrix Lie group. To show that $G_0 = X$ is a normal subgroup of G, take any matrix $x \in X$. Then $Xx^{-1} = R_{x^{-1}}X$ is the diffeomorphic image of the connected set X under the right translation $R_{x^{-1}}$, which is a diffeomorphism. This shows that the set Xx^{-1} is connected and it contains $xx^{-1} = I$. Hence, it follows that $Xx^{-1} \subset X$, since X is the maximal connected subset containing I in G. So, $XX^{-1} = X$ and X is a subgroup of G. Finally, for any $g \in G$, the set $gXg^{-1} = R_{g^{-1}} \circ L_g(X)$ contains I and is also connected, since it is the diffeomorphic image of the connected set X. By the above argument, it follows that $gXg^{-1} \subset X$, which asserts that X is a normal subgroup of G. □

4.3.3 Discontinuous Action of Lie Group

Definition 4.3.30 Let G be a Lie group and M be a smooth manifold. Then G is said to **act properly discontinuously** on M if for any point $x \in M$, the orbit space Gx is a closed discrete subset of M in the sense that any sequence of distinct points $\{g_n x\}$ does not converge to a point in M. In other words, if G acts discontinuously on M, then

(i) every point $x \in M$ has an open nbd U such that

$$gU \cap U = \emptyset, \ \forall g \, (\neq e) \in G$$

and

(ii) for every pair of points $x, y \in M$ with $y \notin Gx$, there exist open nbds U of x and an open V of y such that

$$gU \cap V = \emptyset, \ \forall g \in G, \text{where } gU = \{gx : x \in U\} \subset M.$$

Remark 4.3.31 The **condition (i)** of Definition 4.3.30 asserts that the action of G on M is free, because for any $x \in M$ and $gx = x$ for some $g \in G \implies x \in gU \cap U \, \forall \, nbd \, U \, of \, x \implies gU \cap U \neq \emptyset \implies x = e$. On the other hand, **condition (ii)**

of Definition 4.3.30 is equivalent to the assertion that given any two points $x, y \in M$ that are not equivalent, there exist open nbds U of x and nbds V of y such that

$$gU \cap kV = \emptyset, \ \forall g, k \in G, \text{ since } gU \cap kV = (k^{-1}g)U \cap V.$$

Definition 4.3.32 An action of a Lie group G on a smooth manifold M is said to be **proper** if for each compact subset $C \subset M$, the set

$$G_C = \{g \in G : gC \cap C \neq \emptyset\} \text{ is compact.}$$

Theorem 4.3.33 gives a characterization of proper action by its inverse image of every compact set.

Theorem 4.3.33 *An action of a Lie group G on a smooth manifold M is proper iff the map*

$$\psi : G \times M \to G \times M, (g, x) \mapsto (g, gx)$$

has the property that its inverse image of every compact set is also compact.

Proof First, suppose that ψ has the property that its inverse image of every compact set is also compact. Then for any compact set $C \subset M$,

$$\begin{aligned} G_C &= \{g \in G : gx \in C \ for\ some\ x \in C\} \qquad (4.1) \\ &= \{g \in G : \psi(g, x) \in C \times C \ for\ some\ x \in M\} = p(\psi^{-1}(C \times C)), \end{aligned}$$

where $p : G \times M \to G, \ (g, x) \mapsto g$ is the projection on the first factor, which is continuous. Since $C \times C$ is compact and p is continuous, it follows under the given condition that the set G_C is compact. This proves that the given action is continuous.

Conversely, let G_C be compact for any compact set $C \subset M$ and $K \subset M \times M$ be compact. Let

$$p_1, p_2 : M \times M \to M, \ (x_1, x_2) \mapsto x_1, x_2$$

be the projection maps on the first and second factors, respectively. Then the set

$$A = p_1(K) \cup p_2(K) \subset M$$

is compact and

$$\psi^{-1}(K) \subset \psi^{-1}(A \times A) \subset \{(g, x) \in G \times M : gx \in A, x \in A\} \subset G_A \times A.$$

Since $\psi^{-1}(K)$ is closed in $G_A \times A$ by continuity of ψ and it is also a closed subset of the compact set $G_A \times A$, it follows that the set $\psi^{-1}(K)$ is compact. □

Definition 4.3.34 A topological group is said to be **discrete** if all of its points are open. In other words, G is discrete iff $\{e\}$ is open in G.

Example 4.3.35 The group $\mathbf{Z}^n$ acts smoothly on the Euclidean space $\mathbf{R}^n$ by translation

$$\mathbf{Z}^n \times \mathbf{R}^n \rightarrow \mathbf{R}^n, ((n_1, ..., n_n), (x_1, ..., x_n)) \mapsto (n_1 + x_1, ..., n_n + x_n)$$

(see Exercise 3 of Sect. 4.11).

Proposition 4.3.36 *Let G be a Lie group and K be a closed discrete subgroup of G. Then K acts properly discontinuously on G by left (right) translations*

$$L_k : K \times G \rightarrow G, \ g \mapsto kg \ (R_k : K \times G \rightarrow G, \ g \mapsto gk).$$

Proof Let $e \in G$ be the identity element. Since, by hypothesis, K is a closed discrete subgroup of G, there exists an open nbd V of e in G such that $K \cap V = \{e\}$. Again, since the map

$$\mu : G \times G \rightarrow G, \ (g, h) \mapsto gh^{-1}$$

is continuous, there exists an open nbd U of e in G such that

$$UU^{-1} \subset V \implies \text{only } k = e \in K \text{ has the property that } U \cap kU \neq \emptyset,$$

because

$$U \cap kU \neq \emptyset \implies \text{there exit } u_1, u_2 \in U \text{ such that } u_1 = ku_2$$
$$\implies k = u_1 u_2^{-1} \in UU^{-1} \subset V \implies k = e.$$

Hence, it satisfies condition (i) of Definition 4.3.30. For its condition (ii), consider the projection

$$p : G \rightarrow G/K, x \mapsto gK,$$

which is continuous. By hypothesis, K is a closed subgroup of G. Hence by continuity of μ, it follows that

$$\mu^{-1}(K) = \{(g, k) \in G \times G : g = kh \text{ for some } h \in K\}.$$

Thus $\mu^{-1}(K)$ is a closed subset of $G \times G$. If $Kg \neq Kh$, then $(g, h) \notin \mu^{-1}(K) \implies$ there exist open nbds W of g and S of h such that $(W \times S) \cap \mu^{-1}(K) = \emptyset$. It asserts that the nbds $p(W)$ of Kg and $p(S)$ of Kh are disjoint and hence $kW \cap S = \emptyset, \ \forall k \in K$. □

Theorem 4.3.37 *Let M be a smooth manifold of dimension n and a Lie group act on G properly discontinuously. Then the orbit space M mod G admits the structure of a quotient manifold having dimension n such that the projection*

$$p : M \to M \; mod \; G$$

is a submersion.

Proof Let U be an open nbd of an arbitrary point $x \in M$ such that the canonical surjection $p : M \to M \; mod \; G$ is a homeomorphism. To choose U, take U such that it satisfies condition (i) of Definition 4.3.30. Then for any pair of points $x, y \in U$, the relation $y = gx$ holds iff $g = e$. This implies that the map $p : U \to p(U) = V$ is injective, and hence p is a homeomorphism, since every continuous bijective open map is a homeomorphism. Without any loss of generality, we assume that U is the domain of a chart (ψ, U) in M (by shrinking U, if it becomes necessary). We claim that the pairs (ϕ, V), where $V = p(U)$ and $\phi = \psi \circ (p(U))^{-1}$ form an atlas on the orbit space $M \; mod \; G$. Let (ϕ_1, V_1) and (ϕ_2, V_2) be two such pairs provided by the charts (ψ_1, U_1) and (ψ_2, U_2) on M. Letting $\alpha_i = (p(U_i))^{-1}; i = 1, 2$, the transition map is then given by

$$\phi_2 \circ \phi_1^{-1} = \psi_2 \circ \alpha_2 \circ \alpha_1^{-1} \circ \psi_1^{-1}$$

and the map

$$\alpha_2 \circ \alpha_1^{-1} : \alpha_1(V_1 \cap V_2) \to \alpha_2(V_1 \cap V_2)$$

is smooth. To show it, take any point $g \in G$ and consider the diffeomorphism

$$\delta_g : M \to M, x \mapsto gx.$$

Then every $x \in \alpha_i(V_1 \cap V_2)$ has an open nbd on which the maps $\alpha_2 \circ \alpha_1^{-1}$ agree with δ_g. Now, for any point $x \in \alpha_1(V_1 \cap V_2)$, the point x and the point $y = \alpha_2(\alpha_1^{-1}(x))$ are equivalent and hence $y = gx$ for some $g \in G$. Then for the open nbd $W = U_1 \cap g^{-1}U_2$ of x,

$$W \subset \alpha_1(V_1 \cap V_2), \text{ because, } p(U_1) \cap p(g^{-1}U_g) - V_1 \cap V_2.$$

Hence,

$$p(z) \in V_1 \cap V_2 \text{ for any } z \in W \implies \alpha_2(\alpha_1^{-1}(z)) = t,$$

where $t \in V_2$ is the unique point such that $p(t) = p(z)$. On the other hand,

$$gz \in U_2 \text{ and } p(gz) = p(z) \in V_2 \implies gz = t \implies \alpha_2 \circ \alpha_1^{-1} \text{ and } \delta_g$$

coincide with W. This proves that $\alpha_2 \circ \alpha_1^{-1}$ is smooth. Finally, we claim that the orbit space $(M \; mod \; G, \tau)$ with topology τ is Hausdorff and has a countable open base for τ. By definition, M has a countable open base $\mathcal{B} = \{U_i\}$ for its topology. Since $p : M \to M mod \; G$ is open, it follows that the family of open sets $\{p(U_i)\}$ forms a countable basis for τ. To prove the Hausdorff property of τ, take any pair of elements $z, t \in X \; mod \; G$. Then there exist $x, y \in M$ such that $p(x) = z$ and $p(y) = t$. By using condition (ii) of Definition 4.3.30, it follows that there exist open sets U_1 of x and sets U_2 of y in M such that

$$gU_1 \cap U_2 = \emptyset, \ \forall \, g \in G.$$

This shows that $V_1 = p(U_1)$ and $V_2 = p(U_2)$ are open nbds of z and t in $M \ mod \ G$, respectively, and are such that $V_1 \cap V_2 = \emptyset$, otherwise, we have a contradiction because

$$V_1 \cap V_2 \neq \emptyset \implies p^{-1}(V_1 \cap V_2) = p^{-1}(V_1) \cap p^{-1}(V_2) \neq \emptyset.$$

On the other hand,

$$p^{-1}(V_1) = \bigcup_{g \in G} gU_1 \text{ and } p^{-1}(V_2) = \bigcup_{k \in G} kU_2 \implies gU_1 \cap kU_2 \neq \emptyset \text{ for some } g, k \in K,$$

which contradicts condition (ii) of Definition 4.3.30. This proves that $M \ mod \ G$ is a manifold. Moreover, triviality condition follows and hence p is a submersion. □

Corollary 4.3.43 is similar to a classical result saying that for any closed subgroup K of a Lie group G, the coset space G/K is a manifold.

Corollary 4.3.38 *Let G be a Lie group and K be a discrete subgroup of G, then*

(i) *the coset space G/K is a manifold and*
(ii) $p : G \to G/K, \ g \mapsto gK$ *is a smooth map.*

Proof It follows from Theorem 4.3.37 by using Proposition 4.3.36. □

4.4 Lie Subgroups and Homomorphisms of Lie Groups

4.4.1 Lie Subgroups

This subsection studies Lie subgroups of Lie groups G, which are simultaneously subgroups and submanifolds of G.

Definition 4.4.1 Let G be a Lie group. A Lie subgroup H of G is a subgroup of the group G such that H is also a submanifold of G.

Example 4.4.2 The general linear group $GL(n, \mathbf{R})$ is a Lie subgroup of $GL(n, \mathbf{C})$.

Theorem 4.4.3 *Every subgroup K of a Lie group G is closed in G.*

Proof This is left as an exercise. □

Proposition 4.4.4 *Let G be a Lie group and H be a subgroup of G such that H is a regular submanifold of G. Then H is a Lie subgroup of G.*

Proof By hypothesis, H is a regular submanifold of G. Consider the natural inclusion map $i : H \hookrightarrow G$ and the Lie group multiplication $m : G \times G \to G$. Then the map

$$m \circ (i \times i) : H \times H \to G$$

is a differentiable map having its values in H. Hence, it induces a differentiable map

$$\mu : H \times H \to H$$

making H a Lie subgroup of G under this group function. □

Remark 4.4.5 For More Properties of Subgroups of Lie Groups, see Sect. 4.5.

4.4.2 Properties of Homomorphisms of Lie Groups

Let G and H be two Lie groups. Then $f : G \to H$ is a group homomorphism iff $f \circ L_g = L_{f(g)} \circ f$.

Definition 4.4.6 A group homomorphism $f : G \to GL(n, \mathbf{R})$ (or $f : G \to GL(n, \mathbf{C})$) is called a **real (complex) representation** of G.

Proposition 4.4.7 *Let G and G' be two Lie groups with $e \in G$ and $e' \in G'$ identity elements. If G' is connected and $f : G \to G'$ is a homomorphism of Lie groups such that*

$$f_* : T(G) \to T(G')$$

is surjective, then f is also a surjective.

Proof It follows from the hypothesis that f is surjective onto some open nbd U' of e' by using the inverse function theorem. Since every image under a group homomorphism is a subgroup, f is surjective. □

4.4.3 Action of $O(n+1, \mathbf{R})$ on $\mathbf{R}P^n$

This section considers the action of the Lie group $O(n+1, \mathbf{R})$ on $\mathbf{R}P^n$ by using the usual action of $O(n+1, \mathbf{R})$ on $\mathbf{R}^{n+1}$.

The real projective space $\mathbf{R}^{n+1}$ can be constructed from $\mathbf{R}^{n+1}$ by an equivalence relation $\sim$:
for $x, y \in \mathbf{R}^{n+1}$ and $g \in O(n+1, \mathbf{R})$, if $x \sim y$, (in the sense that $y = tx$ for some

$t \in \mathbf{R} - \{0\}$) then $gx \sim gy$ (in the sense that $gy = tgx$ for some $t \in \mathbf{R} - \{0\}$). This implies that the action

$$\psi : O(n+1, \mathbf{R}) \times \mathbf{R}P^{n+1} \to \mathbf{R}^{n+1}, \ (A, x) \to Ax$$

induces a natural action

$$\psi_* : O(n+1, \mathbf{R}) \times \mathbf{R}P^n \to \mathbf{R}^n, \ (A, [x]) \to [Ax].$$

The action ψ_* is transitive. Let $p \in \mathbf{R}P^n$ be corresponding to the point $x = (1, 0, 0, \ldots, 0) \in \mathbf{R}^{n+1}$ in the sense that $p = [x]$, then the isotropy group at the point p is

$$G_p = \mathbf{Z}_2 \times O(n, \mathbf{R}).$$

Hence, it follows that

$$O(n+1)/(\mathbf{Z}_2 \times O(n, \mathbf{R})) \cong S^n/\mathbf{Z}_2 \cong \mathbf{R}P^n.$$

4.5 More Properties of Lie Subgroups and Connected Lie Groups

This section studies more properties of Lie subgroups of Lie groups G, which are simultaneously subgroups and submanifolds of G. Moreover, it studies connected Lie groups.

Definition 4.5.1 Let G be a Lie group. A Lie subgroup H of G is a subgroup (abstract) of the group G such that H is also a submanifold of G.

Example 4.5.2 The general linear group $GL(n, \mathbf{R})$ is a Lie subgroup of $GL(n, \mathbf{C})$.

Theorem 4.5.3 *Every closed subgroup of a Lie group G is a Lie group.*

Proof Since every Lie group is locally compact, every closed subgroup H of a Lie group G is also locally compact and hence it follows that H is itself a Lie group. □

Example 4.5.4 Theorem 4.5.3 is now applied to prove that each of the following groups is a Lie group.

(i) $SL(n, \mathbf{R})$ is a Lie subgroup of the Lie group $GL(n, \mathbf{R})$. To show it, consider the map

$$f : GL(n, \mathbf{R}) \to \mathbf{R}, A \mapsto det\ A,$$

which is the determinant function and hence f is continuous. Clearly, $SL(n, \mathbf{R}) = f^{-1}(1)$. Since the one-pointic set $\{1\}$ is closed in $\mathbf{R}$, it follows by continuity of f that $f^{-1}(1) = SL(n, \mathbf{R})$ is closed in $GL(n, \mathbf{R})$. By applying Theorem 4.4.3, it follows that $SL(n, \mathbf{R})$ is a Lie subgroup of the Lie group $GL(n, \mathbf{R})$.

(ii) $O(n, \mathbf{R})$ is a Lie subgroup of the Lie group $GL(n, \mathbf{R})$. To show it, proceed as in (i).

(iii) $SO(n, \mathbf{R})$ is a Lie subgroup of the Lie group $GL(n, \mathbf{R})$. To show it, proceed as in (i).

Proposition 4.5.5 *Let a Lie group G act on a differentiable manifold M by an action*

$$\sigma : G \times M, (g, x) \mapsto \sigma(g, x) = gx.$$

Then the isotropy (or stabilizer) group G_{x_0} of G at a point $x_0 \in M$ is a Lie subgroup of G.

Proof By definition,

$$G_x = \{g \in G : \sigma(g, x) = gx = x\}.$$

Then G_x is a subgroup (abstract) of the group G, because

(i) for $g_1, g_2 \in G_x, \sigma(g_1 g_2, x) = \sigma(g_1, \sigma(g_2, x)) = \sigma(g_1, x) = x$;

(ii) $e \in G_x$, since $\sigma(e, x) = x$

(iii) for $g \in G_x$, its inverse $g^{-1} \in G_x$, since $x = \sigma(e, x) = \sigma(g^{-1}g, x) = \sigma(g^{-1}, \sigma(g, x)) = \sigma(g^{-1}, x)$.

To show that the abstract subgroup G_x is a Lie group, keep x fixed and define

$$\psi_x : G \to M, \ g \mapsto gx.$$

Then G_x is the inverse image $\psi_x^{-1}(x)$ of the point x. This implies that G_x is a closed subset of the Lie group G. This proves that G_x is a Lie subgroup of the Lie group G by Theorem 4.5.3. □

Proposition 4.5.6 *Let M be a smooth manifold of dimension n and G be a Lie group of dimension m $(> n)$. If $\psi : G \to M$ is a smooth map having rank n at the identity element $e \in G$, and for $H = \psi^{-1}(\psi(e))$, if*

$$\psi(hg) = \psi(g), \ \forall\, h \in H \text{ and } \forall\, g \in G,$$

then H inherits a Lie subgroup structure of dimension $m - n$ from G.

Proof Since for $h, k \in H$, $\psi(hk) = \psi(k) = \psi(e)$ and $\psi(e) = \psi(hh^{-1}) = \psi(h^{-1})$, it follows that

$$hk \in H, \ h^{-1} \in H, \ \forall\, h, k \in H,$$

hence H is a subgroup of G. Again, since H enjoys a submanifold structure of G, the subgroup H is a Lie subgroup of G. □

Example 4.5.7 gives a direct application of Proposition 4.5.6.

Example 4.5.7 **(i)** The subgroup $H = SL(n, \mathbf{R})$ of the Lie group $GL(n, \mathbf{R})$ is a Lie subgroup of $GL(n, \mathbf{R})$ of dimension $n^2 - 1$ by Proposition 4.5.6 for every integer $n > 1$. It needs verification of the condition that the function

$$\psi : GL(n, \mathbf{R}) \to \mathbf{R}, g \mapsto det\ g$$

satisfies the condition of Proposition 4.5.6. Clearly, for $H = SL(n, \mathbf{R})$,

$$\psi(hg) = det(hg) = det\ h\ det\ g = det\ g = \psi(g),\ \forall\, h \in H,\ and\ \ \forall\, g \in G.$$

For the rest of this example, let x be the standard chart on the Lie group $GL(n, \mathbf{R})$ and 1_d be the identity chart on $\mathbf{R}$.
Then

$$1_d \circ \psi = det(x_{i,j}) \text{ and } \frac{\partial}{\partial(x_{i,j})}(1_d \circ \psi) = C_{i,j} : i, j = 1, 2, \ldots, n,$$

where $C_{i,j}$ stands for the cofactor of $x_{i,j}$ in the matrix $(x_{i,j})$. Since $C_{1,1}e = 1$, it follows that rank of ψ at e is 1.

(ii) The subgroup $H = O(n, \mathbf{R})$ of the Lie group $GL(n, \mathbf{R})$ is a Lie subgroup of $GL(n, \mathbf{R})$ of dimension $\frac{n(n-1)}{2}$ by Proposition 4.5.6 for every integer $n > 1$. Let $Sym\ (n, \mathbf{R})$ be the set of all $n \times n$ real symmetric matrices with its standard C^∞ structure. It needs verification of the condition that the function

$$\psi : GL(n, \mathbf{R}) \to \ Sym\ (n, \mathbf{R}), g \mapsto \ g^t g$$

satisfies the condition of Proposition 4.5.6. Clearly, for $H = O(n, \mathbf{R})$,

$$\psi(hg) = (hg)^t(hg) = (g^t h^t)(hg) = \psi(g),\ \forall\, h, g \in H.$$

Connected Lie Groups

This subsection studies connected Lie groups. It is proved in Proposition 4.5.9 that every nbd of the identity element of a connected Lie group generates the group and other related propositions.

Definition 4.5.8 A Lie G group is said to be **connected** if its underlying manifold is connected.

Proposition 4.5.9 *Let G be a connected Lie group with its identity element e and U be an nbd of e. Then U generates G.*

Proof **Proof I** By hypothesis, U is an nbd of e in G. Let U generate the subgroup K of G. Then K is open in G and for any $k \in K$, the set kU is an open nbd of k in G. It is a submanifold of G, because it is an open subset of the manifold G. Then K is a nonempty Lie subgroup of G, which is closed by Theorem 4.5.3 and hence it follows that $K = G$.

Proof II For any open nbd U of e and an open set $V \subset U$ with VV^{-1}, define an open set

$$K = \bigcup_{n=-\infty}^{\infty} V^n, \text{ with weak topology.}$$

Then K is a subgroup of G. It is closed because its complement is the union of all cosets of K, which are different from K is open. Hence, it follows that $K = G$. □

Proposition 4.5.10 *Let G be a Lie group with identity element e, and G^e be the component containing e in G. Then G^e is a connected Lie subgroup of G.*

Proof By hypothesis, G is a Lie group with identity element e and G^e is the component containing e in G. Let $i : G^e \hookrightarrow G$ be the natural inclusion map. Then the usual multiplication on G,

$$m : G \times G \to G, (x, y) \mapsto xy,$$

is a smooth map and hence the composite map

$$m \circ (i \times i) : G^e \times G^e \to G$$

induces a smooth map

$$\overline{m} : G^e \times G^e \to G^e$$

and it makes G^e a Lie subgroup of G. By the given conditions, it also follows that G^e is connected. □

Proposition 4.5.11 *Let G be a Lie group with identity element e and G^e be the component containing its identity element e. Then the other components of G are cosets of G^e in G.*

Proof Let G be a Lie group with multiplication m and inversion map υ. Since G^e is a connected subset of G, it follows that $G^e \times G^e$ is also connected and

$$m(G^e \times G^e) \subset G^e.$$

Again $\upsilon(G^e)$ is the inverse of G^e, because for any open subset U of G,

$$U^{-1} = \{x^{-1}, \forall x \in G\} = \upsilon(U),$$

and $v(G^e)$ is a connected nbd of e. Consequently,

$$v(G^e) \subset G^e.$$

This asserts that G^e is a subgroup of G. Again, since a translation $T : G \to G$ is a diffeomorphism, it follows that $xG^e x^{-1}$ is also a connected nbd of e for every $x \in G$. This implies that G^e is a normal subgroup of G. Finally let G^x be the component of G containing the element $x \in G$. Then as the above argument, it follows that

$$xG^e \subset G^x \quad \text{and} \quad x^{-1}G^e \subset G^e.$$

This proves that $G^x = xG^e$. □

Proposition 4.5.12 *Let G and $\tilde{G}$ be two connected Lie groups with $e \in G$ and $\tilde{e} \in \tilde{G}$ identity elements. If $p : \tilde{G} \to G$ is a covering of Lie groups, then ker $p = H$ is a discrete normal subgroup of the center $C(\tilde{G})$.*

Proof By the given condition, since p is a local diffeomorphism, there exists an open nbd U of $\tilde{e}$ such that

$$U \cap H = \{\tilde{e}\}.$$

Then

$$hU \cap H = \{h\}, \ \forall\, h \in H,$$

because $hu = h' \implies u = h^{-1}h'$. Since $ker\ p = H$, it is a normal subgroup of $\tilde{G}$. Moreover, it is a discrete normal subgroup of $\tilde{G}$ that is contained in $C(\tilde{G})$, because for any fixed $\tilde{g}_0 \in \tilde{G}$ and any $h \in H$, the path $\alpha : \mathbf{I} \to \tilde{G}$ such that $\alpha(0) = \tilde{e}$ and $\alpha(1) = \tilde{g}_0$ and the element $\alpha(t)\ h\ \alpha(t)^{-1} \in H$. □

Definition 4.5.13 (*Dimension of a Lie group*) Given a Lie group G with identity e, let G^e denote the connected component of G which contains the identity element of G. Then G^e is a connected Lie subgroup of G by Proposition 4.5.10. Any other connected component of G is homeomorphic to G^e. This shows that if G is a Lie group, the (real or complex) dimension of G is well-defined. The dimension of the manifold G^e is called the dimension of the manifold G. Thus the dimension of a Lie group is defined to be its dimension as a manifold.

4.6 Smooth Coverings

This section studies covering projection related to connected Lie groups and proves the existence of a covering projection in Theorem 4.6.5 and gives a sufficient condition for being a connected subgroup of a connected Lie group.

Definition 4.6.1 Given two topological spaces X, B and a continuous onto map $p : X \to B$, an open set V of B is said to be **evenly covered** by p if $p^{-1}(V)$ is a union of disjoint open sets U_i in X, called **sheets** such that

$$p|_{U_i} : U_i \to V$$

is a homeomorphism for each i. Then V is said to be an **admissible open set** in B.

Example 4.6.2 For the exponential map

$$p : \mathbf{R} \to S^1, t \mapsto e^{2\pi it},$$

the open set $U = S^1 - \{1\}$ is evenly covered by p, since $p^{-1}(U) = \bigcup_{n\in\mathbf{Z}} \left(n - \frac{1}{2}, n + \frac{1}{2}\right)$ and corresponding sheets are open intervals in $\mathbf{R}$.

Definition 4.6.3 Let X, B be topological spaces and $p : X \to B$ be an onto map. Then the triplet (X, p, B) is called a covering space if

(i) X is a path-connected topological space;
(ii) the map $p : X \to B$ is continuous;
(iii) each point $b \in B$ has an open nbd U which is evenly covered by p.

The map p is called the covering projection with X its total space and B its base space.

Definition 4.6.3 has an analogue for connected smooth manifolds formulated in Definition 4.6.4.

Definition 4.6.4 Let M and $\tilde{M}$ be two connected smooth manifolds. A covering

$$p : \tilde{M} \to M$$

is said to be a smooth covering if

(i) p is a smooth onto mapping and
(ii) any connected open set U of M is evenly covered by p in the sense that the restriction

$$p|_{\tilde{U}} : \tilde{U} \to U$$

is diffeomorphism for every component $\tilde{U}$ of the set $p^{-1}(U)$.

Theorem 4.6.5 *Let G be a connected Lie group and H be a closed subgroup of G. If H^e is the component of the identity e of H, then the natural projection map*

$$p : G/H^e \to G/H, \ gH^e \to gH$$

is a smooth covering.

Proof Let $\pi : G \to G/H,\ g \mapsto gH$ and $\pi^e : G \to G/H^e$ be the natural projection maps. Then they are both open and continuous. Since the subgroup H^e is open in H, the identity element of G has a connected nbd U such that $U^{-1}U \subset H^e$. To prove the proposition, it is sufficient to prove for any point $\pi(g) = gH \in G/H$, the nbd $U(g) = \pi(gU)$ is evenly covered by p. Take a representative h_i from every component H_i of H and consider open connected sets $\pi^e(gUh_i)$ in G/H^e. The sets $\pi^e(gUh_i)$ have the following properties:

(i) they are disjoint;
(ii) they form together the whole set $p^{-1}(U(g))$;
(iii) p maps each of them homeomorphically onto $U(g)$.

These properties assert that $U(g)$ is evenly covered by p. Hence, it follows from Definition 4.6.3 that p is covering projection. □

Corollary 4.6.6 *Let G be a connected Lie group and H be a closed subgroup of G. If the quotient manifold G/H is simply connected, then H is a connected subgroup of G.*

Proof It follows from Theorem 4.6.5. □

Proposition 4.6.7 *Suppose G is a connected Lie group with multiplication $m : G \times G \to G$ and identity element e. Let $\tilde{G}$ be a connected manifold and $p : \tilde{G} \to G$ be a smooth covering. Then $\tilde{G}$ has the unique Lie group structure such that p is a Lie group homomorphism.*

Proof Let $\tilde{G}$ be a connected manifold and $p : \tilde{G} \to G$ be a smooth covering. Let $\tilde{e} \in p^{-1}(e)$. Consider the composite map

$$\tilde{m} = m \circ (p \times p) : \tilde{G} \times \tilde{G} \to G,\ (\tilde{x}, \tilde{y}) \mapsto m(p(\tilde{x}), p(\tilde{y})).$$

Hence, it follows from the uniqueness property of the lift that m has a unique lifting

$$\tilde{m} : \tilde{G} \times \tilde{G} \to G : (\tilde{e}, \tilde{e}) \mapsto e.$$

Similarly, the inversion map $v : G \to G, x \mapsto x^{-1}$ has a unique lifting

$$\tilde{v} : \tilde{G} \to \tilde{G} : \tilde{e} \mapsto \tilde{e}.$$

Finally, it follows from the uniqueness property of the lift that any two Lie group structures admitted by $\tilde{G}$ such that $p : \tilde{G} \to G$ is a Lie group homomorphism are isomorphic. □

4.6.1 Tangent Bundles of Lie Groups

Recall from Chap. 3 the following definition.

Definition 4.6.8 (*The tangent bundle of a manifold*) The tangent bundle of a smooth manifold M is the bundle $(\mathcal{T}(M), p, M)$, where

(i) the total space $\mathcal{T}(M)$ is the disjoint union of all tangent spaces $T_x(M)$ as x runs over M. This is the set of all ordered pairs (x, v) such that $x \in M$ and $v \in T_x(M)$.
(ii) the map

$$p : \mathcal{T}(M) \to M, \ (x, v) \to x$$

is called the **projection map** of the tangent bundle. $T_x(M)$ is called the **tangent plane** at x and $v \in T_x(M)$ is called the **tangent vector** with initial point x.

Remark 4.6.9 Let M be a smooth manifold. Then the set of all tangent vectors to the manifold M form a bundle, called tangent bundle $\mathcal{T}(M)$ with projection

$$p : \mathcal{T}(M) \to M, v \mapsto x, \ if \ v \in T_x(M).$$

A smooth function $\psi : M \to N$ of smooth manifolds induces a smooth function

$$\psi_* : \mathcal{T}(M) \to \mathcal{T}(N), v \mapsto \psi_{*x}(v), \ if \ x = \psi(v).$$

Proposition 4.6.10 *Let G be a Lie group. Its tangent bundle $\mathcal{T}(G)$ is trivial.*

Proof Consider the right translation for any $g \in G$,

$$R_g : G \to G, x \mapsto xg.$$

Then R_g is a diffeomorphism with its inverse $R_{g^{-1}}$. This implies that

$$dR_g : T_e(G) \to T_g(G)$$

is a linear isomorphism of vector spaces. Let $dim\ G = n$ and $\{v_1, v_2, \dots, v_n\}$ be a basis of the vector space $T_e(G)$. Define vector fields

$$V_i : G \to \mathcal{T}(G), \ g \mapsto d_{v_i} R_g(v_i), \ \forall i = 1, 2, \dots, n.$$

Since the multiplication in G is smooth, it follows that V_i s are smooth vector fields for all $i = 1, 2, \dots, n$. Consider the map

$$\psi : G \times \mathbf{R}^n \to \mathcal{T}(G), \ (g, t_1, t_2, \dots, t_n) \mapsto \ (g, \ \Sigma_i \ t_i \ V_i(g)).$$

Then ψ is an isomorphism of vector bundles (see Chap. 3). This proves that the tangent bundle of any Lie group is trivial. □

Corollary 4.6.11 *Consider S^3 as a Lie group of unit quaternions. Then $\mathcal{T}(S^3)$ is trivial.*

Proof Since S^3 is a Lie group, the corollary follows from Proposition 4.6.10. □

4.6.2 One-Parameter Group of Transformations

This subsection studies one-parameter group of transformations by using the concept of flow in a manifold. A vector field in a manifold generates a flow in the manifold.

Definition 4.6.12 Let G be a Lie group and $f : \mathbf{R} \to G$ be a smooth homomorphism of Lie groups (where $\mathbf{R}$ is treated as a Lie group). Then f is called a one **parameter subgroup of** G**.** In other words, a smooth curve $f : \mathbf{R} \to G$ is said to be a one-parameter subgroup of a Lie group G if

$$f : \mathbf{R} \to G : \ f(t+s) = f(t)f(s), \ \forall\, t, s \in \mathbf{R}.$$

Remark 4.6.13 **(i)** Every one-parameter subgroup of a Lie group G is a map $f : \mathbf{R} \to G$; it is not a subset of G by its definition.

(ii) Every one-parameter subgroup $f : \mathbf{R} \to G$ of a Lie group G passes through the identity element e of G, because $f(0) = f(0+0) = f(0)f(0) = ee = e$.

(iii) Every one-parameter subgroup $f : \mathbf{R} \to G$ of a Lie group G is a differentiable homomorphism of the additive group of real numbers into G.

Definition 4.6.14 For every $v \in T(M)$, there exists a unique $x \in M$. Then the assignment $v \to x$ defines a map

$$p : T(M) \to M, v \to x$$

called a **projection map**. A vector field on M is a map

$$V : M \to T(M) : p \circ V = 1_M.$$

Then $V(x)$ is tangent vector for every $x \in M$. A vector field V on M is said to be smooth if

$$Vf : X \to \mathbf{R}, x \mapsto V(x)f, \ \forall\, f \in C^\infty(M, \mathbf{R}).$$

In other words, a vector field V on M is a section of the tangent bundle $T(M)$ if all such functions Vf are differentiable. Let $p^{-1}(a)$ be denoted by $V_a \in T_a(M)$, $\forall\, a \in M$. For every point $a \in M$, the vector fields $\{V_a : a \in M\}$ form a Linear space of infinite dimension, denoted by $\mathcal{L}(M)$. If $M = G$ is a Lie group, then for any element $a \in G$, a vector field V is said to be **left-invariant** if

$$V_a = (dL_a)_e V_e, \ \ \forall\, a \in G.$$

Definition 4.6.15 Let $f : \mathbf{R} \to G$ be a smooth curve on a Lie group G. Then f is said to be an **integral curve** of a vector field V if the tangent vector to f at each point is equal to the value of V at that point in the sense that

$$\frac{df(t)}{dt} = V_{f(t)}, \ \forall t \in \mathbf{R}.$$

Example 4.6.16 Every one-parameter subgroup $f : \mathbf{R} \to G$ of a Lie group G is an integral curve of some left-invariant vector field.

Definition 4.6.17 A **flow in a manifold** M is a continuous map

$$\psi : \mathbf{R} \times M \to M,$$

where $\mathbf{R}$ acts as an additive group. Given $t \in \mathbf{R}$, a flow in a Lie group M

$$\psi_t : M \to M, \ x \mapsto \psi(t, \ x)$$

is a diffeomorphism such that

(i) $\psi_t(\psi_s(x)) = \psi_{t+s}(x), \forall x \in M$, i.e., $\psi_t \circ \psi_s = \psi_{t+s}$.
(ii) $\psi_0 = 1_M : M \to M$ is the identity map.
(iii) $\psi_{-t} = (\psi_t)^{-1}$.

Remark 4.6.18 The one-parameter group of transformations defined above looks locally like the group $(\mathbf{R}, +)$ but is not necessarily isomorphic to the group $(\mathbf{R}, +)$ globally. For more properties of this group, see Exercise 12 of Sect. 4.11.

4.7 Lie Algebra

This section studies Lie algebra, which is closely related to Lie groups in the sense that every Lie group G defines a Lie algebra, which is the tangent space $T_e(G)$ at its identity element $e \in G$ and conversely, corresponding to every finite-dimensional Lie algebra over $\mathbf{R}$, there exists a connected Lie group uniquely determined up to finite coverings by Lie's third theorem, which asserts that every finite-dimensional Lie algebra over $\mathbf{R}$ is associated with a Lie group G which gives a correspondence between Lie group and Lie algebra. This correspondence facilitates to study the structure and classification of a Lie group in terms of its Lie algebra. For example, a Lie group is found in physics as a group of symmetries of physical systems and its corresponding Lie algebra appears as a tangent vector in the nbd of its identity element, which is considered as an infinitesimal symmetry motion. In particle physics and quantum mechanics, this approach is extensively used.

Definition 4.7.1 Let V be a vector space over the field $\mathbf{F}$ (where F is either $\mathbf{R}$ or $\mathbf{C}$). Then V is said to be a **Lie algebra** over F if there is a composition

$$[\ ,\ ] : V \times V \to V,\ (u, v) \mapsto [u, v]$$

such that

(i) $[v, v] = 0,\ \forall\, v \in V$;

(ii) $[u, v + w] = [u, v] + [u, w],\quad [u, av] = a[u, v]$ and $[v + w, u] = [v, u] + [w, u],\ [au, v] = a[u, v]\ \forall\, u, v, w \in V,\ \forall\, a \in \mathbf{F}$ **(bilinearity)**;

(iii) $[u, v] = -[v, u],\ \forall\, v \in V$; **(skew symmetry)**

(iv) $[u, [v, w]] + [v, [w, u]] + [w, [u, v]] = 0\, \forall\, u, v, w \in V$ **(Jacobi Identity).**
A Lie algebra is said to be real if $\mathbf{F} = \mathbf{R}$, and it is said to be complex if $\mathbf{F} = \mathbf{C}$. For vectors u and v in V, the notation $[u, v]$ is also a vector in V, called their **Lie bracket.**

Example 4.7.2 **(i)** The vector space $\mathbf{R}^3$ is a real Lie algebra under the usual vector product

$$[\ ,\ ] : \mathbf{R}^3 \times \mathbf{R}^3 \to \mathbf{R}^3,\ (\alpha, \beta) \mapsto [\alpha, \beta] = \alpha \times \beta.$$

If $\mathcal{G} = \mathbf{R}^3$ is a Lie algebra of the Lie group of rotations of $\mathbf{R}^3$, then each vector $v \in \mathbf{R}^3$ may be considered as an infinite-dimensional rotation around the axis v, having velocity same as the magnitude of v.

(ii) $M(n, \mathbf{R})$ is a real Lie algebra with $[A, B] = AB - BA,\ \forall\, A, B \in M(n, \mathbf{R})$, where AB denotes the usual matrix multiplication.

Definition 4.7.1 is reformulated in Definition 4.7.3.

Definition 4.7.3 (*Lie algebra over a field*) Let $\mathbf{F}$ be the field either $\mathbf{R}$ or $\mathbf{C}$ and $\mathcal{V}$ be a vector space of finite dimension over $\mathbf{F}$. Moreover, there is given a law of composition

$$[\ ,\ ] : \mathcal{V} \times \mathcal{V} \to \mathcal{V},\ (X, Y) \to [X, Y],$$

called **Lie bracket operation** such that

(i) it is bilinear in the sense

$$[a_1X_1 + a_2X_2,\ Y] = a_1[X_1,\ Y] + a_2[X_2,\ Y],\ \forall\, a_1, a_2 \in \mathbf{F},\ X_1, X_2,\ Y \in \mathcal{V}$$

and

$$[X, a_1Y_1 + a_2Y_2] = a_1[X,\ Y_1] + a_2[X,\ Y_2],\ \forall\, a_1, a_2 \in \mathbf{F},\ X,\ Y_1, Y_2 \in \mathcal{V};$$

(ii)

$$[X, X] = 0,\ \ \forall\, X \in \mathcal{V};$$

(iii)

$$[[X, Y], Z] + [[Y, Z], X] + [[Z, X], Y] = 0,\ \ \forall\, X, Y, Z \in \mathcal{V}.$$

Then the vector space $\mathcal{V}$ equipped with this law of composition is called a **Lie algebra** over **F** and the composition $[\ ,\ ]$ is called the **Lie bracket operation** on $\mathcal{V}$.

Proposition 4.7.4 *Let $\mathcal{V}$ be a Lie algebra over a field F. Then*

$$[X, Y] + [Y, X] = 0, \ \forall X, Y \in \mathcal{V}.$$

Proof Take any $X, Y, Z \in \mathcal{V}$. Then

$$0 = [X + Y, X + Y] = [X, X] + [X, Y] + [Y, X] + [Y, Y]$$

asserts by using $[X, X] = 0, \ \forall X \in \mathcal{V}$ that

$$[X, Y] + [Y, X] = 0, \ \forall X, Y \in \mathcal{V}.$$

□

Definition 4.7.5 (*Direct sum of Lie algebras*) Let $\mathcal{V}_1$ and $\mathcal{V}_2$ be two Lie algebras over the same field **F**. Define the following law of composition in $\mathcal{V}_1 \times \mathcal{V}_2$,

$$[(X_1, Y_1), (X_2, Y_2)] = ([X_1, X_2], [Y_1, Y_2]).$$

With this law of composition, a Lie algebra is obtained whose underlying vector space is the product of the underlying vector spaces of $\mathcal{V}_1$ and $\mathcal{V}_2$. This Lie algebra is called the direct sum (or product) of the Lie algebras $\mathcal{V}_1$ and $\mathcal{V}_2$, and it is written as $\mathcal{V}_1 \oplus \mathcal{V}_2$ (or $\mathcal{V}_1 \times \mathcal{V}_2$). If $\mathcal{V}$ is a Lie algebra and $\mathcal{V}_1, \ \mathcal{V}_2$ are subalgebras of $\mathcal{V}$, it is said that $\mathcal{V}$ decomposes as the Lie algebra of direct sum of $\mathcal{V}_1$ and $\mathcal{V}_2$ as vector spaces denoted by

$$\mathcal{V} = \mathcal{V}_1 \oplus \mathcal{V}_2$$

if $\mathcal{V}$ is the direct sum of $\mathcal{V}_1$ and $\mathcal{V}_2$ as vector spaces and $[X_1, Y_2] = 0, \ \forall X_1 \in \mathcal{V}_1$ and $Y_2 \in \mathcal{V}_2$.

4.7.1 Lie Algebra of Vector Fields

This subsection proves in Theorem 4.7.7 that the set of all vector fields on a smooth manifold forms a real Lie algebra under the Lie bracket composition and studies the Lie algebra of vector fields.

Let M be a smooth manifold, $\mathcal{F}_U$ be the set of all differentiable functions $f : M \to \mathbf{R}$ whose domains intersect a given open subset $U \subset M$ and $\mathcal{V}(M)$ be the real vector space of all vector fields on M. Thus, $\mathcal{F}_U$ denotes the set of all differentiable functions $f \in C^\infty(M, \mathbf{R})$ whose domains intersect a given open subset U of M. Recall the concept of a vector field on a smooth manifold.

Definition 4.7.6 Let M be a smooth manifold of dimension n. A **vector field** on a smooth manifold M of dimension n is a smooth map X, which assigns to every point $p \in M$ a tangent vector $X_p \in T_p(M)$. Given local coordinates $x_1, x_2, \dots, x_n$ near a point $p \in M$, the vector field is formulated by

$$X : M \to \mathcal{T}(M),\ \ p \mapsto \Sigma_{i=1}^{n} X_i(p) \frac{\partial}{\partial x_i}.$$

A vector field X is said to be a **smooth vector field** if every component function $X_i(p)$ of X is smooth.

Theorem 4.7.7 *Let M be a real smooth manifold of dimension n and $\mathcal{V}(M)$ be the set of all vector fields on M. Then $\mathcal{V}(M)$ forms a real Lie algebra under the Lie bracket composition*

$$[\ ,\ ] : \mathcal{V}(M) \times \mathcal{V}(M) \to \mathcal{V}(M), (X, Y) \mapsto [X, Y] \text{ where } [X, Y](f) = X(Yf) - Y(Xf), \forall f \in C^{\infty}(M, \mathbf{R}).$$

Proof The composition [,] is well-defined, because $[X, Y] \in \mathcal{V}(M)$. To prove it, let (ψ, U) be a coordinate system with coordinate functions $(x_1, x_2, \dots, x_n)$. Then there exist b_i, c_i in $C^{\infty}(U, \mathbf{R})$ such that

$$X|_U = \Sigma_{i=1}^{n} b_i (\frac{\partial}{\partial x_i}),\ \ Y|_U = \Sigma_{i=1}^{n} c_i (\frac{\partial}{\partial x_i}).$$

For $X = b(\frac{\partial}{\partial x_i}),\ Y = c(\frac{\partial}{\partial x_i})$,

$$[X, Y](f) = b\frac{\partial}{\partial x_i}(c\ \frac{\partial}{\partial x_j}(f)) - c\frac{\partial}{\partial x_j}(b\ \frac{\partial}{\partial x_i}(f))$$

$$= b\frac{\partial}{\partial x_i}(c\ \frac{\partial}{\partial x_j}(f)) - c\frac{\partial}{\partial x_j}\ (b)\ \frac{\partial}{\partial x_i}(f) + bc\frac{\partial}{\partial x_i}\frac{\partial}{\partial x_j}(f) - bc\frac{\partial}{\partial x_j}\frac{\partial}{\partial x_i}(f)$$

$$= [b\frac{\partial}{\partial x_i}(c)\ \frac{\partial}{\partial x_j} - c\frac{\partial}{\partial x_j}\ (b)\ \frac{\partial}{\partial x_i}](f), \forall f \in C^{\infty}(M, \mathbf{R})$$

asserts that

$$[X,\ Y] = b\frac{\partial}{\partial x_i}(c)\ \frac{\partial}{\partial x_i} - c\frac{\partial}{\partial x_i}\ (b)\ \frac{\partial}{\partial x_i},$$

which is a vector field, since $b\frac{\partial}{\partial x_i}(c)$ and $c\frac{\partial}{\partial x_i}\ (b) \in C^{\infty}(M, \mathbf{R})$.

Jacobi identity: For any $X, Y, Z \in \mathcal{V}(M)$, let their domains intersect on an open set U. Then for any $f \in \mathcal{F}_U$,

$$[X, [Y, Z]](f) = X(Y(Z(f)) - X(Z(Y(f)) - Y(Z(X(f)) + Z(Y(X(f)).$$

Similarly, find $[Y, [Z, X](f)$ and $[Z, [X, Y](f)$ and add them. This asserts that

$$([X, [Y, Z]] + [Y, [Z, X] + [Z, [X, Y])(f) = 0, \ \forall f \in \mathcal{F}_U$$

and hence

$$[X, [Y, Z]] + [Y, [Z, X] + [Z, [X, Y] = 0.$$

Since [,] is well-defined and satisfies all the conditions of a Lie algebra, it follows that $\mathcal{V}(M)$ forms a real Lie algebra under this Lie bracket composition. □

Corollary 4.7.8 *Let M be a smooth manifold of dimension n. If $X = \Sigma_{i=1}^n X_i(\frac{\partial}{\partial x_i})$ and $Y = \Sigma_{i=1}^n Y_i(\frac{\partial}{\partial x_i})$ are two vector fields on M in local coordinates in M, then*

$$\Sigma_{i,j=1}^n \left(X_i \frac{\partial Y_j}{\partial x_i} - Y_i \frac{\partial X_j}{\partial x_i}\right) \frac{\partial}{\partial x_j}$$

is the Lie bracket $[X, Y]$.

Proof It follows from the proof of Theorem 4.7.7. □

4.7.2 Lie Algebra of a Lie Group

A Lie group enjoys many spatial properties of manifolds, because of the existence of its smooth multiplication structure. For example, given a Lie group G, for any $g \in G$, the left translation

$$L_g : G \to G, x \mapsto gx$$

is a diffeomorphism with its inverse $L_{g^{-1}}$. If e is the identity element of G, then $L_g(e) = ge = g$ implies that the diffeomorphism L_g sends the identity element e to the element g and induces an isomorphism of vector spaces

$$L_{g^*} = (L_g)_* : T_e(G) \to T_g(G).$$

This asserts that for the study of tangent space $T_g(G)$ at any point, it is sufficient to study the tangent space $T_e(G)$. For example, we can identify the tangent space $T_g(GL(n, \mathbf{R}))$ at any point $g \in T_g(GL(n, \mathbf{R}))$ with the tangent space $T_I(GL(n, \mathbf{R}))$ at I (identity matrix) by the isomorphism

$$L_{g^*} : T_I(GL(n, \mathbf{R})) \to T_g(GL(n, \mathbf{R}))$$

induced by the left translation $L_g : GL(n, \mathbf{R})) \to GL(n, \mathbf{R})), A \mapsto gA$.

Definition 4.7.9 Let G be a Lie group and $V : G \to T(G)$ be a vector field on G. Then V is said to be **left-invariant** if it is invariant under all left translations of G:

$$L_g : G \to G, \ x \mapsto gx, \ \forall g \in G.$$

Let $g_* : T(G) \to T(G)$ denote the differential dL_g of L_g. Then V is said to be a left-invariant vector field of G if

$$V(gx) = g_*(V(g)), \ \forall g, x \in G.$$

Proposition 4.7.10 *Let G be a Lie group with identity element e and V be a left-invariant vector field on G. Then V is determined by its value at e.*

Proof Let $g \in G$ be an arbitrary element. Then

$$V(g) = V(ge) = g_*(V(e)), \ \forall g \in G$$

shows that V is determined by its value at e. □

Remark 4.7.11 Converse of Proposition 4.7.10 saying that for any element (vector) $v \in T_e(G)$, the correspondence

$$g \to g_* v$$

defines a left-invariant vector field having its value v at e, which is available in Theorem 4.7.13.

Remark 4.7.12 Definition 4.7.3 shows that the left-invariant infinitesimal transformations of an analytic group $\mathcal{G}$ constitute a Lie algebra over the field $\mathbf{R}$ having dimension the same as the dimension of the analytic group $\mathcal{G}$.

Theorem 4.7.13 *Let G be a Lie group with identity element e and $\mathcal{L}G$ be the vector space of all left-invariant vector fields on G. Then it admits a Lie group structure such that the map*

$$\psi : \mathcal{L}G \to T_e(G), X \mapsto Xe$$

is an isomorphism of vector spaces.

Proof The map

$$\psi : \mathcal{L}G \to T_e(G), \ X \mapsto Xe$$

is well-defined, since any left-invariant vector field X on G is determined by its value at e. It is a monomorphism of vector spaces. To show that it is surjective, take an arbitrary element $v \in T_e(G)$. Since the function $X : g \mapsto g_* v$ gives a left-invariant vector field having its value v at e, it follows that ψ is an epimorphism of vector spaces and hence it is proved that ψ is an isomorphism of vector spaces. □

Theorem 4.7.14 *Let G be a Lie group with identity e. Then the vector space $T_e(G)$ admits the structure of a Lie algebra.*

Proof To prove the theorem, we utilize the isomorphism

$$\psi : \mathcal{L}G \to T_e(G), X \mapsto Xe$$

of vector spaces given in Theorem 4.7.13. Consider a given basis $\{v_1, v_2, \ldots, v_n\}$ for the vector space $T_e(G)$. Then the corresponding collection of vector fields

$$\{X_1, X_2, \ldots, X_n \text{ where } X_i : g \mapsto g_* v, i = 1, 2, \ldots, n\}$$

forms a basis for $\mathcal{L}G$. Consequently,

$$[X_i, X_k] = \Sigma_j \, a_{ik}^j X_j : a_{ik}^j \in \mathbf{R} \text{ and } i, j, k = 1, 2, \ldots, n.$$

The real numbers a_{ik}^j are called **structure constants** of G. Using the given basis $\{v_1, v_2, \ldots, v_n\}$ for the vector space $T_e(G)$, the multiplication in $LG \in \mathcal{L}G$ is represented by

$$[v_i, v_k] = \Sigma_j a_{ik}^j v_j : a_{ik}^j \in \mathbf{R} \text{ and } i, j, k = 1, 2, \ldots, n.$$

□

Definition 4.7.15 The vector space $\mathcal{G} = T_e(G)$ is called the **Lie algebra of the Lie group** G.

Example 4.7.16 If $G = S^1 = \{z \in \mathbf{C} : |z| = 1\} = \mathbf{R}/\mathbf{Z}$, then $\mathcal{G} = \mathbf{R}$.

$$exp : \mathcal{G} \to G, x \mapsto \alpha_x(e) = r.$$

Definition 4.7.17 Let G be a Lie group with its associated Lie algebra $\mathcal{G} = T_e(G)$. Consider the exponential map

$$exp : \mathcal{G} \to G, x \mapsto \alpha_x(e),$$

where $\alpha_x(t)$ is the one-parameter subgroup with tangent vector x at the point e.

Example 4.7.18 If $G = \mathbf{R}$, then $\mathcal{G} = \mathbf{R}$ and for any $r \in \mathcal{G}$, its associated one-parameter subgroup $\alpha_r(t) = tr$ with exponential map

$$exp : \mathcal{G} \to G, x \mapsto \alpha_x(e) = r.$$

Theorem 4.7.19 *The Lie algebra $\mathcal{G}L(n, \mathbf{R})$ of $GL(n, \mathbf{R})$ and the Lie algebra $\mathcal{M}(n, \mathbf{R})$ of $M(n, \mathbf{R})$ are isomorphic.*

Proof Consider the left-invariant vector fields on the Lie group $GL(n, \mathbf{R})$. Let x be the standard chart on $GL(n, \mathbf{R})$. Then the vectors $e_{i,j} = (\partial/\partial x_{ij})_e$ constitute a basis for the vector space $T_e(GL(n, \mathbf{R}))$. Hence, the left-invariant vector fields $X_{i,j}$ obtained by $e_{i,j}$ are represented by

$$X_{i,j} : g \mapsto g_*(\partial/\partial x_{ij})_e = \Sigma_{k,t,s}(\partial g_{ks}x_{st})/\partial x_{ij})_e(\partial/\partial x_{kt})_g = \Sigma_k\, g_{ki}(\partial/\partial x_{kj})_g.$$

It asserts that

$$X_{i,j} = \Sigma_k x_{ki}\partial/\partial x_{kj}.$$

This implies that

$$[X_{i,j}, X_{m,l}] = \delta_{mj}X_{i,l} - \delta_{il}X_{m,j}.$$

Hence, the multiplication in the Lie algebra of $GL(n, \mathbf{R})$ can be expressed in terms of the basis $\{e_{ij}\}$ as

$$[e_{ij}, e_{ml}] = \delta_{mj}e_{il} - \delta_{il}e_{mj}.$$

Next, consider the left-invariant vector fields on $M(n, \mathbf{R})$. Let $\{E_{ij}\}$ be a basis for the vector space $M(n, \mathbf{R})$, where E_{ij} denotes the matrix in $M(n, \mathbf{R})$ having 1 in the ijth position and zeros elsewhere. Then

$$[E_{ij}, E_{ml}] = \delta_{mj}E_{il} - \delta_{il}E_{mj}$$

gives the Lie algebra structure on $M(n, \mathbf{R})$. Then the map

$$\psi : T_e(GL(n, \mathbf{R})) \to M(n, \mathbf{R}),\ e_{ij} \to E_{ij}$$

is a vector space isomorphism and hence it gives an isomorphism

$$\psi_* : \mathcal{G}L(n, \mathbf{R}) \to \mathcal{M}(n, \mathbf{R}), v \mapsto [vx_{ij}]$$

of the corresponding Lie algebras. This proves that the Lie algebras $\mathcal{G}L(n, \mathbf{R})$ of $GL(n, \mathbf{R})$ and the Lie algebra $\mathcal{M}(n, \mathbf{R})$ of $M(n, \mathbf{R})$ are isomorphic. □

4.7.3 Homomorphism of Lie Algebras and Adjoint Representation

For every Lie group G, the vector space $\mathcal{G} = T_e(G)$ has a canonical structure of a Lie algebra, and every morphism of Lie groups gives rise to a morphism of Lie algebras.

Definition 4.7.20 Let $\mathcal{G}$ and $\mathcal{H}$ be two Lie algebras. A linear map $f : \mathcal{G} \to \mathcal{H}$ is said to be

(i) a **Lie algebra homomorphism** if it preserves the Lie bracket in the sense that

$$f([A, B]) = [f(A), f(B)], \ \forall A, B \in \mathcal{G};$$

(ii) If in addition, f is a bijection, then f is said to be a **Lie algebra isomorphism;**
(iii) A Lie algebra isomorphism of a Lie algebra onto itself is called a **Lie algebra automorphism;**
(iv) A Lie algebra homomorphism of a Lie algebra into itself is called a **Lie algebra endomorphism.**

Definition 4.7.21 Let G be a Lie group with corresponding Lie algebra $\mathcal{G}$. For every $X \in \mathcal{G}$, define a linear map

$$ad_X : \mathcal{G} \to \mathcal{G}, Y \mapsto [X, Y].$$

The map $\psi : G \to G, X \mapsto ad_X$ is called the **Adjoint map or adjoint representation**.

Remark 4.7.22 (*Properties of Adjoint maps or adjoint representations*) Let G be a Lie group and let

$$\sigma : G \times G \to G, (g, k) \mapsto gkg^{-1} = \sigma_g(k)$$

act on G from the left by inner automorphism. Then its identity element is fixed by $\sigma_g : G \to G, k \mapsto gkg^{-1}$. Moreover,

$$ad(g) = d(\sigma_g)_e : \mathcal{G} \to \mathcal{G}$$

is a Lie algebra homomorphism. This implies that for every $g \in G, ad(g) \in End\,(\mathcal{G})$, this gives rise to a map

$$ad : G \to End(\mathcal{G}), \ g \mapsto ad(g),$$

which is a representation or an adjoint map of G into the vector space $End(\mathcal{G})$ of linear operators on $\mathcal{G}$. On the other hand, given an element $X \in \mathcal{G}$, there is a linear map

$$ad_X : \mathcal{G} \to \mathcal{G}, \ Y \mapsto [X, Y]$$

and the assignment $X \to ad_X$ is the adjoint map or adjoint representation. The map $X \to ad_X$ is a linear map of $\mathcal{G}$ into the vector space $End(\mathcal{G})$ of linear operators on $\mathcal{G}$. The Jacobi identity asserts that ad_X is the derivation of the bracket in the sense that

$$ad_X([Y, Z]) = [ad_X(Y), Z] + [Y, ad_X(Z)].$$

Proposition 4.7.23 *Let $\mathcal{G}$ be a Lie algebra. Then*

$$ad : \mathcal{G} \to End(\mathcal{G}), g \mapsto ad(g)$$

is a Lie algebra homomorphism.

Proof For any elements $X, Y, Z \in \mathcal{G}$, it follows from the definition of $ad : \mathcal{G} \to End(\mathcal{G})$ that

$$ad_{[X,Y]}(Z) = [[X, Y], Z]$$

and

$$[ad_X, ad_Y](Z) = [X, [Y, Z]] - [Y, [X, Z]].$$

This implies that $ad : \mathcal{G} \to End(\mathcal{G})$ satisfies the property

$$ad_{[X,Y]} = ad_X ad_Y - ad_Y ad_X = [ad_X, ad_Y], \forall X, Y \in \mathcal{G}.$$

Hence the proposition follows. □

Summarizing the discussion in the above sections, a result is formulated in Theorem 4.7.24.

Theorem 4.7.24 *For every Lie group G, the vector space $\mathcal{G} = T_e(G)$ has a canonical structure of a Lie algebra, and every morphism of Lie groups gives rise to a morphism of Lie algebras.*

4.8 Subalgebra, Ideal, Stabilizer and Center

This section continues the study of Lie groups and associated algebras by defining the subalgebra of a Lie algebra and associated ideals and centers and studies them. This study is based on Theorem 4.7.14 that for every Lie group G with identity e, the vector space $\mathcal{G} = T_e(G)$ is the Lie algebra of the Lie group G.

Subalgebra and Ideals of a Lie Algebra

This subsection introduces the concepts of subalgebra and ideals of a lie algebra.

Definition 4.8.1 Let $\mathcal{G}$ be a Lie algebra. A subspace $\mathcal{H}$ of the vector space $\mathcal{G}$ is said to be

(i) a **Lie subalgebra** of $\mathcal{G}$ if for every pair of elements $x, y \in \mathcal{H}$, $[x, y] \in \mathcal{H}$ and
(ii) an **ideal** of $\mathcal{G}$ if for any element $x \in \mathcal{G}$ and any element $y \in \mathcal{H}$, $[x, y] \in \mathcal{H}$. In other words, an ideal $\mathcal{H}$ is a subalgebra of a Lie algebra $\mathcal{G}$ if

$$[x, y] \in \mathcal{H}, \forall x \in \mathcal{G}, y \in \mathcal{H}.$$

Proposition 4.8.2 *Let $\mathcal{G}$ be the Lie algebra of a Lie group G.*

(i) *If H is a Lie subgroup of G, then $\mathcal{H} = T_e(H)$ is a Lie subalgebra of $T_e(G) = \mathcal{G}$.*
(ii) *If $\mathcal{H}$ is a subalgebra of $\mathcal{G}$, then there exists a unique connected Lie subgroup H of G with $\mathcal{H}$ its Lie algebra.*
(iii) *If H is a normal subgroup of G, then $\mathcal{H} = T_e(H)$ is an ideal of $\mathcal{G}$ and conversely, if H is a connected Lie subgroup of G and $\mathcal{H} = T_e(H)$ is an ideal of $\mathcal{G}$, then H is a normal subgroup of G.*

Proof It is left as an exercise. □

4.9 Some Special Lie Algebras

This section turns to the study of the general theory of Lie algebra instead of the study of Lie algebra associated with a Lie group initiating the description of some special classes of Lie algebras such as simple, solvable and nilpotent Lie algebras. One of the basic theorems in general theory of Lie algebra is **Ado theorem** (see Execise 41) saying that every finite-dimensional Lie algebra (over a field F of characteristic 0) is isomorphic to some matrix Lie algebra. This theory is vast. For its detailed study, see Postnikov (1986).

4.9.1 Simple Lie Algebra

Definition 4.9.1 Let $\mathcal{G}$ be a Lie algebra. It is said to be

(i) **irreducible** if only ideals of $\mathcal{G}$ are $\mathcal{G}$ and $\{0\}$.
(ii) **simple** if it is irreducible and its dimension is ≥ 2.

Remark 4.9.2 Every 1-dimensional Lie algebra $\mathcal{G}$ is irreducible, because it has no nontrivial subspaces and hence it has no nontrivial subalgebras and no nontrivial ideals. Since its dimension is one, it is not considered as a simple algebra. An analogy between Lie algebras and Lie groups may be drawn by considering the role of subalgebras in Lie algebra played by Lie subgroups in Lie groups and the role of ideals in Lie algebra played by normal Lie subgroups in Lie groups. For example, the kernel of every Lie algebra homomorphism is an ideal of the Lie algebra, which is analogous to the result that the kernel of every Lie group homomorphism is a normal subgroup of the Lie group.

4.9.2 Solvable Lie Algebra

Let F be a field of characteristic 0. All algebras over F are assumed to be finite-dimensional.

Definition 4.9.3 Let $\mathcal{G}$ be a Lie algebra over F. For any two subspaces $\mathcal{A}$ and $\mathcal{B}$ of $\mathcal{G}$, let $[\mathcal{A}, \mathcal{B}]$ denote the subspace generated by the elements of the form $[a, b]$, where $a \in \mathcal{A}, b \in \mathcal{B}$. Define

$$\mathcal{G}^{(1)} = \mathcal{G},\ \mathcal{G}^{(2)} = [\mathcal{G}, \mathcal{G}],\ \ldots, \mathcal{G}^{(n)} = [\mathcal{G}^{n-1}, \mathcal{G}^{n-1}], \ldots,$$

and construct a descending chain of ideals:

$$\mathcal{G} = \mathcal{G}^{(1)} \supset \mathcal{G}^{(2)} \supset \cdots \supset \mathcal{G}^{(n)} \supset \cdots .$$

(i) A Lie algebra $\mathcal{G}$ is said to be **solvable** if there exists some $n \geq 0$ such that $\mathcal{G}^{(n)} = 0$.

(ii) If $dim\ \mathcal{G} = n$, then a descending chain of subspaces

$$\mathcal{G} = \mathcal{G}_0 \supset \mathcal{G}_1 \supset \mathcal{G}_2 \supset \cdots \supset \mathcal{G}_i \supset \cdots \supset \mathcal{G}_n = 0$$

is said to be a **flag** if

$$dim\ \mathcal{G}_i = n - i,\ \forall i = 0, 1, 2, \ldots, n.$$

A flag consisting of subalgebras is said to be a **flag of subalgebras.**

Theorem 4.9.4 characterizes a Lie algebra over a field in terms of a flag of subalgebras.

Theorem 4.9.4 *Let $\mathcal{G}$ be a Lie algebra over F. It is solvable iff it has a flag of subalgebras*

$$\mathcal{G} = \mathcal{G}_0 \supset \mathcal{G}_1 \supset \mathcal{G}_2 \supset \cdots \supset \mathcal{G}_i \supset \cdots \supset \mathcal{G}_n = 0,$$

such that for all $i = 1, 2, \ldots, n$,

(i) *every subalgebra $\mathcal{G}_i$ is an ideal of the preceeding subalgebra $\mathcal{G}_{i-1}$ and*
(ii) *the relation*

$$[\mathcal{G}_{i-1}, \mathcal{G}_i] \subset \mathcal{G}_i$$

holds.

Proof First, suppose that the Lie algebra $\mathcal{G}$ has the given flag of subalgebras. Claim that $\mathcal{G}$ is solvable. By the given condition, since $dim\ \mathcal{G}_i = n - i$ for all $i = 0, 1, 2, \ldots, n$ it follows that

$$dim\ \mathcal{G}_{i-1} = dim\ \mathcal{G}_i + 1.$$

This implies that every element of $\mathcal{G}_{i-1}$ is of the form $x + \alpha v_0$, where $x \in \mathcal{G}_i$, $\alpha \in F$ and $v_0 \in \mathcal{G}_i$ is a fixed element. Since

$$[x + \alpha v_0, y + \beta v_0] = [x, y] + \alpha[x, v_0] - \beta[y, v_0] \ \ where\ [x, y], \alpha[x, v_0], [y, v_0] \in \mathcal{G}_i$$

and $\mathcal{G}_i$ is an ideal of $\mathcal{G}_{i-1}$, it follows that $[\mathcal{G}_{i-1}, \mathcal{G}_{i-1}] \subset \mathcal{G}_i$. Thus, if $\mathcal{G}^{(i)} \subset \mathcal{G}_{i-1}$, then

$$\mathcal{G}^{(i+1)} = [\mathcal{G}^{(i)}, \mathcal{G}^{(i)}] \subset [\mathcal{G}_{i-1}, \mathcal{G}_{i-1}] \subset \mathcal{G}_i.$$

Since the inclusion relation $\mathcal{G}^i \hookrightarrow \mathcal{G}_{i-1}$ holds for $i = 1$, it is proved that this inclusion relation holds for all $i = 1, 2, \ldots, n+1$. This gives in particular,

$$\mathcal{G}^{(n+1)} \subset \mathcal{G}_n = 0$$

and hence it follows $\mathcal{G}^{(n+1)} = 0$. Conversely, suppose that the given Lie algebra $\mathcal{G}$ is solvable and $dim\ \mathcal{G} = n$. Then $\mathcal{G}^{(n)} = 0$ but $\mathcal{G}^{(n-1)} \neq 0$. Consider the strictly descending chain of ideals

$$\mathcal{G} = \mathcal{G}^{(0)} \supset \mathcal{G}^{(1)} \supset \mathcal{G}^{(2)} \supset \cdots \supset \mathcal{G}^{(n)} = 0.$$

Then this chain can be embedded into some flag of subalgebras

$$\mathcal{G} = \mathcal{G}_0 \supset \mathcal{G}_1 \supset \mathcal{G}_2 \supset \cdots \supset \mathcal{G}_i \supset \cdots \supset \mathcal{G}_n = 0.$$

For $i = 0, 1, 2, \ldots, n$, let k be the greatest index such that $\mathcal{G}_i \subset \mathcal{G}^{(k)}$. Consider the two cases:

Case I If $\mathcal{G}_i \neq \mathcal{G}^{(k)}$, then $\mathcal{G}_{i-1} \subset \mathcal{G}^{(k)}$ and hence

$$[\mathcal{G}_{i-1}, \mathcal{G}_i] \subset [\mathcal{G}^{(k)}, \mathcal{G}^{(k)}] = \mathcal{G}^{(k+1)} \subset \mathcal{G}_i.$$

Case II If $\mathcal{G}_i = \mathcal{G}^{(k)}$, then $\mathcal{G}_{i-1} \subset \mathcal{G}^{(k-1)}$ and hence

$$[\mathcal{G}_{i-1}, \mathcal{G}_i] \subset [\mathcal{G}^{(k-1)}, \mathcal{G}^{(k)}] \subset \mathcal{G}^{(k)} = \mathcal{G}_i.$$

Hence, it follows that in either case, $[\mathcal{G}_{i-1}, \mathcal{G}_i] \subset \mathcal{G}_i$. This proves that the above flag is a flag of subalgebras such that every subalgebra is an ideal of the previous one, because $[\mathcal{G}_i, \mathcal{G}_i] \subset [\mathcal{G}_{i-1}, \mathcal{G}_i] \subset \mathcal{G}_i$. □

4.9.3 Nilpotent Lie Algebra

This subsection continues the study of Lie algebra by defining nilpotent Lie algebra. Let $\mathcal{G}$ be a Lie algebra. With the help of any ideal $\mathcal{G}^{(k)}$ of $\mathcal{G}$, one can define ideals $\mathcal{G}^k$ recursively as

$$\mathcal{G}^1 = \mathcal{G} \text{ and } \mathcal{G}^k = [\mathcal{G}, \mathcal{G}^{k-1}].$$

Then $\mathcal{G}^2 = \mathcal{G}^{(2)}$.

Definition 4.9.5 Let $\mathcal{G}$ be a Lie algebra.

(i) $\mathcal{G}$ is said to be **nilpotent** if there is an integer $k \geq 1$ such that $\mathcal{G}^k = 0$.
(ii) Lie algebra $\mathcal{G}$ is said to be **abelian** if $\mathcal{G}^2 = 0$, which means that

$$[X, Y] = 0, \ \forall\, X, Y \in \mathcal{G}.$$

Example 4.9.6 Every abelian Lie algebra is nilpotent.

Proposition 4.9.7 *Every nilpotent Lie algebra is solvable.*

Proof Let $\mathcal{G}$ be a nilpotent Lie algebra. Since by recursion on k, the inclusion $\mathcal{G}^{(k)} \hookrightarrow \mathcal{G}^k$ holds, the proposition follows. □

Definition 4.9.8 $\mathcal{G}$ be the Lie algebra of a Lie group G.

(i) The center $C(\mathcal{G})$ of the Lie algebra $\mathcal{G}$ is

$$C(\mathcal{G}) = \{X \in \mathcal{G} : [X, Y] = 0, \ \ \forall\, Y \in \mathcal{G}\}.$$

(ii) The center $C(G)$ of the Lie group G is

$$C(G) = \{x \in G : xg = gx \ \ \forall\, g \in G\}.$$

Proposition 4.9.9 *$\mathcal{G}$ is the nilpotent Lie algebra of a Lie group G. Then the center $C(\mathcal{G})$ of $\mathcal{G}$ is nonzero.*

Proof Suppose that $\mathcal{G}^{k-1} \neq 0$ but $\mathcal{G}^k = 0$. Then the nonzero ideal $\mathcal{I} = \mathcal{G}^{k-1}$ satisfies the condition that $[\mathcal{I}, \mathcal{G}] = 0$ and hence it follows that $\mathcal{I} \in C(\mathcal{G})$. It proves that center $C(\mathcal{G})$ of $\mathcal{G}$ is nonzero. □

Theorem 4.9.10 characterizes the nilpotency of a Lie algebra of a Lie group in terms of a flag of subalgebras.

Theorem 4.9.10 *$\mathcal{G}$ is a Lie algebra of a Lie group G. Then $\mathcal{G}$ is nilpotent iff it has a flag of subalgebras*

$$\mathcal{G} = \mathcal{G}_0 \supset \mathcal{G}_1 \supset \mathcal{G}_2 \supset \cdots \supset \mathcal{G}_i \supset \cdots \supset \mathcal{G}_n = 0$$

such that

$$[\mathcal{G}, \mathcal{G}_{i-1}] \subset \mathcal{G}_i, \; \forall i = 1, 2, \ldots, n$$

and in particular, every subalgebra $\mathcal{G}_{i-1}$ is an ideal of $\mathcal{G}$.

Proof Proceed as in the proof of Theorem 4.9.4. □

4.10 Applications

Proposition 4.10.1 *Given two connected Lie groups G and H, if $f, k : G \to H$ are two Lie group homomorphisms such that $df = dk$, then $f = k$.*

Proof By hypothesis, G and H are two connected Lie groups. Hence under multiplication defined component-wise, $G \times H$ is a Lie group with corresponding Lie algebra $\mathcal{G} \oplus \mathcal{H}$ under bracket operation defined component-wise. The map

$$f : G \to H$$

is a homomorphism iff its graph

$$G_f = \{(g, f(g)) : g \in G\} \subset G \times H$$

is a Lie subgroup of $G \times H$. Then its Lie subalgebra is the graph

$$G_{df} = \{(v, df(v)) : v \in \mathcal{G}\}.$$

Hence, the hypothesis $df = dk$ implies that $G_{df} = G_{dk}$. This implies by Exercise 4.11 that $f = k$. □

Proposition 4.10.2 *The Lie algebra of the Lie group $SL(2, \mathbf{C})$ is simple.*

Proof With respect to the basis

$$\mathcal{B} = \{X = \begin{pmatrix} 0 & 1 \\ 0 & 0 \end{pmatrix}, Y = \begin{pmatrix} 0 & 0 \\ 1 & 0 \end{pmatrix}, Z = \begin{pmatrix} 1 & 0 \\ 0 & -1 \end{pmatrix}\}$$

for the vector space $\mathcal{SL}(2, \mathbf{C})$, it follows that

$$[X, Y] = Z,\ [Z, X] = 2X,\ [Z, Y] = -2Y.$$

Let $\mathcal{H}$ be an ideal of $\mathcal{SL}(2, \mathbf{C})$ and $H = \alpha X + \beta Z + \gamma Y \in \mathcal{H}$ be an arbitrary element, where α, β, γ (are not all zero). If $\gamma \neq 0$, then the element

$$[X, [X, H]] = [X, -2\beta X + \gamma Z] = -2\gamma X$$

is a nonzero multiple of X and hence it follows that $X \in \mathcal{H}$, since $\mathcal{H}$ is an ideal of $\mathcal{SL}(2, \mathbf{C})$ by hypothesis. On the other hand, $[Y, X]$ is a nonzero multiple of Z and $[Y, [Y, X]]$ is a nonzero multiple of Y and hence $Y \in \mathcal{H}$ and $Z \in \mathcal{H}$. This implies that $\mathcal{H} = \mathcal{SL}(2, \mathbf{C})$. If $\gamma = 0$ but $\beta \neq 0$, then $[X, H]$ is a nonzero multiple of X and hence by earlier argument, it follows that $\mathcal{H} = \mathcal{SL}(2, \mathbf{C})$. For the other possibility, when $\alpha \neq 0$ but both β and γ are zero, then H is a nonzero multiple of X and hence it follows that $\mathcal{H} = \mathcal{SL}(2, \mathbf{C})$. Considering all the cases, it is proved that the Lie algebra $\mathcal{SL}(2, \mathbf{C})$ of the Lie group $SL(2, \mathbf{C})$ is simple. □

Definition 4.10.3 A matrix Lie group T is said to be an m-**torus** if T is isomorphic to the direct product of m copies of the circle group $S^1 \cong U(1, \mathbf{C})$ for some positive integer m.

Example 4.10.4 Let T be the group of $n \times n$ diagonal, unitary matrices with determinant 1. Then T is isomorphic to $n - 1$ copies of S^1, because every element of T can be expressed uniquely as

$$diag(z_1, z_2, \ldots, z_{n-1}, (z_1 z_2 \cdots z_{n-1})^{-1}),$$

where $z_1, z_2, \ldots, z_{n-1}$ are complex numbers with absolute value 1. This implies that T is isomorphic to the direct product of $n - 1$ copies of the circle group $S^1 \cong U(1, \mathbf{C})$.

Definition 4.10.5 A subgroup T of a Lie group G is a torus if T is isomorphic to $(S^1)^m$ for some positive integer m, and T is said to be a **maximal torus** in G if T is a torus and it is not properly contained in any other torus in G.

Theorem 4.10.6 *Let G be a compact connected abelian Lie group of dimension n. Then as a topological group, G is isomorphic to the n-dimensional torus $\mathbf{T}^n$.*

Proof By hypothesis, G is a compact connected abelian Lie group of dimension n. Let $\mathcal{G}$ be the Lie algebra of Lie group of G. Since G is abelian, $[X, Y] = 0,\ \forall\, X, Y \in \mathcal{G}$. This implies that the Lie algebra $\mathcal{G}$ is the same as the Lie algebra of $\mathbf{R}^n$. Using the result that a compact connected group which is locally isomorphic to $\mathbf{R}^n$ is isomorphic to $\mathbf{T}^n$, the theorem follows. □

Definition 4.10.7 (*Matrix exponential map*) The map

$$exp : M(n, \mathbf{C}) \to M(n, \mathbf{C}),\ A \mapsto exp\ A = \Sigma_{k=0}^{\infty} \frac{1}{k!} A^k$$

is called the matrix exponential map. This power series converges and defines an analytic map.

Remark 4.10.8 **(i)** For arbitrary matrices $A, B \in \mathcal{M}(n, \mathbf{R})$, it is not necessarily true that

$$e^A e^B = e^{A+B}.$$

(ii) But for commutating matrices $A, B \in \mathcal{M}(n, \mathbf{R})$ in the sense that $AB = BA$, the relation

$$e^A e^B = e^{A+B}$$

is true.

We consider $M(n, \mathbf{R})$ identified with $\mathbf{R}^{n^2}$. So an element $A \in \mathbf{R}^{n^2}$ is identified with a matrix in $M(n, \mathbf{R})$. Let $tr\ A$ denote the trace of A.

Proposition 4.10.9 *For any $A \in M(n,\ \mathbf{R})$,*

$$\frac{d}{dt} e^{tA} = Ae^{tA} = e^{tA}\ A.$$

Proof Consider the power series for the exponential function e^{tA}. Its every (i, j)-entry is a power in t. Differentiating term by term, it follows that

$$\frac{d}{dt} e^{tA} = \frac{d}{dt}(I + tA + \frac{1}{2!}t^2 A^2 + \frac{1}{3!}t^3 A^3 + \frac{1}{4!}t^4 A^4 \cdots) = A + tA^2 + \frac{1}{2!}t^2 A^3 + \frac{1}{3!}t^3 A^4 + \cdots$$

$$= A(I + tA + \frac{1}{2!}t^2 A^2 + \frac{1}{3!}t^3 A^3 + \cdots) = Ae^{tA}.$$

Taking the second equality from the above equalities, it shows that

$$\frac{d}{dt} e^{tA} = A + tA^2 + \frac{1}{2!}t^2 A^3 + \frac{1}{3!}t^3 A^4 + \cdots.$$

$$= (I + tA + \frac{1}{2!}t^2 A^2 + \frac{1}{3!}t^3 A^3 + \cdots)A = e^{tA} A.$$

By identifying $M(n, \mathbf{R})$ with $\mathbf{R}^{n^2}$, this also proves that for any $A \in \mathbf{R}^{n^2}$,

$$\frac{d}{dt} e^{tA} = Ae^{tA} = e^{tA}\ A.$$

□

Proposition 4.10.10 *For any $A \in M(n, \mathbf{R})$, the determinant function det has the property*

$$det(e^A) = e^{tr\ A}.$$

Proof **Case I** Take A first as an upper triangular matrix of the form

$$A = \begin{pmatrix} \tau_1 & \cdots & * \\ 0 & \tau_2 & \\ 0 & \cdots & \ddots\ \tau_n \end{pmatrix}.$$

Then it follows that

$$e^A = \Sigma_{k=0}^{\infty}(1/k!)A^k = \Sigma_{k=0}^{\infty}(1/k!)\begin{pmatrix} \tau_1 & \cdots & * \\ 0 & \tau_2 & \\ 0 & \cdots & \ddots\ \tau_n \end{pmatrix} = \begin{pmatrix} e^{\tau_1} & \cdots & * \\ 0 & e^{\tau_2} & \\ 0 & \cdots & \ddots\ e^{\tau_n} \end{pmatrix}.$$

This implies that

$$det\ e^A = \Pi\ e^{\tau_i} = e^{\sum \tau_i} = e^{trA}.$$

Case II Let $A \in M(n, \mathbf{R})$ be a matrix having eigenvalues $\lambda_1, \lambda_2, \ldots, \lambda_n$. Then there exists a nonsingular complex matrix X such that

$$XAX^{-1} = \begin{pmatrix} \lambda_1 & \cdots & * \\ 0 & \lambda_2 & \\ 0 & \cdots & \ddots\ \lambda_n \end{pmatrix}$$

is an upper triangular matrix . This implies that

$$e^{XAX^{-1}} = I + XAX^{-1} + \frac{1}{2!}(XAX^{-1})^2 + \frac{1}{3!}(XAX^{-1})^3 + \cdots$$

$$= I + XAX^{-1} + X(\frac{1}{2!}(A^2)X^{-1} + X\frac{1}{3!}(A^3)X^{-1} + \cdots = Xe^AX^{-1}.$$

Hence, it follows that

$$det\ e^A = det\ (Xe^AX^{-1}) = det\ (e^{XAX^{-1}}) = e^{tr\ (XAX^{-1})}$$

(because XAX^{-1} is an upper triangular matrix). This asserts that

$$det(e^A) = e^{tr\ A}.$$

□

Remark 4.10.11 Proposition 4.10.10 displays the importance of e^A, because det $(e^A) = e^{tr\ A} \neq 0$ and hence e^A is always nonsingular. This property permits to define a curve on $GL(n, \mathbf{R})$ with a given initial point and having a given initial velocity.

Example 4.10.12 Consider the curves α and β on $GL(n, \mathbf{R})$:

(i) $\alpha : \mathbf{R} \to GL(n, \mathbf{R}),\ t \mapsto e^{tA}$. Since $\alpha(0) = I$ and $\frac{d}{dt}\ e^{tA}|_{t=0} = Ae^{tA}|_{t=0} = A$, it follows that α is a curve on $GL(n, \mathbf{R})$ with the initial point I and initial velocity A.

(ii) $\beta : \mathbf{R} \to GL(n, \mathbf{R}),\ t \mapsto Xe^{tA}$ is a curve on $GL(n, \mathbf{R})$ with the initial point X and initial velocity XA.

Definition 4.10.13 (*Matrix logarithm map*) The map

$$log : M(n, \mathbf{C}) \to M(n, \mathbf{C}),\ A \mapsto exp\ A = \Sigma_{k=0}^{\infty}(1/k!)A^k$$

is called the **matrix exponential map.** This power series converges and defines an analytic map. $exp\ A$ is also written as e^A.

Theorem 4.10.14 *The matrix exponential map*

$$log : M(n, \mathbf{C}) \to M(n, \mathbf{C}),\ A \mapsto exp\ A$$

has the following properties:

(i) *log is analytic;*

(ii) *$log\ (M(n, \mathbf{C})) \subset GL(n, \mathbf{C})$;*

(iii) *log maps an nbd of $0 \in M(n, \mathbf{C})$ diffeomorphically onto an nbd of $I_n \in GL(n, \mathbf{C})$;*

(iv) *if $A,\ B \in M(n, \mathbf{C})$ commute, then*

$$log\ (\alpha A + \beta B) = log\ (\alpha A)\ log\ (\beta B),\ \ \forall \alpha, \beta \in \mathbf{R}.$$

Proof It follows from the definition of the matrix exponential map that

$$log : M(n, \mathbf{C}) \to M(n, \mathbf{C}),\ A \mapsto exp\ A = \Sigma_{k=0}^{\infty}(1/k!)A^k.$$

□

4.11 Exercises

1. The circle group S^1 is a Lie group under group multiplication.
   [Hint: $e^{\theta}e^{\alpha} = e^{\theta+\alpha}$ *and* $(e^{i\theta})^{-1} = e^{-i\theta}$ are differentiable.]
2. Let H be a subgroup of the Lie group $G = GL(n, \mathbf{R})$. If H is also a manifold in $\mathbf{R}^{n^2}$, show that H is a Lie group.

3. Let G be a discrete (Lie) group and M be a smooth manifold. Show that the action of G on M is smooth iff for every g, the map

$$\psi_g : X \to X, x \mapsto gx$$

is smooth. Use this result to show that the action

$$\psi : \mathbf{Z}^n \times \mathbf{R}^n \to \mathbf{R}^n \ ((n_1, n_2, \ldots, n_k), (x_1, x_2, \ldots, x_k)) \mapsto (n_1 + x_1, n_2 + x_2, \ldots, n_k + x_k)$$

is smooth.

4. Let G be a Lie group and H be a discrete subgroup of G. Show that the factor space G/H is a manifold and the natural projection

$$p : G \to G/H, g \mapsto gH$$

is smooth.

5. Let G be a Lie group and H be a discrete subgroup of G. Show that the factor space G/H is a manifold and the natural projection

$$p : G \to G/H, g \mapsto gH$$

is smooth.

6. Show that in a Lie group G with identity element e, the component C_e containing e is a connected Lie subgroup of G.

7. Show that the topological group $SL(n, \mathbf{R})$ is a Lie subgroup of $GL(n, \mathbf{R})$. [Hint: Consider the determinant function $det : GL(n, \mathbf{R}) \to \mathbf{R},\ M \mapsto det\ M$. It is continuous and hence $det^{-1}(1) = SL(n, \mathbf{R})$ is closed in $GL(n, \mathbf{R})$, since $\{1\}$ is closed in $\mathbf{R}$. This shows that $SL(n, \mathbf{R})$ is a closed subgroup of $GL(n, \mathbf{R})$. Finally, use Theorem 4.5.3].

8. Let X be and a submanifold of G. A Lie subgroup H of a Lie group G is said to be **connected** if its underlying manifold is connected to the set of all matrices $M \in M(3, \mathbf{R})$ of the form

$$M = \begin{pmatrix} 1 & x & y \\ 0 & 1 & t \\ 0 & 0 & 1 \end{pmatrix}.$$

Show that

(i) X is a differentiable manifold of dimension 3 under the global chart

$$\psi : X \to \mathbf{R}^3,\ M \mapsto (x, y, t).$$

(ii) X forms a Lie group.
(iii) X forms a Lie subgroup of $GL(3, \mathbf{R})$.

9. Show that

(i) $O(n, \mathbf{R})$ is a Lie subgroup of $GL(n, \mathbf{R})$;
(ii) $SO(n, \mathbf{R})$ is a Lie subgroup of $GL(n, \mathbf{R})$;
(iii) $SL(n, \mathbf{R})$ is a Lie subgroup of $GL(n, \mathbf{R})$.

[Hint: Use the result that every closed subgroup of a Lie group is a Lie subgroup].

10. Find all finite-dimensional Lie groups of transformations of the real line space $\mathbf{R} = \mathbf{R}^1$.

11. Let G be a finite-dimensional Lie group and H be a Lie subgroup of G. Show that

(i) the coset space G/H is a manifold having

$$dim\ G/H = dim\ G - dim\ H;$$

(ii) the coset space G/H is a Lie group if H is a normal subgroup of G.

12. Let $X(x, y) = -y\ \partial/\partial x + x\ \partial/\partial y$ be a vector field in the Euclidean plane $\mathbf{R}^2$. Show that

(i) the action

$$\psi : \mathbf{R} \times \mathbf{R}^2 \to \mathbf{R}^2,\ (t, (x, y)) \mapsto (xcos\ t - y\ sint,\ xsin\ t + ycos\ t)$$

is a flow generated by the vector field X satisfying the following property:
(ii) the flow through (x, y) is the circle having the center at the origin.

13. Let $X(x, y) = y\ \partial/\partial x + x\ \partial/\partial y$ be a vector field in the Euclidean plane $\mathbf{R}^2$. Find the flow generated by the vector field $X(x, y)$.

14. **(Existence of Finite Partition of Unity** Let (X, τ) be a normal space and $\mathcal{C} = \{V_1, V_2, \ldots, V_n\}$ be an open covering of X. Show that there exists a partition of unity subordinate to the covering $\mathcal{C}$.

15. Let M be a C^r-manifold of dimension n for $r = 1, 2, \ldots, \infty$ and $\mathcal{C} = \{U_i\}$ be an open covering of M. Show that there exists a partition of unity of C^r-class subordinate to $\mathcal{C}$.

16. Let $\mathbf{H}^*$ be the multiplicative group of quaternions.

(i) Show that $\mathbf{H}^*$ admits an open submanifold structure of $\mathbf{R}^4$;
(ii) Find the Lie algebra of the group $\mathbf{H}^*$.

[Hint: Identify the set $\mathbf{H}$ with $\mathbf{R}^4$ under the bijection

$$f : \mathbf{R}^4 \to \mathbf{H},\ (x, y, z, t) \mapsto x + yi + zj + tk$$

]

17. Show that

(i) $M(n, \mathbf{C})$ considered as a real vector space admits the structure of a Lie algebra under

$$[X, Y] = XY - YX.$$

(ii) This Lie algebra is isomorphic to the Lie algebra of $GL(n, \mathbf{C})$.

18. Given a one-parameter subgroup $\alpha(t)$ of a Lie group G, if α intersects with itself, show that there exists a positive real number K such that

$$\alpha(t + K) = \alpha(t), \ \forall t \in \mathbf{R}.$$

19. If G is a compact connected Lie group, show that every point $x \in G$ is in a certain one-parameter subgroup of G.

20. Let G be a finite-dimensional commutative and connected Lie group. Show that G is locally isomorphic to a finite-dimensional vector space.

21. Let G be a finite-dimensional compact, commutative and connected Lie group. Show that G is isomorphic to the torus.

22. Let G be a finite-dimensional commutative and connected Lie group. Show that G is isomorphic to the product of the torus and a vector space.

23. **(Matrix exponential map)** The map

$$exp : M(n, \mathbf{C}) \to M(n, \mathbf{C}), \ A \mapsto exp\ A = \Sigma_{k=0}^{\infty}(1/k!)A^k$$

is called the matrix exponential map.
Prove the following statements:

(i) exp is analytic;
(ii) $exp\ (M(n, \mathbf{C})) \subset GL(n, \mathbf{C})$;
(iii) exp maps an nbd of $0 \in M(n, \mathbf{C})$ diffeomorphically onto an nbd of $I_n \in GL(n, \mathbf{C})$;
(iv) If $A, \ B \in M(n, \mathbf{C})$ commute, then

$$exp\ (\alpha A + \beta B) = exp\ (\alpha A)\ exp\ (\beta B), \ \ \forall \alpha, \beta \in \mathbf{R}.$$

24. Identify the Lie algebra of $GL(n, \mathbf{R})$ with the Lie algebra $M(n, \mathbf{R})$. Prove that the subalgebra $\mathcal{SL}(n, \mathbf{R})$ corresponding to $SL(n, \mathbf{R})$ arises from the matrix set $\mathcal{M} = \{A \in M(n, \mathbf{R}) : trac\ A = 0\}$.

25. Prove that the Lie algebra $\mathcal{SP}(n, \mathbf{R})$ of the Lie group $Sp(n, \mathbf{R}$ and the Lie algebra $\mathcal{LA}$ of the real $2n \times 2n$ matrices of the form

$\begin{pmatrix} M & N \\ P & -M^t \end{pmatrix}$ are isomorphic, where M, N and P are real $n \times n$ matrices such that both N and P are symmetric matrices.

26. Let $\mathcal{A}$ be an associative algebra and $\mathcal{G}$ be a subspace of $\mathcal{A}$ such that $XY - YX \in \mathcal{G}, \forall X, Y \in \mathcal{G}$. Show that $\mathcal{G}$ endowed with bracket operation

$$[\ ,\ ] : \mathcal{G} \times \mathcal{G},\ (X, Y) \mapsto XY - YX, i.e., [X, Y] = XY - YX$$

forms a Lie algebra. Hence, show that if $\mathcal{S}L(n, \mathbf{C}) = \{X \in M(n, \mathbf{C}) :\ \text{trace } X = 0\}$, then $\mathcal{S}L(n, \mathbf{C})$ is a Lie algebra with bracket operation

$$[X, Y] = XY - YX.$$

27. Let G be a Lie group. Show that

(i) a Lie subgroup H of G is embedded in G, iff H is closed in G;
(ii) every closed abstract subgroup H of G is a Lie subgroup of G.
[Hint: Use Theorem 4.5.3]

28. Let G be a matrix Lie group and $\mathcal{G}$ be its Lie algebra. If $\mathcal{H}$ is a Lie subalgebra of $\mathcal{G}$, show that there exists a unique connected Lie subgroup H of G with Lie algebra $\mathcal{H}$.

29. Let G be a connected Lie group with identity e. Show that

(i) if U is an nbd of e, then the nbd U generates the Lie group G.
(ii) G is abelian iff its Lie algebra $\mathcal{G}$ is abelian.
(iii) if G is abelian, then it is isomorphic to the product Lie group $T^n \times \mathbf{R}^m$ for $T = \mathbf{R}/\mathbf{Z}$.
(iv) the center $C(G)$ of G is a closed subgroup of G with Lie algebra the center $C(\mathcal{G})$ of the Lie algebra $\mathcal{G}$.
(v) the center $C(G)$ of G is the kernel of the adjoint map

$$ad : G \to \mathcal{A}ut(\mathcal{G})$$

torus.
(vi) a connected Lie subgroup H of the Lie G is a normal subgroup of G, iff its Lie algebra $\mathcal{H}$ is an ideal of $\mathcal{G}$.

30. **(Isomorphism theorem of Lie groups)** Let G and H be two Lie groups and $f : G \to H$ be a homomorphism of Lie groups. Prove that

(i) $K = ker\ f$ is a normal Lie subgroup of G;
(ii) f determines an injective homomorphism $\tilde{f} : G/K \to H,\ gK \mapsto f(g)$, which is an immersion of differentiable manifolds;
(iii) if $im\ f$ is closed, then it is a Lie subgroup of H and $\tilde{f} : G/K \to im\ f$ is an isomorphism of Lie groups, denoted by $\tilde{f} : G/K \cong im\ f$.

31. Let G and H be two Lie groups with their corresponding Lie algebras $\mathcal{G}$ and $\mathcal{H}$. If G is simply connected and $\psi : G \to \mathcal{H}$ is a Lie group homomorphism in the sense of Definition 4.7.20, show that there exists a unique Lie group homomorphism

$$f : G \to H \text{ such that } df = \psi.$$

32. Prove the following statements:

(i) Let G and H be two connected Lie groups such that the corresponding Lie algebras $\mathcal{G}$ and $\mathcal{H}$ are isomorphic. Then the Lie groups G and H are isomorphic.

(ii) Given any Lie algebra $\mathcal{A}$, there exists a unique simply connected Lie group G such that its Lie algebra $\mathcal{G}$ is isomorphic to the Lie algebra $\mathcal{A}$.

33. **(Representations of compact Lie groups)** Let G be topological group. A continuous homomorphism

$$\theta : G \to GL(n, \mathbf{R}) \; (\; or \; GL(n, \mathbf{C}))$$

is called a real (or complex) representation of G. If G is any compact Lie group, prove the following statements:

(i) every real representation of the compact Lie group G is equivalent to a representation by orthogonal matrices;

(ii) every complex representation of the compact Lie group G is equivalent to a representation by unitary matrices;

(iii) every representation of the compact Lie group G is semisimple.

34. Show that the Lie algebra $\mathcal{G}$ of the Heisenberg group G (formulated in Definition 4.1.26) is the space of all matrices of the form given by

$$M = \begin{pmatrix} 1 & x & y \\ 0 & 1 & z \\ 0 & 0 & 1 \end{pmatrix},$$

where $x, y, z \in \mathbf{R}$. Prove that the Lie algebra $\mathcal{H}$ is the space of all real matrices of the form

$$M = \begin{pmatrix} 0 & x & y \\ 0 & 0 & z \\ 0 & 0 & 0 \end{pmatrix}.$$

35. Let G be the Heisenberg group and $\mathcal{G}$ be its Lie algebra. Show that the exponential map

$$exp : \mathcal{G} \to G$$

is bijective.

36. Show that the action of the Lie group $O(n, \mathbf{R})$ on the n-dimensional Euclidean plane $\mathbf{R}^n$

$$O(n, \mathbf{R}) \times \mathbf{R}^n, (A, x) \mapsto Ax$$

is not transitive but its action on S^{n-1} is transitive having orbit space of the point x is the spheres of radius $||x||$.

37. Show that every connected, compact and commutative matrix Lie group is a torus.

38. Let $T = (S^1)^m$ be the torus and $\alpha = (e^{2\pi\theta_1}, e^{2\pi\theta_2}, \ldots, e^{2\pi\theta_m}) \in T$. Show that α generates a dense subgroup of T iff the set

$$\{1, \theta_1, \theta_2, \ldots, \theta_m\}$$

is linearly independent over the field $\mathbf{Q}$ of rational numbers.

39. Let T be a torus and $\alpha \in T$. Show that the subgroup H of T generated by α is not dense in T iff there is a nonconstant homomorphism

$$\psi : T \to S^1 : \psi(\alpha) = 1.$$

40. Prove the following statements on maximal torus.

(i) If T is a maximal torus, then its Lie algebra $\mathcal{T}$ is a maximal commutative subalgera of a Lie algebra $\mathcal{G}$ of a matrix group G.

(ii) Conversely, if $\mathcal{T}$ is a maximal commutative subalgera of a Lie algebra $\mathcal{G}$, then the connected Lie subgroup T of G with Lie algebra $\mathcal{T}$, is a maximal torus.

41. **(Ado theorem)** Show that every finite-dimensional Lie algebra (over a field F of characteristic 0) is isomorphic to some matrix Lie algebra.

42. **(Realizable Lie algebra)** A Lie algebra $\mathcal{G}$ is said to be **realizable** if there is a Lie group G such that its Lie algebra is $\mathcal{G}$. If the center $\mathcal{C}$ of a Lie algebra $\mathcal{G}$

$$\mathcal{C} = \{X \in \mathcal{G} : [X, Y] = 0, \ \forall X, Y \in \mathcal{G}\}.$$

is zero, show that $\mathcal{G}$ is realizable.

43. Let M be a smooth manifold and $\psi : U \to \mathbf{R}^n$ be a coordinate system on M, with coordinate functions $(x_1, x_2, \ldots, x_n)$. Show that a smooth curve $\alpha : \mathbf{R} \to U$ is an integral curve of a vector field $V = \Sigma a_i(\partial/\partial x_i)$ iff

$$\frac{d\ (x_i \circ \alpha)}{dt} = a_i, \ \ \forall i = 1, 2, \ldots, n.$$

44. Show that every one-parameter subgroup of Lie group G is an integral curve of some left-invariant vector field on G.

45. Consider the Lie groups $Sp(1, \mathbf{H})$ (symplectic group) and $SU(2, \mathbf{C})$ (unitary group) formulated in Definition 4.2.3. Show that

(i) the Lie groups $Sp(1, \mathbf{H})$ and $SU(2, \mathbf{C})$ are isomorphic as Lie groups;
(ii) each of the Lie groups $Sp(1, \mathbf{H})$ and $SU(2, \mathbf{C})$ is diffeomorphic to the sphere S^3.

46. Show that

(i) each of the Lie groups $SO(n, \mathbf{R})$, $U(n, \mathbf{C})$, $SU(n, \mathbf{C})$ and $Sp(n, \mathbf{H})$ is connected;
(ii) there are only two connected components in the group $O(n, \mathbf{R})$.

47. If a one-parameter subgroup $\alpha : \mathbf{R} \to G$ of a Lie group G intersects itself, show that there exists a real number $k > 0$ such that

$$\alpha(t + k) = \alpha(t), \quad \forall t \in \mathbf{R}.$$

48. Given a compact, connected Lie group G, show that every point $g \in G$, is a point of some one-parameter subgroup of G.
49. Show that every connected commutative Lie group G is locally isomorphic to a finite-dimensional vector space.
50. **(Closed subgroup theorem)** An abstract subgroup of a Lie group G which is a closed subset in the topology in G, is called a **closed subgroup** of G. Prove that every closed subgroup of a Lie group is an embedded Lie subgroup of $GL(n, \mathbf{R})$. [Hint : See [Warner 1983].]
51. Show that the Lie subgroups $SL(n, \mathbf{R})$ and $O(n, \mathbf{R})$ are embedded Lie subgroups of $GL(n, \mathbf{R})$.
[Hint: Use closed subgroup theorem (see Exercise 50).]

Multiple Choice Exercises

Identify the correct alternative(s) (there may be more than one) from the following list of exercises:

1. **(i)** The Euclidean plane $\mathbf{R}^2$ is a manifold of dimension 2.
(ii) The standard sphere $S^2 \subset \mathbf{R}^3$ is a manifold of dimension 3.
(iii) The real projective space $\mathbf{RP}^3$ is a manifold of dimension 3.
2. Let $V_r(\mathbf{R}^n)$ denote the Stiefel manifold of (orthogonal) r-frames in $\mathbf{R}^n$ and $G_r(\mathbf{R}^n)$ denote the Grassmann manifold of r-planes of $\mathbf{R}^n$ through the origin. Then

(i) $V_1(\mathbf{R}^n) = S^n$.
(ii) $G_1(\mathbf{R}^n) = \mathbf{RP}^n$.
(iii) the orthogonal group $O(n, \mathbf{R})$ acts transitively on $G_r(\mathbf{R}^n)$.

3. Let $\mathbf{R}^n$ be the Euclidean n-space.

(i) Let X be the circle at which the standard sphere S^2 in $\mathbf{R}^3$ intersects the plane $z = 0$. Then X is a submanifold of S^2.
(ii) The Euclidean subspace $X = \{x = (x_1, x_2, x_3, x_4, x_5) \in \mathbf{R}^5 : x_4 = x_5 = 0\} \subset \mathbf{R}^5$ is a submanifold of $\mathbf{R}^5$.
(iii) Let $X = \{M \in M(n, \mathbf{R}) : M\ is\ symmetric\} \subset M(n, \mathbf{R})$. It is not a submanifold of $M(n, \mathbf{R})$.

4. **(i)** The space X of nonzero real numbers (with usual topology) is a 1-dimensional Lie group under usual multiplication of real numbers.
(ii) The space X of nonzero complex numbers (with usual topology) is a 1-dimensional (real) Lie group under usual multiplication of complex numbers.
(iii) The orthogonal group $O(2, \mathbf{R})$ is a 4-dimensional Lie group under usual multiplication of matrices.
(iv) The real projective plane $\mathbf{RP}^2$ is a surface.

5. Let G be a Lie group and $a \in G$ be a given point.

(i) The map $f_1 : G \to G,\ g \mapsto ag$ is a diffeomorphism.
(ii) The map $f_2 : G \to G,\ g \mapsto ga$ is a homeomorphism but not a diffeomorphism.
(iii) The map $f_3 : G \to G,\ g \mapsto g^{-1}$ is a diffeomorphism.
(iv) If X is a closed subgroup of the Lie group G, then X is a Lie subgroup of G.

References

Adhikari, M.R.: Basic Algebraic Topolgy and its Applications. Springer, India (2016)
Adhikari, M.R.: Basic Topology, Volume 3: Algebraic Topology and Topology of Fiber Bundles. Springer, India (2022)
Adhikari, M.R., Adhikari, A.: Groups. Rings and Modules with Applications. Universities Press, Hyderabad (2003)
Adhikari, M.R., Adhikari, A.: Textbook of Linear Algebra: An Introduction to Modern Algebra. Allied Publishers, New Delhi (2006)
Adhikari,M.R., Adhikari, A.: Basic Modern Algebra with Applications. Springer, New Delhi, New York, Heidelberg (2014)
Adhikari, A., Adhikari, M.R.: Basic Topology, Volume 1: Metric Spaces and General Topology. Springer, India (2022)
Baker, Andrew: Matrix Groups, an Introduction to Lie Group Theory. Springer, London (2002)
Bredon, G.E.: Topology and Geometry. Springer, New York (1983)
Brickell, F., Clark, R.S.: Differentiable Manifolds. Van Nostrand Reinhold, London (1970)
Chevelly, C.: Theory of Lie Groups. Princeton University Press, Princeton (1957)
Hall Brian, C., Hall, A.: Lie Groups. Lie Algebras and Representations. Springer, Switzerland (2015)
Hirsch, M.W.: Differential Topology. Springer, New York (1976)

Lie, S.: Theorie der transformations gruppen. Math Ann **16**, 441–528 (1880)
Moise, E.E.: Geometric Topology in Dimensions 2 and 3, Graduate Texts in Mathematics, vol. 47. Springer (1977)
Nakahara, M.: Geometry. Topology and Physics. Institute of Physics Publishing, Taylor and Francis, Bristol (2003)
Pontragin, L.: Topological Groups. Princeton University Press, Princeton (1939)
Postnikov, L.: Lectures in Geometry, Semester V: Lie Groups and Lie Algebras. Mir Publishers, Moscow (1986)
Singer, I.M., Thorpe, J.A.: Lecture Notes on Elementary Topology and Geometry. Springer, New York (1967)
Sorani, G.: An Introduction to Real and Complex Manifolds. Gordan and Breech. Science Pub, New York (1969)
Warner, F.W.: Foundation of Differential Manifold and Lie Group. Springer (1983)

Chapter 5
Brief History of Topological Groups, Manifolds and Lie Groups

This chapter conveys **the history of emergence** of the concepts leading to the development of topological groups, manifolds and also Lie groups as mathematical topics with their motivations. They are specialized topological spaces having additional structures other than topological structure which are interlinked. **Historically,** the word "topology" comes from the Greek words "$\lambda \acute{o} \gamma o \zeta$" and "$\tau \acute{o} \pi o \varsigma$" with an alternative name "analysis situs" aiming at the study of situations. This subject arising as a branch of geometry plays a key role in modern mathematics, because of its study of continuous deformations such as stretching, twisting, crumpling and bending, which are allowed, whereas tearing or gluing are not allowed.

For additional reading the readers may refer to Adhikari (2016, 2022), Adhikari and Adhikari (2003, 2006, 2014, 2022), Bredon (1983), Dieudonné (1989), Freedman (1982), Milnor (1956), Poincaré (2010), Weyl (1913).

Recall a statement of M. Morse

"Any problem which is non-linear in character, which involves more than one coordinate system or more than one variable or where structure is initially defined in the large, is likely to require considerations of topology and group theory for its solution. In the solution of such problems classical analysis will be frequently appear as an instrument in the small, integrated over the whole problem with the aid of topology or group theory". M. Morse, 1965.

5.1 A Brief History of Topological Groups

This section communicates a brief history of topological groups with an emphasis on motivation of its study and historical development. **Topological algebra** studies topological groups and topological vector spaces and provides many interesting

A. Adhikari and M. R. Adhikari, *Basic Topology 2*,
https://doi.org/10.1007/978-981-16-6577-6_5

geometrical objects of our interest. It relates algebra with geometry and analysis. The concept of topological groups was born through an exercise to connect two branches of mathematics such as abstract group theory and topology. This unification is an outcome of a great influence of Lie groups and direct consequence of various types of transformation groups. The basic concept of a topological group is that it is an abstract group endowed with a topology such that the multiplication and inverse operations are both continuous. This concept was accepted by mathematicians in the early 1930s. For example, the papers of

(i) A. Harr (1885–1933) published in 1933;
(ii) L. S. Pontragin (1908–1988) published in 1932;
(iii) H. Freudenthal (1905–1990) published in 1936 and
(iv) A. Weil (1906–1998) published in 1936

carry the early work of topological group theory, which witnessed the early development of the theory of topological groups in 1930s.

5.1.1 Motivation of the Study of Topological Groups and Topological Vector Spaces

Basic Topology, Volume 1 of the present series of the books studies the general properties of topological spaces and their continuous maps, but there are many important topological spaces having other structures such as S^1, S^3, $GL(n, \mathbf{R})$, $GL(n, \mathbf{C})$, etc., each of them admits a natural group structure under usual multiplication having their topological and group structures compatible in the sense that the corresponding group operations are continuous. This asserts that in a topological group, the algebraic and topological structures are united and interrelated. This phenomenon leads to the concept of topological groups which is studied in Chap. 2. On the other hand, a topological vector space (linear space) is a vector space endowed with a topology such that the scalar multiplication and addition are both continuous. Topological vector spaces also provide a rich supply of topological groups. Most of the interesting concepts that arise in the analysis are studied through vector spaces (real or complex), instead of single object such as a function, measure or operator. The concepts of topological groups and topological vector spaces facilitate a deep study of topological algebra and establish a key link between topology and algebra. Lie groups studied in Chap. 4 include a special family of topological groups.

5.1.2 Historical Development of Topological Groups and Vector Spaces

The concept of a topological group was born through the work of Felix Klein (1849–1925) in 1872 and Marius Sophus Lie (1842–1899) in 1873. The concepts of topological groups and topological vector spaces display an interplay between topology and algebra, which are two basic areas of mathematics. The subject topology studies continuity and convergence of functions as well as topological properties such as compactness, connectedness, etc., for classification of topological spaces. On the other hand, algebra studies operations to endow different algebraic structures as well as functions, preserving their algebraic structures called homomorphisms to obtain interesting algebraic results. In a topological group, the rules that govern the relationship between a topological structure and an algebraic operation are that the (algebraic) group operations are to be continuous, jointly or separately under the given topological structure. Topological groups provide a rich supply of important topological spaces. The important class of topological groups include the classical topological groups of matrices such as $GL(n, \mathbf{R})$, $GL(n, \mathbf{C})$, $GL(n, \mathbf{H})$, $SL(n, \mathbf{R})$, $SL(n, \mathbf{C})$, $O(n, \mathbf{R})$, $U(n, \mathbf{C})$ *and* $SL\ (n, \mathbf{H})$.

5.2 A Brief History of Manifolds

This section communicates a brief history of topological manifolds with an emphasis on motivation of its study and historical development. A manifold is a special topological space such that each point has an open nbd, i.e., near every point, it looks like a Euclidean space. The plane, the sphere and the torus are common examples of manifolds. They can be realized in $\mathbf{R}^3$, but the Klein bottle and real projective plane are also manifolds that cannot be realized in $\mathbf{R}^3$.

5.2.1 Motivation of the Study of Manifolds

There are many branches of mathematics where topological spaces can be described locally by n-tuples $(x_1, x_2, \dots, x_n) \in \mathbf{R}^n$ of real numbers. Such a space which is locally homeomorphic to the Euclidean space $\mathbf{R}^n$ is called a (topological) manifold of dimension n. Manifolds as mathematical objects provide generalizations of curves and surfaces to arbitrary dimensions invite a study of their topological properties with geometric intuition needed for a deep study of manifold theory related to differential topology and the topology from a differential view-point.

Manifolds are locally Euclidean in the sense that every point has a neighborhood (nbd), called a chart, homeomorphic to an open subset of a Euclidean space $\mathbf{R}^n$. The coordinates on a chart facilitate to study manifolds through many concepts born in $\mathbf{R}^n$ such as differentiability, tangent spaces and differential forms, etc. Thus, the study of manifolds is facilitated by introducing a coordinate system in each of these Euclidean neighborhoods (nbds) and changing of coordinates, which are continuous real-valued functions of several variables. A (topological) manifold of dimension n may be considered as being made of pieces of the Euclidean space $\mathbf{R}^n$ glued together by homeomorphisms. For example, the soccer ball is a manifold of dimension 2. If in particular, these homeomorphisms are differentiable, then the concept of differentiable or smooth manifold (or C^∞-manifold) is introduced.

Geometers working on differential geometry discuss curves and surfaces in ordinary space, i.e., in Euclidean plane and Euclidean space through local concepts such as curvature, etc. But Riemann introduced a concept, known as Riemann surface in an abstract setting in the sense that these surfaces are not defined as subsets of Euclidean space, which led to the concept of topological manifolds. This concept plays a key role in the study of modern research in topology.

5.2.2 Motivation of the Study of Differentiable Manifolds

A differentiable manifold is a topological manifold endowed with a differentiable structure, which is an additional structure other than its topological structure. A differentiable manifolds is studied based on the standard differentiable structure on a Euclidean space $\mathbf{R}^n$. Motivation of the study of differential manifold comes from the following observation: in a topological manifold of dimension n, every point admits a nbd homeomorphic to $\mathbf{R}^n$, and hence in each of these Euclidean nbds, a coordinate system can be introduced. An n-dimensional topological manifold has a nbd homeomorphic to $\mathbf{R}^n$ and thereby a coordinate system is introduced in each of such Euclidean nbds. The change of coordinate is given by continuous functions of several real variables. By using the standard differentiable structure on a Euclidean space, such changes of coordinates may be differentiable. This leads to the concept of differentiable manifolds (also called smooth or C^∞ -manifold).

5.2.3 Historical Development of Manifolds

The concept of manifolds can be traced to the work of B. Riemann (1826–1866) on differential and multi-valued functions. Riemann introduced a concept, known as Riemann surface, in an abstract setting. **Historically,** B. Riemann (1826–1866) made an extensive work generalizing the concept of a surface to higher dimensions. But before him, Carl Friedrich Gauss published a paper in 1827, where he used local coordinates on a surface carrying the concept of charts. The term manifold is derived from the German word Mannigfaltigkeit, which Riemann first used. Henri Poincaré (1854–1912) used the concept of locally Euclidean spaces (defining the concept of topological manifold) in his monumental work when Henri Poincaré published his Analysis Situs, in 1895, and its subsequent complements on homotopy and homology in the late nineteenth century.

A rapid development of manifold theory occured from 1930s following the definition of a manifold based on general topology. But its deep study was born through the search of solution of **Poincaré conjecture** posed in 1904. This conjecture asks: is a compact simply connected n-dimensional topological manifold with the same homology groups as S^n homeomorphic to S^n ? Its equivalent statement says: **is a compact n-manifold homotopically equivalent to S^n homeomorphic to S^n?** (see

Sect. 5.4). This conjecture is very significant in the study of the general theory of topological manifolds and is the most important topological problem of the twentieth century. For dimension $n \geq 3$ the classification problems become more complicated. Since 1895, it took more than century till for $n = 3$. G. Perelman (1966–) proved this conjecture in 2003 by using Ricci flow. For other values of n, it was solved by others before 1994. To be specific, for 5 -dimensional or its higher dimensional manifolds, the problem was solved by Stephen Smale in 1961 and for 4 -dimensional manifolds, the problem was solved by Michael Freedman in 1982. It is remarkable that William Thurston formulated, in 1970, a more powerful conjecture, now known as the **Thurston geometrization conjecture** saying that every compact 3-manifold has a geometric decomposition, in the sense that it can be cut along specific surfaces into finitely many pieces, admitting each of them one of eight highly uniform (but mostly non-Euclidean) geometric structures. Unfortunately we fail to study them here, because of the techniques used by them need far more groundwork than we cover in this book.

Historically, the approach of study a differentiable manifold based on the standard differentiable structure on a Euclidean space $\mathbf{R}^n$ was formally given

(i) by Hermn Weyl (1885–1955) in 1912 and
(ii) by H Whitney (1907–1989) during 1930's.

Remark 5.2.1 Prior to them, Dini (1845–1918), a student of E. Betti (1823–1892) published his first paper in 1864 on differential geometry with comments on his own thesis as suggested by Betti. The theory on implicit functions developed by Dini is closely related to the basic work of H. Whitney on embedding of manifolds in Euclidean space $\mathbf{R}^k$. The foundation of modern theory of differentiable manifolds was laid by the work of Dini, Hermn Weyl and H Whitney. This theory witnessed its rapid development in the second half of the nineteenth century through differential geometry and Lie group theory.

The most basic results embedded in the theory of differential manifolds is essentially due to the work of a single man, H. Whitney (1907–1989). For example, the basic property of a manifold which plays the key role toward various developments of manifold is the **Whitney's Embedding Theorem** saying that every manifold can be embedded in a Euclidean space as a closed subspace (see Chap. 3). This theorem implies that any manifold may be considered as a submanifold of a Euclidean space. It reconciles this earlier concept of manifolds with the modern abstract concept of manifolds and facilitates various development of manifold theory. This result broke new ground. His concept of regular maps are now called immersions as C^r- maps such that the tangent map at every point of the manifold is injective and injective immersions are now called embeddings. The main aim of differential topology is to study interactions between topology and calculus as well as analysis and differential equations. The concept of the set of critical values of a smooth map plays a key role in **differential topology.** This set is not necessarily finite but A. Sard proved in 1942 based on earlier work by A. P. Morse that this set has measure zero.

It was believed before 1956 that a topological space may admit only one differentiable structure. Examples show that differentiable structure of a topological space may not be unique. For example, John Willard Milnor (1931–) proved in 1956 that **the 7-sphere S^7 admits 28 different differentiable structures** [Milnor, 1956]. Milnor was awarded the Fields Medal in 1962 for his work in differential topology. He was also awarded the Abel Prize in 2011. Sir Simon Kirwan Donaldson (1957–) proved in 1983 that **$\mathbf{R}^4$ admits an infinite number of different differentiable structures** [Donaldson, 1983]. He was awarded a Fields Medal in 1986.

Remark 5.2.2 The work of H. Whitney, H. Hopf and E. Stienfel established the importance of fiber bundles for applications of topology to geometry It also makes a return of algebraic topology to its origin and revitalized this topic from its origin in the study of classical manifolds (see Basic Topology, Volume 3 of the present series of books).

Remark 5.2.3 Theorem 5.2.4 proves that the property of being a set of measure zero is invariant under smooth map. Theorem 5.2.4 was proved by Arthur B. Brown in 1935, which rediscovered by Dubovickil in 1953 and by Thorn in 1954.

Theorem 5.2.4 *Let the subset $A \subset \mathbf{R}^n$ have measure zero. If $f : A \to \mathbf{R}^m$ is an arbitrary smooth map, then $f(A)$ is a set of measure zero in $\mathbf{R}^m$.*

5.3 Classification of Surfaces and 1-Dimensional Manifolds

This section communicates motivation of classification of surfaces and 1-dimensional manifolds with a brief history of their classification. These classification theorems are simple and excellent results of geometric topology proved in Chap. 3.

Motivation of the Classification

A basic problem in topology is the classification problem of manifolds up to topological equivalence. More attention has been paid for classification of compact manifolds in the sense that they are considered homeomorphic to closed and bounded subsets of some Euclidean space $\mathbf{R}^n$. A complete classification of closed surfaces (2-manifold) is known. This **classification theorem** is a beautiful example of geometric topology. It says that any compact connected surface S is either homeomorphic to a sphere, or to a connected sum of tori (with $n \geq 1$ holes), or to a connected sum of $n(\geq 1)$ projective planes. No two of these three types of manifolds are homeomorphic to each other. The classification theorem for closed surfaces is a landmark theorem in topology. It exhibits the elegance of statement, the power of combinatorial or geometric topology and asserts its central importance to the subject.

The proof of **classification theorem of surfaces** was first given by M Dehn (1878–1952) and P Heegaard (1871–1948) in 1907.

Theorem 5.3.1 *(Classification theorem of surfaces) Any compact connected surface S is either homeomorphic to a sphere, or to a connected sum of tori, or to a connected sum of projective planes.*

Theorem 5.3.2 gives another form of complete classification of compact surfaces.

Theorem 5.3.2 *(An alternative form of classification theorem of surfaces) Any compact surface is homeomorphic either*

(i) *to the sphere S^2,*
(ii) *or to the sphere S^2 with a finite number of handles glued to the sphere, called a connected sum of tori,*
(iii) *or to the sphere S^2 with a finite number of Möbius strips glued to the sphere obtained by a finite number of disks removed and replaced by Möbius strips, called a connected sum of projective planes. No two of these surfaces are homeomorphic.*

Remark 5.3.3 The classification theorem of compact connected 1-dimensional manifolds is formulated in Theorem 5.3.4 and in Theorem 5.3.5.

Theorem 5.3.4 *(Classification theorem of **1-dimensional compact connected manifold**) Any compact connected 1-dimensional smooth manifold is diffeomorphic to either $[0, 1]$ or S^1.*

Theorem 5.3.5 *(Complete Classification of **1-manifolds**) Any connected 1-dimensional smooth manifold is diffeomorphic to either some interval of real numbers or to the circle S^1.*

5.4 Poincaré Conjecture on Compact Smooth n-Dimensional Manifold

While investigating the 3-dimensional and higher dimensional manifolds in 1895, in his "Analysis Situs", Henri Poincaré (1854–1912) formally introduced the concepts of homotopy, fundamental group, homology groups and Betti numbers (studied in **Basic Topology: Volume 3** of the present book series) for further development of the theories of functions and differential equations. He posed a conjecture in 1894, known as **Poincaré's conjecture** asking that a compact smooth n-dimensional manifold, which is homotopy equivalent to the n-sphere S^n is homeomorphic to S^n.

Its equivalent statement says: is a compact n-manifold homotopically equivalent to S^n homeomorphic to S^n? For $n = 3$, G. Perelman (1966–) proved this conjecture in 2003 by using Ricci flow. For other values of n, it was solved by others before 1994 as stated below

(i) For $n = 4$, **M. Freedman** (1951–) proved that the conjecture is true and wins Fields medal for this proof,
(ii) For $n = 5$, **C. Zeeman** (1925–2016) demonstrated the conjecture in 1961,
(iii) For $n = 6$, **J.R. Stallings** (1935–2008) proved in 1961 that the conjecture is true,
(iv) For $n \geq 7$, **S. Smale** (1930–) proved that the conjecture is true and also extended his proof for all $n \geq 5$. He wins the Fields medal in 1966 for this work.

5.5 Beginning of Differential Topology

Differential topology is the study of those properties of differentiable manifolds which are invariant under diffeomorphism, i.e., invariant under differentiable homeomorphism. This subject considers problems that arise from an interplay between different strucures of a manifold such as its topological, combinatorial and differentiable structures. But it does not consider the notions such as connections, geodesics, curvature, etc., which are considered in differential geometry. This makes a difference between the sujects' differential topology and differential geometry. Many concepts and results now found in differential topology were used in the work of mathematicians of the eighteenth and nineteenth centuries, but they do not constitute a mathematical branch of their own rights in the this period. Many tools of algebraic topology are well-suited to the study of manifolds.

5.6 A Brief History on Lie Groups

5.6.1 Motivation of the Study of Lie Groups and Lie Algebra

Lie groups, named after Sophus Lie (1842–1899), was born through his study of continuous transformations. In a natural way, Lie groups form topological manifolds and provide a rich supply of important topological groups. A Lie group G is algebraically a group and is also a differentiable manifold satisfying certain axioms compatible with the group structure and the manifold structure on G. The most important Lie groups are the classical groups of matrices such as $GL(n, \mathbf{R})$, $GL(n, \mathbf{C})$, $GL(n, \mathbf{H})$, $SL(n, \mathbf{R})$, $SL(n, \mathbf{C})$, $O(n, \mathbf{R})$, $U(n, \mathbf{C})$ and $SL(n, \mathbf{H})$.

Lie groups and Lie algebra provide a rich supply of manifolds. The topological structure and algebraic structure of a Lie group are sometimes used to study the topology of manifolds. For example, the Lie group $SO(2, \mathbf{R})$ is topologically equivalent to the circle S^1 and the Lie group $SU(2, \mathbf{C})$ is topologically equivalent to the 3-sphere S^3. Lie theory gives a rich supply of tools to modern geometry. For example, Felix Klein investigated various "geometries" by using specified suitable transformation groups having certain geometric properties invariant in his Erlangen program. Accordingly, Euclidean geometry corresponds to the particular choice of the group $\mathbf{R}^3$ of distance-preserving transformations of the Euclidean space $\mathbf{R}^3$. On the other hand, projective geometry corresponds to the specific choice of the properties invariant under the projective group. Moreover, in Lie theory, continuous symmetries represented through an action of a Lie group on a manifold facilitates a deep study in representation theory.

5.6.2 Historical Development of Lie Groups and Lie Algebra

Lie groups named after Sophus Lie, born through his study of continuous transformations, laid the foundation of the theory of continuous transformation groups. His paper published in **Math. Ann., Vol 16, 1880 [Lie, 1880]** provides useful tools for the deep study of different areas in contemporary mathematics and modern theoretical physics. Lie groups are smooth differentiable manifolds as well as topological spaces with specific interrelations. They are investigated by using differential calculus but this technique does not work for the study of more general topological groups. The main study is based on replacement of the global object (topological group) by its local or linearized version, which is now called its Lie algebra. Sophus Lie himself termed it "infinitesimal group"

Hermn Weyl inaugurated, in 1925, the global theory of Lie groups. He proved a basic result asserting that the fundamental group of a compact Lie group is finite. This important results is applied in the theory of integration for any compact Lie group. Cartan's work through publication of a series of papers on global theory of Lie groups and homogeouos spaces during the perid 1927–1935 carries a mile-stone in the history of Lie theory.

5.6.3 Hilbert's Fifth Problem and Lie Group

D. Hilbert poised 23 problems at ICM 1900, of which the fifth problem made a revolution in the deep study of Lie theory. His fifth problem asked whether each locally Euclidean topological group admits a Lie group structure. The first half of the twentieth century saw a positive answer of this problem through the work of Gleason, Iwasawa, Montgomery, Yamabe and Zippin, which determined the structure of almost connected locally compact groups. The importance of this family of groups lies in the result that this family contains every compact group and every connected locally compact groups.

Lie theory provides enormous tools to modern geometry. Felix Klein studied different "geometries" by specifying an appropriate transformation group, which leaves certain geometric properties invariant in his Erlangen program, says in his Erlangen program that one can consider various "geometries" by specifying an appropriate transformation group that leaves certain geometric properties invariant. For example, Euclidean geometry corresponds to the choice of the group $\mathbf{R}^3$ of distance-preserving transformations of the Euclidean space $\mathbf{R}^3$. On the other hand, projective geometry corresponds to the choice of the properties invariant under the projective group. Moreover, in Lie theory, continuous symmetries represented through an action of a Lie group on a manifold facilitates a deep study in representation theory.

References

Adhikari, M.R.: Basic Algebraic Topolgy and its Applications. Springer, India (2016)
Adhikari, M.R.: Basic Topology, Volume 3: Algebraic Topology and Topology of Fiber Bundles. Springer, India (2022)
Adhikari, M.R., Adhikari, A.: Groups. Rings and Modules with Applications. Universities Press, Hyderabad (2003)
Adhikari, M.R., Adhikari, A.: Textbook of Linear Algebra: An Introduction to Modern Algebra. Allied Publishers, New Delhi (2006)
Adhikari, M.R., Adhikari, A.: Basic Modern Algebra with Applications. Springer, New Delhi, New York, Heidelberg (2014)
Adhikari, A., Adhikari, M.R.: Basic Topology, Volume 1: Metric Spaces and General Topology. Springer, India (2022)
Bredon, G.E.: Topology and Geometry. Springer-Verlag, New York (1983)
Dieudonné, J.: A History of Algebraic and Differential Topology, 1900–1960. Modern Birkhäuser (1989)
Freedman, M.: The topology of four-dimensional manifolds. J Differ Geom **17**, 357–453 (1982)
Milnor, J.W.: On Manifolds Homeomorphic to the 7-Sphere. Ann Math Princeton University Press **64**(2), 399–405 (1956)
Poincaré, H.: Papers on Topology: Analysis Situs and its Five Supplements, Translated by Stillwell, J., History of Mathematics, **37**. Amer Math Soc (2010)
Weyl, H.: Die Idee der Riemannschen Flche, Druck und Verlag von B.G. Teubner, (English translation: The Concept of a Riemann Surface, Addison-Wesley Publishing Company, Inc., Reading, Mas-sachusetts, 1955). Leipzig und Berlin (1913)

Index

A. Adhikari and M. R. Adhikari, *Basic Topology 2*,
https://doi.org/10.1007/978-981-16-6577-6

Printed in the United States
by Baker & Taylor Publisher Services